ANCIENT SEDIMENTARY ENVIRONMENTS

Ancient Sedimentary Environments

and their sub-surface diagnosis

Fourth edition

RICHARD C. SELLEY

CHAPMAN & HALL

London · Glasgow · Weinheim · New York · Tokyo · Melbourne · Madras

Published by Chapman & Hall, 2–6 Boundary Row, London SE1 8HN, UK

Chapman & Hall, 2–6 Boundary Row, London SE1 8HN, UK

Blackie Academic & Professional, Wester Cleddens Road, Bishopbriggs, Glasgow G64 2NZ, UK

Chapman & Hall GmbH, Pappelallee 3, 69469 Weinheim, Germany

Chapman & Hall USA, 115 Fifth Avenue, New York, NY 10003, USA

Chapman & Hall Japan, ITP-Japan, Kyowa Building, 3F, 2-2-1 Hirakawacho, Chiyoda-ku, Tokyo 102, Japan

Chapman & Hall Australia, 102 Dodds Street, South Melbourne, Victoria 3205, Australia

Chapman & Hall India, R. Seshadri, 32 Second Main Road, CIT East, Madras 600 035, India

First edition 1970
Reprinted 1972, 1973, 1976
Second edition 1978
Third edition 1985
Reprinted 1992, 1994
Fourth edition 1996

Typeset in 10.5/12 pt Times by Saxon Graphics Ltd, Derby
Printed in Great Britain by Clays Lts, St. Ives plc., Bungay, Suffolk

ISBN 0 412 57970 7

A catalogue record for this book is available from the British Library

Library of Congress Catalog Card Number: 95-71083

∞ Printed on permanent acid-free text paper, manufactured in accordance with ANSI/NISO Z39.48–1992 and ANSI/NISO Z39.48–1984 (Permanence of Paper).

Contents

Preface

When I wrote the first edition of this little book in 1969 I little thought that it would run to four editions and be translated into five languages, one of them actually authorized.

Geology has changed much in the intervening years. Twenty-five years ago geologists were hammer-wielding, bearded pipesmokers, dressed in boots and tweeds. Today geology is practised by technicians in white-coats, and by computer-buffs in city suits and high heels. Twenty-five years ago it first became possible to produce reasoned diagnoses of the depositional environments of sedimentary rocks. This technique was both intellectually satisfying, and also of inestimable commercial value as a tool for predicting the geometry and trend of petroleum reservoirs. Today 3D seismic can often now directly image the geometry of petroleum reservoirs, so the application of environmental interpretation to petroleum exploration declines. It continues to be invaluable, however, to predict the smaller scale heterogeneities of petroleum reservoirs that are below the limits of resolution of the seismic method. Here it is now helped immensely by a new range of borehole logging tools which can image sedimentary structures, accurately calculate their orientation, and thus indicate palaeoflow direction, and predict reservoir trends.

The rapid advance of technology in the last quarter century has also stimulated the intellectual approach to stratigraphy in general and facies analysis in particular. The advent of modern high resolution seismic has encouraged geophysicists to study geology, and, often ignoring traditional geological terminology and concepts, advance the new discipline of seismic sequence stratigraphy.

The various editions of this book have reflected the evolution of technology on the environmental interpretation of sedimentary rocks. The first edition dealt solely with outcrop geology. The second edition introduced the study of borehole logs as a means of diagnosing depositional environments in the subsurface. The third edition embraced the seismic method. This, the fourth edition, includes examples of 3D seismic, and of the borehole imaging tools which produce facsimiles of sedimentary structures. This edition also endeavours to evaluate critically the impact of seismic sequence stratigraphy on environmental interpretation.

Despite the evolution of technology and ideas, the object of *Ancient Sedimentary Environments* remains unchanged, namely to show how to diagnose the depositional environment of sedimentary rocks, at the surface, and in the sub-surface, and to show how this information may be applied in the exploration for, and exploitation of, petroleum and other natural resources. It seeks to achieve these objectives by examining one or more case histories of each of the main environments, and by reviewing their diagnostic geological and geophysical characteristics.

Similarly *Ancient Sedimentary Environments* is designed to be read by geologists and geophysicists who already have a basic grounding in geology, including some basic sedimentology. Hopefully, with the aid of a geological dictionary, readers with little prior knowledge of geology may find the book entertaining and instructive. Even petroleum reservoir engineers have been known to find it useful.

25 December 1994

Richard C Selley
Royal School of Mines
Imperial College
London

Preface to the third edition

I wrote the first edition of this book at a very impressionable age. I had just completed nine years of university research on the environmental interpretation of sediments, and had started work for an oil company. The first edition was thus concerned with the interpretation of sedimentary environments from outcrop, from an aesthetic stance, with no thought of vulgar commercial applications.

I wrote the second edition after spending several years learning how to interpret the depositional environment of sediments in the subsurface, and in trying to apply this knowledge to petroleum exploration. Thus the second edition was expanded to include sections of the use of geophysical well logs in subsurface facies analysis.

Within the last few years seismic geophysical surveying has improved tremendously. A whole new field has developed, loosely termed seismic stratigraphy, which applies sedimentary concepts to the interpretation of seismic data. Today seismic surveys may delineate channels, deltas, reefs, submarine fans, and other deposits. In this edition I have included sections on the seismic characteristics of the various sedimentary facies. I have also expanded sections on metalliferous sedimentary deposits, and on recent environments. New case histories have been introduced and I have attempted the impossible task of updating the bibliographies.

September, 1984

Richard C Selley
Imperial College, London

Preface to the second edition

I began to write the first edition of this book after spending nearly ten years at university studying and teaching how to diagnose the depositional environments of sedimentary rocks where they crop out at the earth's surface.

The first edition was written during the first three months of my five-year sabbatical in the oil industry. From then on a very large part of my time has been spent learning how to diagnose the environments of sediments from bore holes in the sub-surface. This is a far more challenging occupation for there are fewer data, the techniques are quite different, and the economic implications may be immense.

The new edition reflects this experience. The introductory chapter includes a discussion of the techniques of sub-surface facies analysis, and subsequent chapters discuss the criteria by which each environment may be recognized in the sub-surface.

Most of the chapters have been modified in one way or another; sections on modern environments have been expanded and some of the case histories have been extensively modified and, for the Capitan 'reef', completely rewritten.

I have updated the bibliographies, but since the sedimentology part of GeoAbstracts show that some 20 000 papers have been published in this field since the first edition, I may well have missed one or two important papers. I apologize to the reviewers, for experience has taught me that the unquoted seminal papers are inevitably their own.

August, 1977

Richard C Selley
Imperial College, London

Preface to the first edition

In this book I have attempted to show how the depositional environments of sedimentary rocks can be recognized. This is not a work for the specialist sedimentologist, but an introductory survey for readers with a basic knowledge of geology.

Within the last few years environmental analysis of ancient sediments has been enhanced by intensive studies of their modern counterparts. Thus Geikie's dictum 'the present is the key to the past' can now be applied with increasing accuracy. While an understanding of Recent processes and environments is critical to the interpretation of their ancient analogues, it is beyond the scope of this book to describe these studies in detail. I have, however, attempted to summarize the results of this work; detail being sacrificed in the interest of brevity. Inevitably, this has led to a tendency to generalize; I have tried to counteract this by providing bibliographies of studies describing Recent sediments.

The economic importance of environmental analysis of ancient sediments is increasing. With the world expected to consume as much oil and gas in the decade 1970–80 as has been produced to date, the search extends increasingly to the more elusive stratigraphically controlled accumulations.

Similarly environmental analysis has a part to play in locating metallic ores in sediments whose geometries are facies controlled.

The book begins with a discussion of the classification of sedimentary environments and an evaluation of the methods which may be used to identify them in ancient deposits.

Each subsequent chapter describes a particular depositional environment, beginning with a summary of its characteristics as seen on the earth's surface at the present time. This is followed by a description of an ancient case history whose origin is then deduced. A general discussion of the problems of identifying the environment in ancient sediments comes next, and each chapter concludes with a brief review of its economic significance.

Neither the selection of environments nor the discussion of economic aspects is intended to be comprehensive. I hope, however, that there are sufficient examples to show how the environment of a sedimentary rock can be determined and some of the ways in which sedimentology can be used in the search for economic materials. The discussions of the economic aspects of the various environments are heavily biased in favour of the oil industry at the expense of mining, hydrogeology, and engineering geology. This is no accident. The oil industry is the largest employer of sedimentologically oriented geologists and has done more to advance and apply sedimentology than any other branch of industrial geology.

Critical readers will notice that metres, feet, kilometres and miles are used indiscriminately throughout the book. Since the oil industry refuses to go metric, the student must quickly learn to correlate the two systems. A conversion scale is included in the first figure.

January, 1970

Richard C Selley
Tripoli, Libya

Acknowledgements

I am very grateful to the many geologists who have aided me in the preparation of this, and earlier editions of this book. This gratitude extends not only to colleagues and students at Imperial College, but to the multitude of oil company staff who have been exposed to my 'Subsurface Facies Analysis' course around the globe in the last 20 years. This course is, to a very large extent, the oral version of the book. The present edition has been tried and tested before a diverse audience, and there is no doubt that it has greatly benefited from their comments over the years.

Specific thanks are due to Schlumberger for permission to use examples of their art (Figures 1.8, 2.6, 2.16, 2.17, 2.18 and 2.22 and Plates 1–4), and particularly to O. Serra, J. Roestenburg and I. Stowe. To the Geological Society of London (Figures 1.11, 3.8, 3.10, and 8.3–8.10), to British Petroleum (Figures 1.15 and 6.17 and Plate 5), to Longman Technical and Scientific (Figure 2.14), to Academic Press Inc. (Figures 2.15, 3.3, 4.4 and 7.1), to the Society of Economic Paleontologists and Mineralogists (Figures 3.2 and 5.5), to the Scottish Journal of Geology (Figures 3.4 and 13.2), to the American Association of Petroleum Geologists (Figures 3.17, 3.19, 3.22, 6.18, 6.21, 7.4, 7.9, 9.9, 11.15, 11.16 and 12.23 and Plate 7), to Springer-Verlag (Figure 3.18), to the Institute of Petroleum (Figures 6.13 and 12.20), to Gebruder Borntraeger (Figure 6.16), to Leading Edge (Figure 6.19 and Plate 6), to Oasis Oil Company of Libya Inc. (Figures 10.2, 10.4 and 10.5), to the Society of Petroleum Engineers (Figure 11.5), to W H Freeman (Figure 1.13) and to the Norwegian Petroleum Society (Figure 12.15).

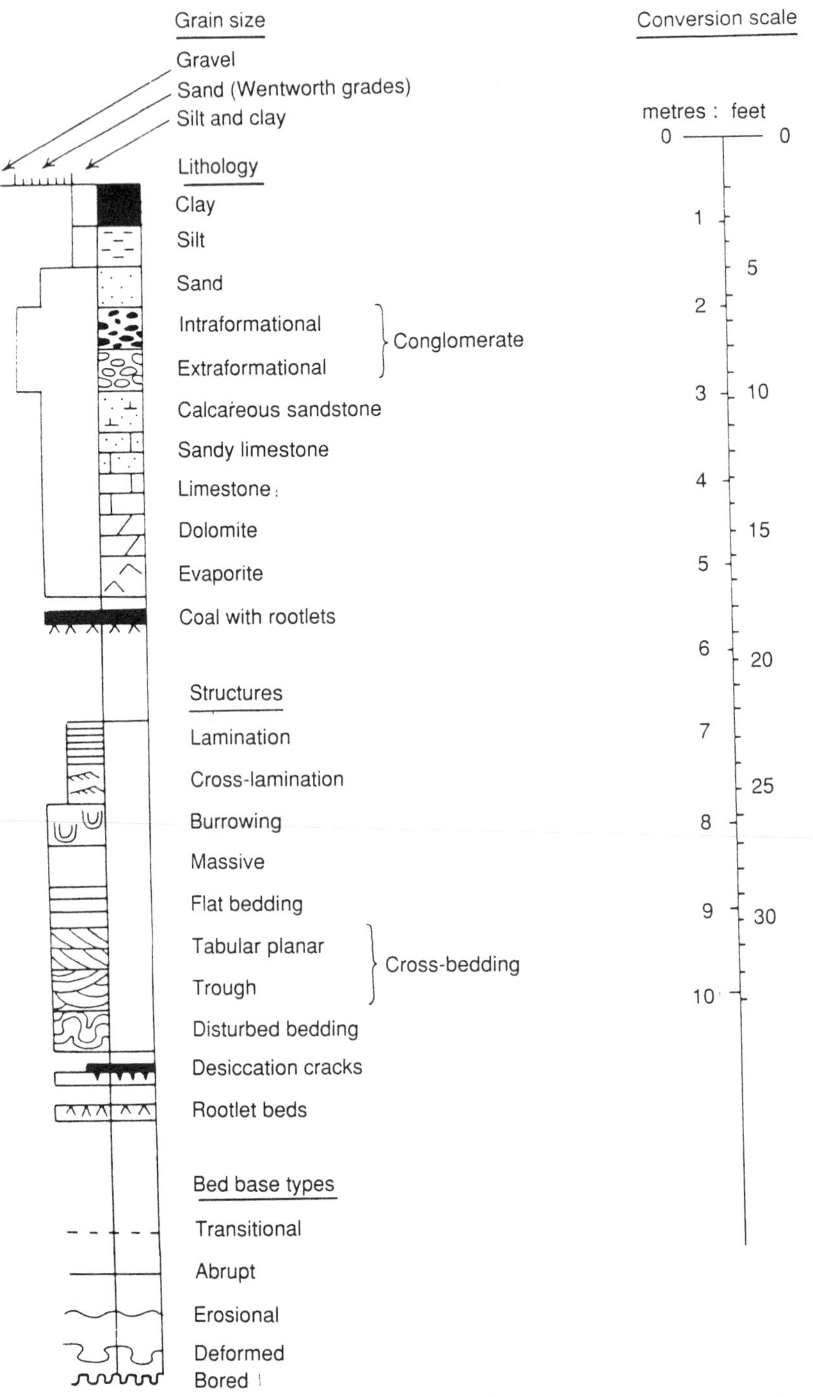

Detailed measured rock sections, illustrated throughout the book, are redrawn using the above key. Note that grain size is drawn increasing to the left so as to correlate with wireline logs.

A note about references

In the old days scientists only published the results of their research when they had discovered something worthy of report. Thus one of the most valuable parts of any paper or book was the bibliography. Nowadays academics are pressured to publish, no matter what, not only for their own careers, but also for the financial solvency and continued existence of their department or university. Thus publication accelerates at an alarming rate, and one is reminded of the words of the poet Waller:

> We write in sand,
> our language grows,
> and like the tide,
> our work o'er flows

It is estimated that some 70 000 papers on geology are published annually. In this book I have tried to select three categories of references: papers that document the case histories, review papers, and papers that first introduce seminal concepts or terms. I make no claim for completeness. Like any book the bibliography is out of date as soon as the manuscript is sent to the publisher. As the great Dr Johnson remarked in 1755, however:

> Knowledge is of two kinds. We know a subject ourselves, or we know where to find information about it.

With the advent of on-line electronic retrieval databases the importance of a bibliography appended to a book or paper is diminished. The reader is urged to consult an on-line retrieval system when wishing to obtain the most up-to-date literature on a particular topic.

1 Concepts and methods

INTRODUCTION

This chapter introduces and discusses the various concepts that are employed in the interpretation of the depositional environments of ancient sedimentary rocks. It includes a review of how environments may be classified, a discussion of the concepts of facies and models, and an account of how facies analysis relates to stratigraphy in general, and to seismic sequence stratigraphy in particular.

SEDIMENTARY ENVIRONMENTS

A sedimentary environment may be defined as a part of the earth's surface that is physically, chemically and biologically distinct from adjacent areas. Examples include deltas, reefs and submarine fans (Figure 1.1).

The physical variables that affect a sedimentary environment include temperature, water depth, current velocity, sunlight, wind speed and direction, and so forth. The chemical variables that affect a sedimentary environment include water composition (salinity), and the mineralogy of both the autochthonous and allochthonous sediment. The biological variables that affect a sedimentary environment include, on land, soil, grass or forest cover, and the life styles of land animals, especially the number and amount of herbivores. Under water the type and abundance of animals and plants exert a major influence on the sedimentary environment, particularly in generating carbonate material. These physical, chemical and biological variables form an intricately interconnected system. A small change in one of the parameters may have a significant effect on one or more of the others. The effect of mankind on modern environments is an obvious instance.

A cursory view of the globe shows that there are many different types of sedimentary environment, and that these occur repeatedly around the world. There are many schemes that attempt to classify sedimentary environments. Table 1.1 illustrates what may best be termed the 'classic' scheme.Though hallowed by time this classification has two main limitations for the purpose of environmental interpretation. The scheme contains several environments which are either erosional, equilibrial, or of only limited significance as depositors of sediments. More seriously, marine environments are differentiated by water depth.

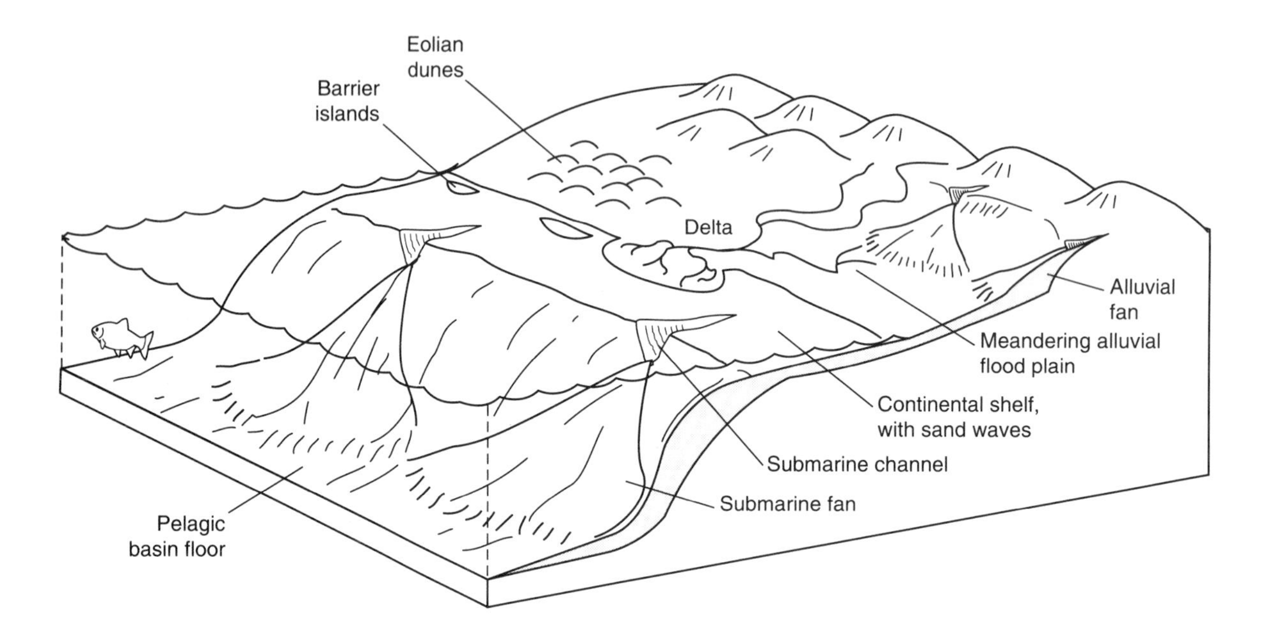

Figure 1.1 Geophantasmogram to illustrate the various sedimentary environments. Note how, in addition to groupings into continental, shoreline and marine environments, it is also possible to differentiate environments of erosion, environments of non-deposition, or equilibrium, and environments of deposition.

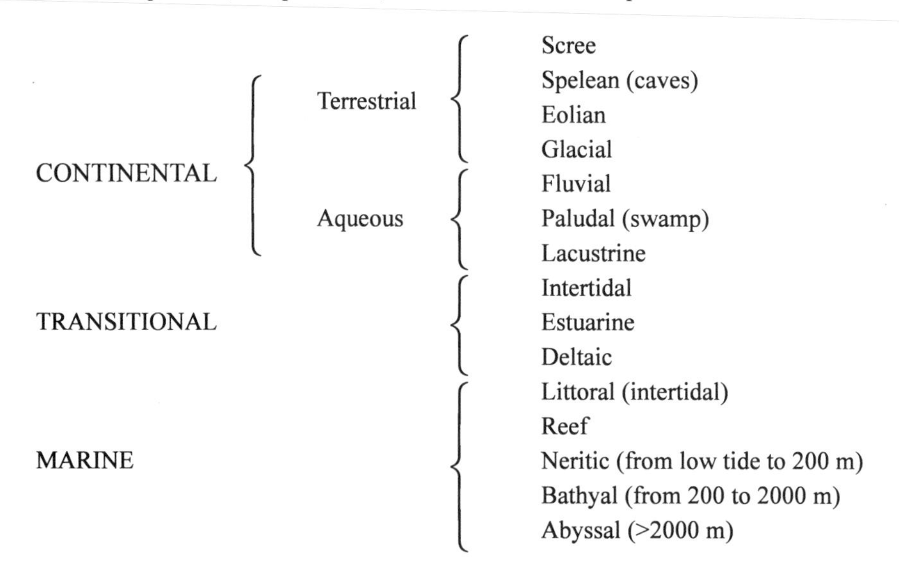

CONTINENTAL	Terrestrial	Scree
		Spelean (caves)
		Eolian
		Glacial
	Aqueous	Fluvial
		Paludal (swamp)
		Lacustrine
TRANSITIONAL		Intertidal
		Estuarine
		Deltaic
MARINE		Littoral (intertidal)
		Reef
		Neritic (from low tide to 200 m)
		Bathyal (from 200 to 2000 m)
		Abyssal (>2000 m)

Table 1.1 A typical classification of sedimentary environments. This has two limitations as a basis for interpreting ancient sedimentary environments. It contains several environments that are essentially erosional or equilibrial, and do not actually deposit sediment. Furthermore the division of marine environments by water depth is both difficult to do with any great accuracy, and is often of little significance as a controlling factor on sedimentation.

Water depth is a very obvious and easily measured parameter of modern environments. It is, however, not a major controlling factor on sedimentary processes. Furthermore, it is very difficult to determine the water depths in which ancient sediments were deposited, beyond such broad zones as intertidal, shelf and 'pelagic', meaning of the open sea. Interpretations of ancient water depths are largely based on palaeontology, particularly using foraminifera. It is widely recognized, however, that foraminiferal assemblages may correlate with temperature as much as depth. Thus changes in ancient faunal assemblages may indicate climatic fluctuations as much as depth changes.

Look at Figure 1.1 again. Note that in addition to the obvious groupings of the various environments in to continental shoreline and marine, there is another trinity of environmental types, namely environments of erosion, environments of non-deposition, or equilibrium, and environments of deposition. These are worthy of consideration.

Environments of erosion are areas that are predominantly continental and terrestrial, apart from cliffed coastlines. They are thus only of concern in so far as they are the sources for terrigenous detritus. The detrital mineralogy of a sand exerts a strong control on its diagenesis and porosity evolution, but is of little significance in terms of its environment of deposition.

Environments of non-deposition, or equilibrium, are areas where there is neither erosion nor net deposition taking place, though sediment may be transported across them. Such equilibrial environments may be found on land, on continental shelves, and even on ocean floors. Terrestrial equilibrial surfaces characterize many deserts. For example, travellers crossing the vast desert plains of the Sahara sometimes encounter the stone circles of old camp fires. Closer inspection may reveal the beer cans of a geological field party, the land mines of a World War II patrol, or the flint instruments of the Stone Age. The last of these indicate the equilibrial nature of the desert surface, where neither deposition nor erosion has occurred for tens of thousands of years.

Terrestrial equilibrial surfaces may be marked by palaeosols that may include cemented horizons of caliche, ferrocrete, or silcrete. As discussed in Chapter 9 many continental shelves are equilibrial surfaces, across which sand may be driven by tidal, wind or storm generated currents, leaving little trace of their passage. Shelf environments may thus leave nothing in the stratigraphic record beyond an unconformity, a lag gravel and, in carbonate sediments, a hard ground. Surfaces of non-deposition or equilibrium also occur in deep marine environments. Some modern ocean floors have sedimentation rates so negligible that the teeth of long extinct species of shark lie unburied on the ocean bed. Environments of equilibrium or non-deposition are represented in the stratigraphic record by unconformities. Furthermore, because they are commonly marked by a cemented layer overlain by less cemented sediment, they often show up on seismic sections as prominent reflectors and sequence boundaries.

Lastly, there is the third type of environment, the environment of deposition. This type occurs in continental, shoreline and open marine settings. It is the depositional environment that is of principal concern and interest to sedimentologists, because this is the environment that actually generates sedimentary sequences that are preserved in the stratigraphic column.

Table 1.2 A classification of depositional sedimentary environments. Equilibrial and erosional environments are omitted. Note that water depth is not considered. The marine environments may include both terrigenous and carbonate deposits. Like all classifications this one breaks down under close scrutiny. For example, lakes may contain deltas, linear coasts, reefs and subaqueous channel and fan systems. None the less this classification provides a useful framework for further study and discussion.

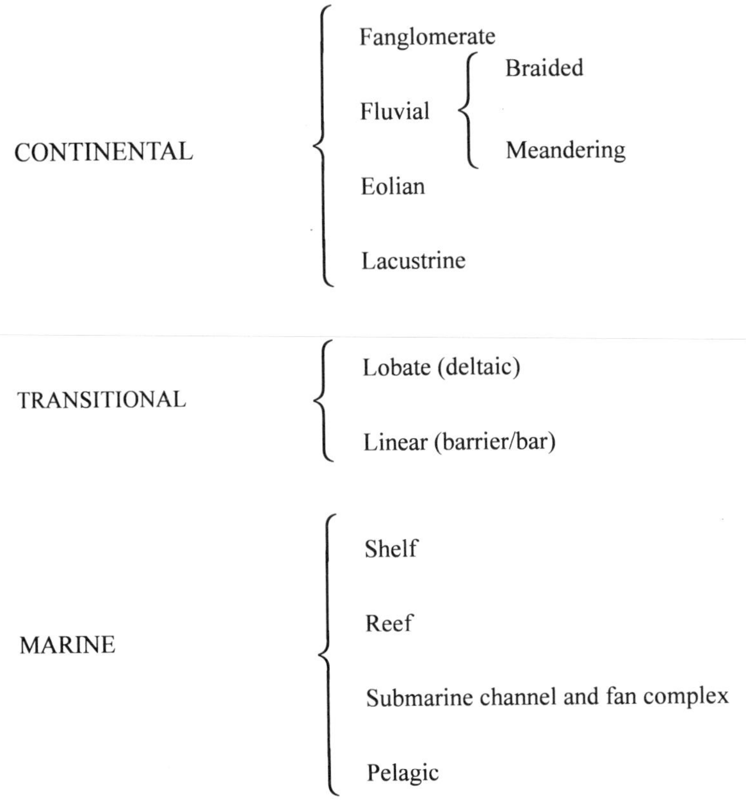

CONTINENTAL	Fanglomerate	
	Fluvial	Braided
		Meandering
	Eolian	
	Lacustrine	
TRANSITIONAL	Lobate (deltaic)	
	Linear (barrier/bar)	
MARINE	Shelf	
	Reef	
	Submarine channel and fan complex	
	Pelagic	

This recognition of environments of erosion, equilibrium and deposition may be used to modify the classic environmental classification shown in Table 1.1 and discussed earlier. Table 1.2 shows the result. This scheme retains the broad grouping into continental, shoreline and marine environments, but now omits the rare, obscure, erosional and equilibrial environments. This scheme is, therefore, not a classification of sedimentary environments. It is a classification of depositional sedimentary environments. Note also that the marine environmental grouping avoids reference to water depth. This scheme is much more appropriate for geologists to use when studying ancient sedimentary environments. It will therefore be adopted in this book, and forms the framework for the subsequent chapters.

When classifying sedimentary environments it is often hard to define their boundaries precisely. Rivers merge gradually in to deltas, and so forth. Furthermore most environments can be divided in to sub-environments. For instance fluvial environments contain active and abandoned channels, levees, crevasse splay and flood basin sub-environments. This hierarchical system of environments and sub-environments will be illustrated in subsequent chapters.

SEDIMENTARY FACIES

A second important aspect of environmental interpretation is appreciation of what is meant by the term sedimentary facies. The regional geological mapping of sedimentary rocks began in the early part of the nineteenth century. The first significant observations were made by men such as William Smith who published his *Geological Map of England and Wales, with Part of Scotland* in 1815. By this time two significant observations had been made. First, that sedimentary rocks occur in layers that can be mapped on the surface across the countryside, and underground in mine workings. Furthermore the sediment layers are characterized by different assemblages of fossils. These assemblages occur in a regular vertical sequence.

The term 'facies', from the German word for face or aspect, was first applied to geology by Steno in 1669. Gressly (1838), working in the Alps, developed the term 'facies' for units of rock that shared similar lithological and palaeontological features. His definition, translated from the original French by Teichert (1958), reads:

> To begin with, two principal facts characterize the sum total of the modifications which I call facies or aspects of a stratigraphic unit: one is that a certain lithological aspect of a stratigraphic unit is linked everywhere with the same palaeontological assemblage; the other is that from such an assemblage fossil genera and species common in other facies are invariably excluded.

In the same year Prevost (1838) applied the term 'formation' to mappable lithostratigraphic units. He noted that different formations could contain the same fossils. He interpreted this to indicate that different formations could be deposited at the same time.

For much of the nineteenth century the terms facies and formation were used interchangeably. Thus Lyell (1865) wrote:

> The term 'formation' expresses in geology any assemblage of rocks which have some character in common, whether of origin, age or composition. Thus we speak of stratified and unstratified, fresh-water and marine, aqueous and volcanic, ancient and modern, metalliferous and non-metalliferous formations.

This is much more what would now be termed facies. By the twentieth century, however, the term formation was firmly embalmed as the term applied to a mappable rock unit. The Geological Society of London took the view:

> The formation is the basic practical division in lithostratigraphical classification. It should possess some degree of internal lithological homogeneity, or distinctive lithological features that constitute a form of unity in comparison with adjacent strata.

The North American Stratigraphic Code definition states:

> The formation is the fundamental unit in lithostratigraphic classification. A formation is a body of rock identified by lithic characteristics and stratigraphic position; it is prevailingly, but not necessarily tabular, and is mappable at the Earth's surface or traceable in the subsurface. (Article 24 of the North American Stratigraphic Code, 1983)

By contrast a twentieth-century definition of the term facies proposed by Moore (1949) states:

> A sedimentary facies is defined as an areally restricted part of a designated stratigraphic unit which exhibits characters significantly different from those of other parts of the unit

or, as was proposed in the first edition of this book in 1970:

> A sedimentary facies is a mass of rock which can be defined and distinguished from others by its geometry, lithology, sedimentary structures, palaeocurrent pattern and fossils.

This is the definition that will be used in this book, naturally, and the five defining parameters provide a coherent framework not only for defining facies, but also for interpreting their environment of deposition in an organized and logical way.

There is a further complication to the use of the term facies. As early as 1879 the term facies had begun to develop an environmental connotation. Thus Mojsisovics (1879) wrote:

> Following Gressly and Oppel, one now customarily applies the term facies to deposits formed under different environmental conditions (translation in Teichert, 1958).

This use of the term facies in an environmental sense continues to the present day (see, for example Reading, 1986; Anderton, 1985). This is dangerous because there is a confusion between observations and interpretations. Facies are observed, and their environment of deposition is interpreted from them. Observations may be checked, and are seldom disputed. Environmental interpretations may well be challenged, disputed and changed. Consider one simple example. Once upon a time there was a cliff of graded greywacke sandstones and interbedded shales. Generations of students have acquired incipient hypothermia standing in the pouring rain listening to professors expound.

In the 1960s the turbidity current process for depositing graded beds was very fashionable. Then professors told their students that this outcrop was of turbidite facies, and waxed lyrical about turbidity flows and submarine fans. In the 1970s oceanographers publicized the importance of contour currents in depositing deep sea sands, so professors told their students that this outcrop was of contourite facies.

In the 1980s storm processes became the fashion, and the outcrop was described as a sequence of tempestites. Surely it is safer to describe the outcrop as an example of graded greywacke facies, what in the old days might have been termed flysch facies. Then one may discuss the possible interpretation of depositional environments and processes. These may range from submarine fan turbidites to storm-dominated shelf tempestites. Facies definitions should be objective and immutable, environmental interpretations are subjective and may evolve. Figure 1.2 shows the relationship between environments and facies.

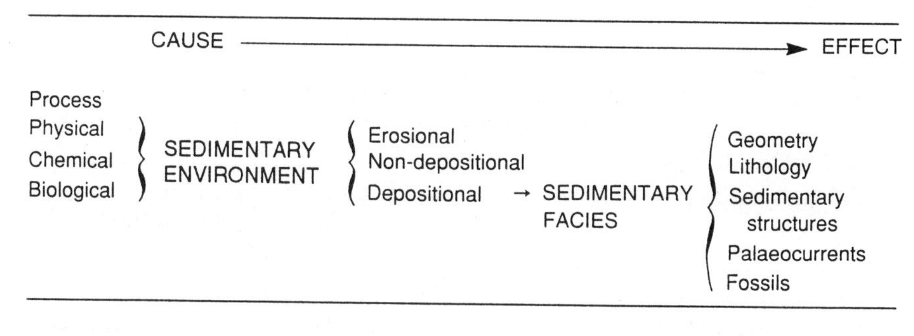

Figure 1.2 The relationship between sedimentary environments and sedimentary facies. Note that sedimentary facies are the product of a depositional environment, a particular type of sedimentary environment.

Initially facies were defined qualitatively, as in ‘red pebbly sandstone facies’, or ‘graded greywacke facies’. In the same way that environments can be divided into sub-environments, so may facies be hierarchically arranged into sub-facies. Then it may be more appropriate to combine letters and numbers, defining facies 1, with sub-facies 1A, lB, 1C; facies 2, with sub-facies 2A, 2B and so on. The characterization of petroleum reservoirs, in particular, requires objective facies definition. Engineers have great difficulty grasping abstract concepts like beauty, God and skeletal grainstone facies. They require objective facies definitions whose parameters may be measured, digitized and fed into a computer. This can now be done for subsurface facies analysis using geophysical well logs. When a borehole is drilled divers sondes are lowered down it with the ability to measure many of the physical properties of the strata penetrated. These variables include resistivity, radioactivity, acoustic velocity, density and so forth. Thus, for any rock unit, at any depth, there are measurements of many of its physical properties.

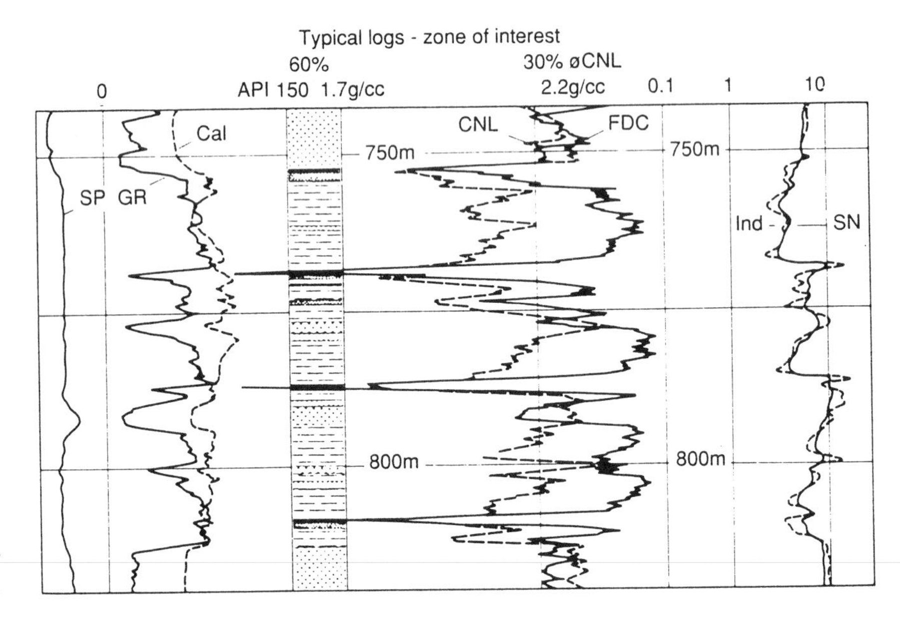

Figure 1.3 Representative suite of geophysical logs through a deltaic sequence. From Rider and Laurier (1979).

A first attempt to use these to define facies objectively was made by Rider and Laurier (1979). They studied a petroleum reservoir that generated typical suites of logs as shown in Figure 1.3. Core and log studies suggested to the authors that the reservoir consisted of repeated sequences of upward-coarsening shale to sand, capped by coal beds, intermittently cut through by channel sands. They interpreted the sedimentary facies as having been deposited on a delta plain, in which bay mud, crevasse splay and swamp deposits were cut through by delta distributaries. They defined an 'ideal cycle' and handcrafted the log responses through it. They then, however, set about objectively defining the reservoir facies. They did this using a conventional log analysis technique, namely the cross-plot, in which the values of two logs are plotted on a graph to calculate porosity accurately. They cross-plotted values from the density and neutron tools, two conventional porosity logs. Instead of just plotting values for reservoir sands, however, they extended this to plot, initially, one whole genetic cycle (Figure 1.4). They then did this for 100 metres of well log, again, irrespective of lithology (Figure 1.5). They were then able to interprete the results geologically (Figure 1.6).

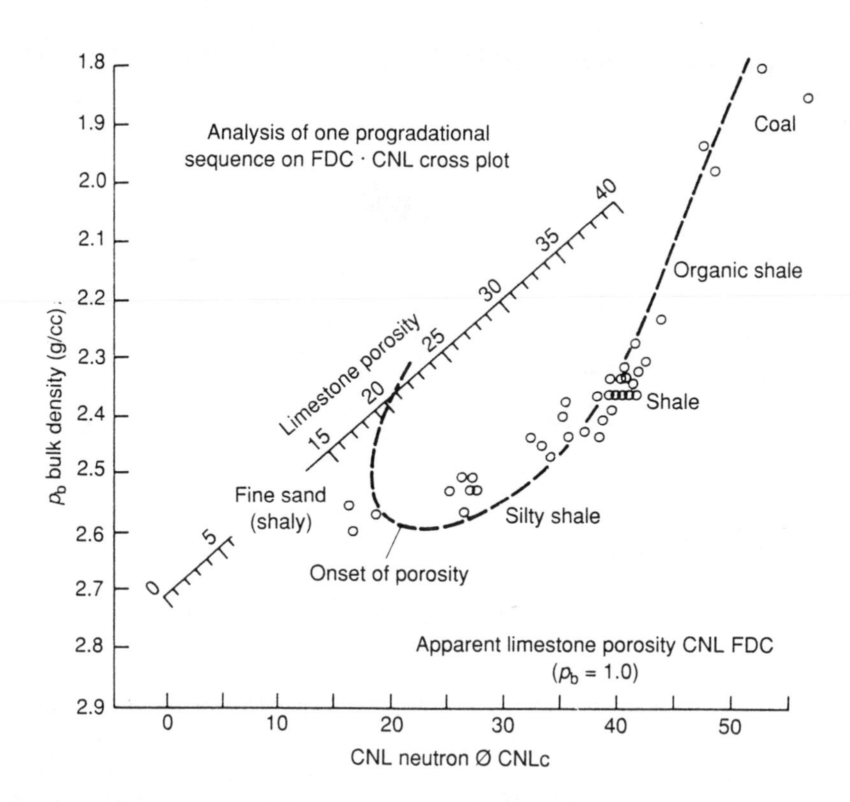

Figure 1.4 Cross-plot of bulk density and neutron log readings for a single cyclic sequence of the deltaic deposits illustrated in Figure 1.3. From Rider and Laurier (1979).

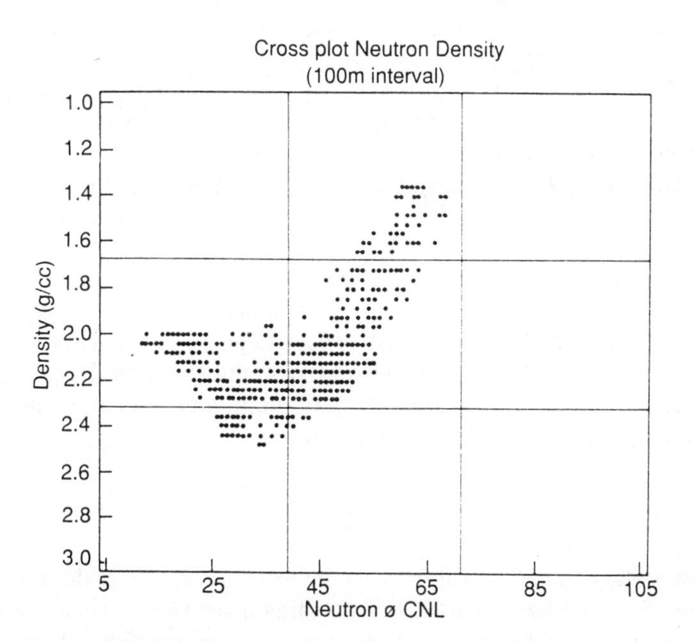

Figure 1.5 Density-neutron cross-plot through 100 metres of the deltaic sediments characterized in Figure 1.3. From Rider and Laurier (1979).

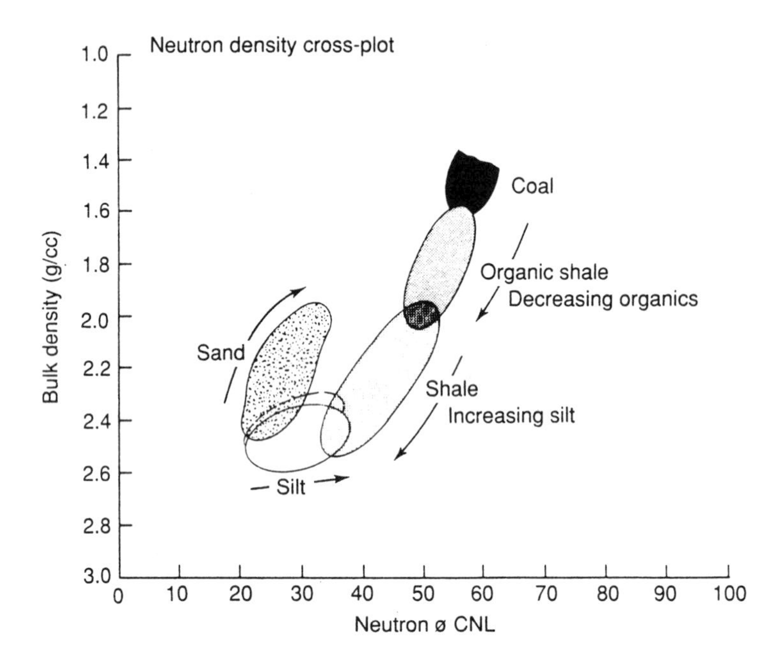

Figure 1.6 Geological interpretation of the cross-plot illustrated in Figure 1.5. From Rider and Laurier (1979).

Now the sedimentary facies of the reservoir are objectively defined. Comparisons can be made between facies of adjacent intervals of the same well, of adjacent wells, or even of different reservoirs in other fields.Thus it became possible to objectively define what are now termed **electrofacies**, though more than just the electrical properties of the rocks are used. Serra and Abbott (1980) developed this approach, using initially three (Figure 1.7), but subsequently many different log values, and manipulating them using cluster analysis and other tricks of statistical gymnastics to produce what are termed electrofacies logs (Figure 1.8).

The scale on the log is measured in cluster numbers. But note that these are dimensionless numbers. In other words the electrofacies with cluster number 2 may be no more similar to electrofacies 1 than electrofacies 5. The concept of the electrofacies is a great step forward because it enables sedimentary facies to be numerically and hence objectively defined. But note that electrofacies will not just reflect depositional environment, but may also include the effects of diagenesis, porosity and pore fluid chemistry. Nonetheless bold enthusiasts are now attributing depositional environments to log defined facies (Bolviken, *et al.*, 1992).

THE SEDIMENTARY MODEL

The sedimentary model is a third very important concept in the interpretation of the depositional environment of sedimentary facies. This is based on two observations, and one important conclusion drawn from them. If one

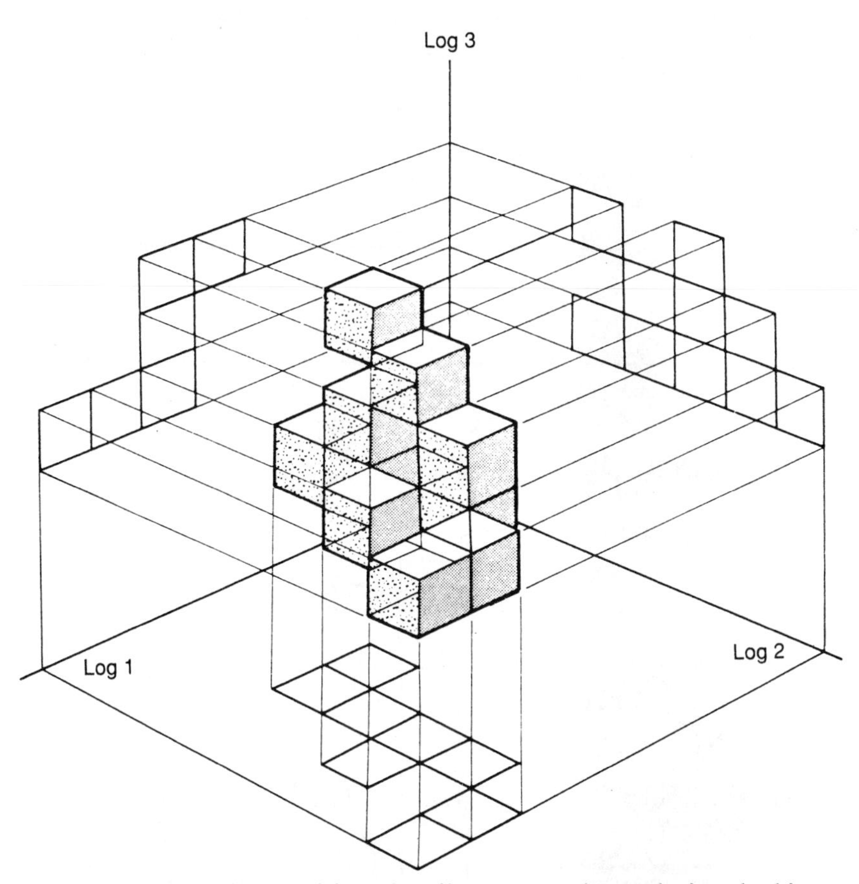

Figure 1.7 Three dimensional display of 'Electrofacies' defined by three geophysical well logs. From Serra and Abbott (1980).

flies around the globe acquiring air miles, or spends much time looking at satellite data, one notes that there are sedimentary environments that occur repeatedly. There are many alluvial floodplains, deltas, deserts, and so forth. One may conclude, therefore, that there is on the earth's surface today a finite number of sedimentary environments.

This observation must be qualified to some degree. Detailed examination shows that no two similar environments are exactly identical. There will always be differences, albeit minor, in grainsize, biota, or whatever. Furthermore the lateral boundaries between environments may often be gradational. It is, for example, difficult to define the boundary between an alluvial flood plain and a delta.

The second observation on which the concept of the sedimentary model is based is that there is a limited number of sedimentary facies that re-occur in time and place throughout the geological record. This becomes apparent to geologists as, over the years, they travel around the globe. You will keep on seeing coarse cross-bedded pebbly channeled red beds, and keep on seeing upward-coarsening shale to sandstone coal-capped cycles in rocks of many different ages on many different continents.

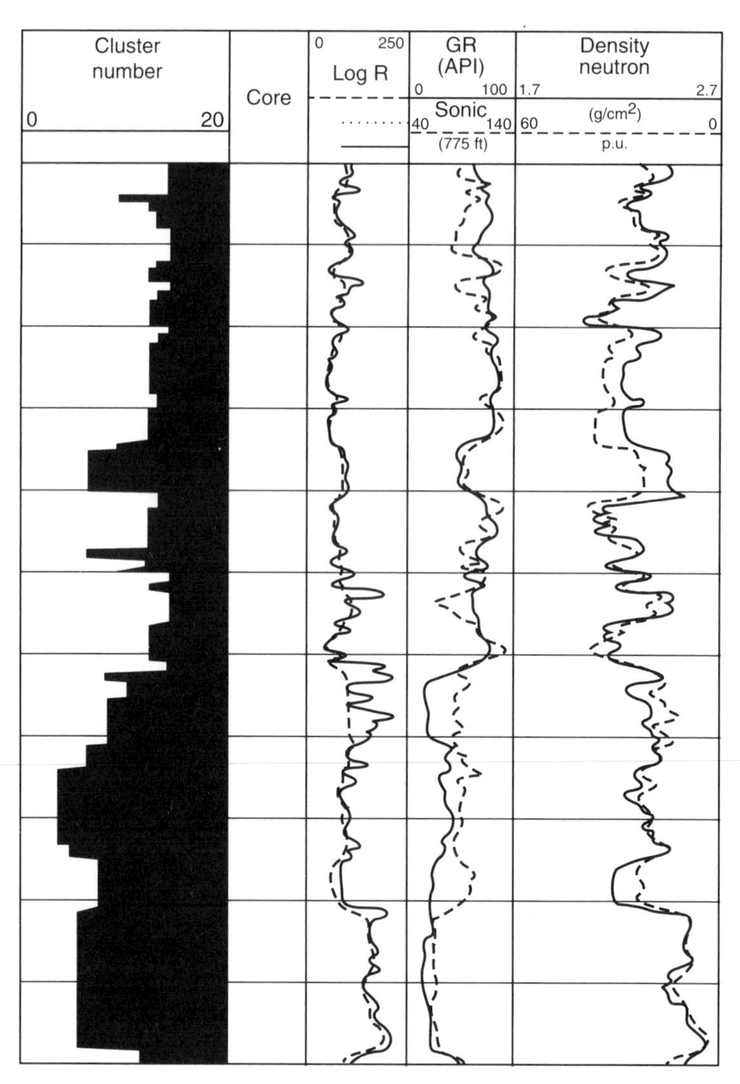

Figure 1.8 An example of a Schlumberger electrofacies log. This permits sedimentary facies to be objectively defined and compared, one with another. Note that the scale on the Cluster log is non-linear. Cluster numbers 1, 2 and 3 could equally be termed apples, pears and bananas.

One concludes therefore that there is a finite number of sedimentary facies that occur repeatedly in time and space. As with sedimentary environments, however, this statement needs qualification. Some geopedant will always be able to discover that no two similar facies are ever exactly identical. There will always be some slight difference in lithology, palaeontology, or whatever. Furthermore, though facies boundaries may be abrupt, or even erosional, many facies grade laterally and sometimes vertically, into different adjacent facies.

The observations that there is a finite number of depositional environments, and a finite number of facies form the basis of the sedimentary model concept. Using the old adage that 'the key to the past lies in the present' the parameters of ancient sedimentary facies can be matched with those of modern deposits whose environments are known. This is how the depositional environments of ancient sedimentary facies may be diagnosed. Thus the conclusion is reached that there is, and always has been, a finite number of sedimentary environments which deposit characteristic facies. These may be classified into various ideal depositional systems or sedimentary models. The term was first defined by Potter and Pettijohn (1977, p. 314) as:

> A sedimentary model in essence describes a recurring pattern of sedimentation.

The model concept is now deemed to be essential to the interpretation of ancient sedimentary environments (Walker, 1984; Anderton, 1985). Sedimentary models serve many functions. Not least they may be used to predict the distribution of porosity and permeability within petroleum reservoirs, and the distribution of both syngenetic and epigenic ores within sedimentary rocks. Subsequent chapters of the book review the various models proposed for the main sedimentary environments and their resultant facies.

SIDEWAYS SEDIMENTATION VERSUS EVENT SEDIMENTATION

Early geologists thought that sedimentary formations were laid down not only one on top of the other, but also one after the other. They might interpret a sedimentary sequence thus: 'Once upon a time a mudstone unit was laid down. Then the sea shallowed and quartz sand was deposited. The sediment supply was then cut off and an oolitic lime sand was deposited, and so on ...'. In some instances it may be possible to demonstrate that sedimentation did indeed take place by the synchronous deposition of bed after bed over large areas. This is often referred to as 'event sedimentation', resulting in 'event stratigraphy' (Goldring, 1991). This is where a geological event, normally of a catastrophic kind, deposits a clearly identifiable bed that can be traced over a considerable area. Turbidites would be a case in point, though, because of their tendency to be gregarious, it is difficult to unequivocally identify one specific bed from outcrop to outcrop, or well to well.

Storms may also deposit event beds and erode scour surfaces. As with turbidites, however, the problem of correlating individual events in a sequence of storm deposits may be difficult. Volcanic tuffs, and the bentonites into which they degrade, may be identifiable over a large area. The volcanic ash layer in eastern Mediterranean sediments attributed to the eruption of Santorini on the island of Thera *c.* 1628 BC, is a celebrated example; since it may have caused, or at least contributed to, the destruction of the Mycenean civilization.

Such demonstrable events are rare in the rock record, however, and most stratigraphy is based on palaeontology. As biostratigraphy enabled ever shorter time units to be defined it became possible to demonstrate that the boundaries between successive formations were not always synchronous, but were sometimes diachronous. Biostratigraphically defined time units cross-cut lithological boundaries when traced over considerable distances. In other words it became possible to show that, as time passed, sedimentary environments moved laterally across the face of the earth. The commonplace nature of what may be termed 'sideways sedimentation' was firmly established by studies of Alpine stratigraphy by the German geologist Johannes Walther. He proposed the term '*faciesbezirk*' (Walther, 1893). This translates literally as 'facies tract', but is now termed a 'systems tract' (Brown and Fisher, 1977). Walther also defined his eponymous law which states:

> The various deposits of the same facies area and similarly the sum of the rocks of different facies areas were formed beside each other in space, but in a crustal profile we see them lying on top of each other. It is a basic statement of far-reaching significance that only those facies and facies areas can be superimposed, primarily, that can be observed beside each other at the present time. (Walther, 1893, p. 979, translation from Blatt *et al.*, 1980, p. 187)

More succinctly:

> A conformable vertical sequence of facies was generated by a lateral sequence of environments (Middleton, 1973).

Walther's Law is one of the basic assumptions of facies analysis. Subsequently studies of modern sedimentary environments confirmed the ubiquity of 'sideways sedimentation', showing how several sub-environments prograde laterally across the surface of the earth depositing sub-facies that lie diachronously one above the other.

These observations from sediments ancient and modern show that it is dangerous to assume that lithological boundaries are time lines.

Ager (1993) whimsically contrasts event sedimentation and sideways sedimentation as the 'gentle rain from heaven' type (Shakespeare, 1623), versus the 'moving finger writes' type (Khayyam, *c.* 1100). Both occur. Their relative importance becomes significant, however, when addressing the dogma of seismic sequence stratigraphy discussed later in this chapter.

SEQUENCES AND CYCLES

From the earliest days of stratigraphy it was recognized that sediments were often laid down in predictable sequences, and that these were often repeated vertically. These sequences were given a number of terms. 'Cyclothem' was widely used. Busch (1971) proposed two fundamental definitions:

> A **genetic increment** of strata is a mass of sedimentary rock in which the facies or sub-facies are genetically related to one another. E.g. a single prograding delta prism of delta platform delta front and prodelta deposits

and:

> A **genetic sequence** of strata includes more than one increment of the same genetic type.

More recently it has been suggested that the terms 'parasequence' and 'parasequence set' replace the established terms 'genetic increments' and 'genetic sequences' (Van Wagoner *et al.*, 1988; 1990).

Figure 1.9 demonstrates the relationships between facies, environments, genetic increments and sequences. Genetic increments and sequences can normally be defined from wireline logs, calibrated with seismic sections, and thus mapped therefrom. One important feature to emerge from studies of modern sedimentary environments is that most sedimentary models generate a characteristic vertical profile of grainsize that can variously be termed a genetic increment or parasequence. This can often be imaged by geophysical well logs, and is commonly an important diagnostic feature of the environment (Figure 1.10). In many stratigraphic sections genetic increments occur in genetic sequences (parasequences occur in parasequence sets). This phenomenon is termed cyclic sedimentation.

Statisticians frequently tease geologists for our ability to find patterns in random data. This is particularly true when it come to recognizing cyclicity in sedimentary sequences. In one celebrated instance geologists confidently

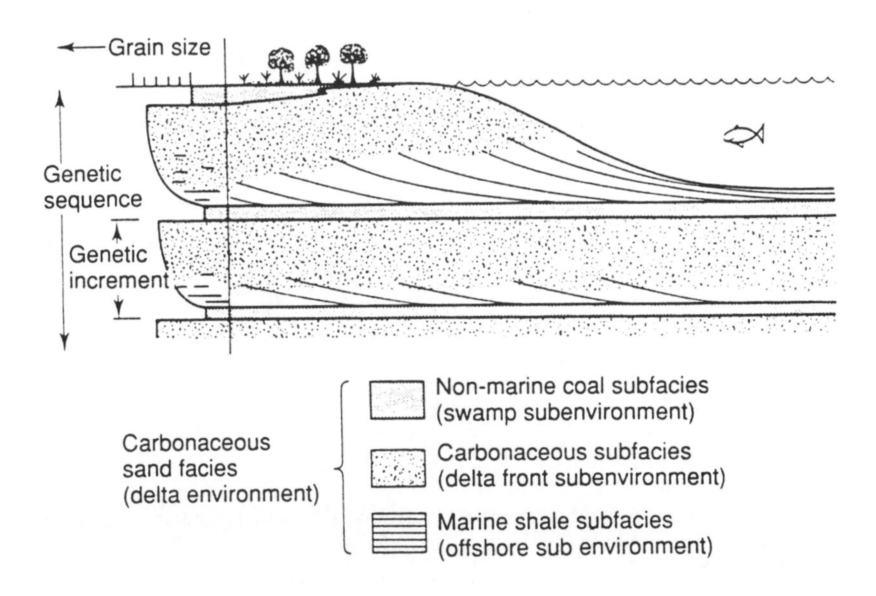

Figure 1.9
Geophantasmogram to display many of the basic concepts of facies analysis. Note environments deposit facies, and how they may be divided into sub-environments and sub-facies respectively. See how genetic increments form by the lateral migration of a suite of environments (Walther's Law) and how genetic sequences are produced by the repeated migration of the same depositional system.

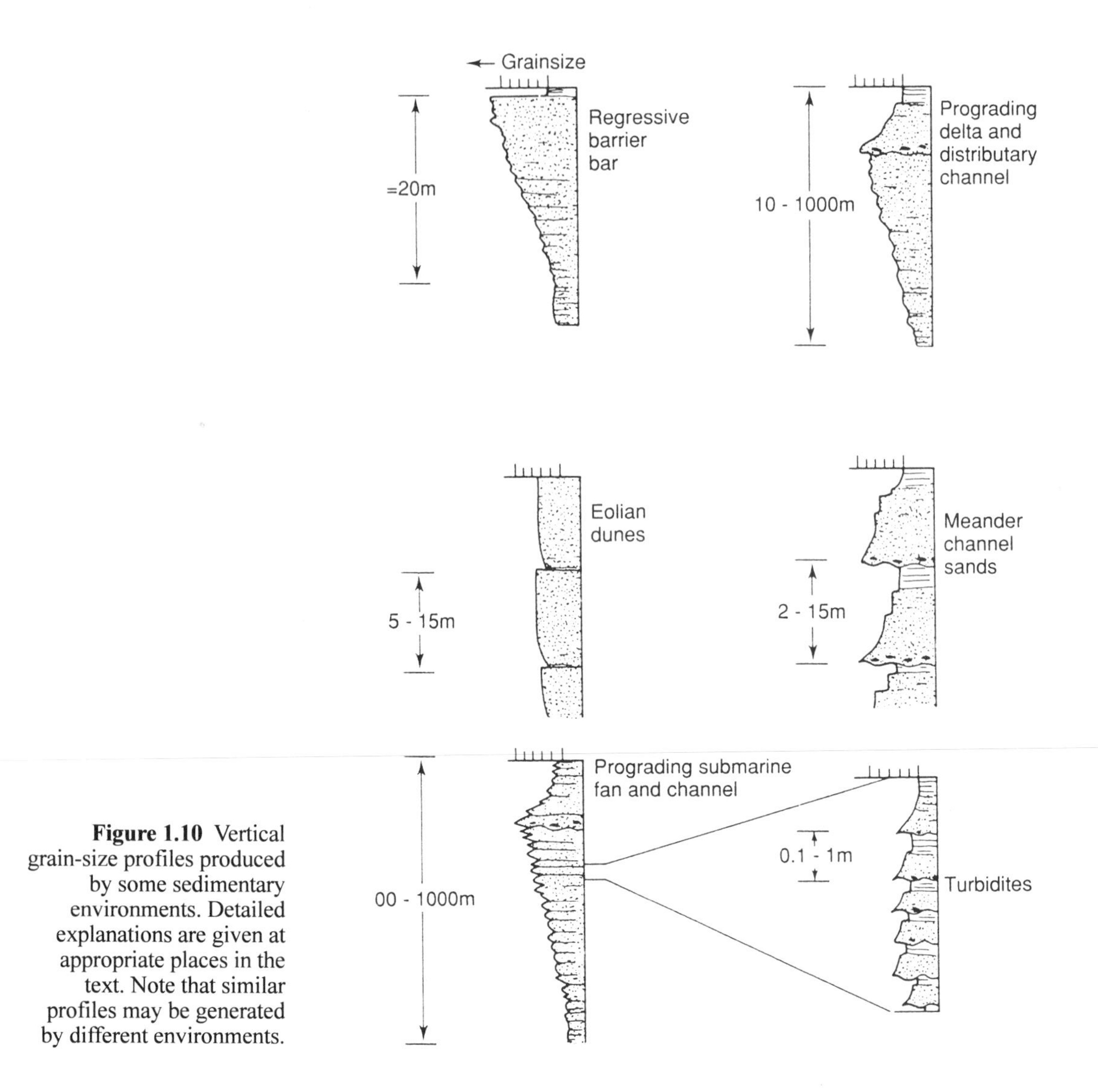

Figure 1.10 Vertical grain-size profiles produced by some sedimentary environments. Detailed explanations are given at appropriate places in the text. Note that similar profiles may be generated by different environments.

Figure 1.11 (opposite) Diagram to illustrate a test for cyclicity in a sedimentary sequence (a) using a simple mathematical device; a transition matrix (b) shows the numbers of transitions from one lithology down to another; (d) shows the number of transitions from one lithology to another expected, were the lithologies randomly deposited; (e) shows the differences between the observed and predicted lithology transitions; (c) and (f) show how lithological transitions may be graphically displayed (from Selley, 1970).

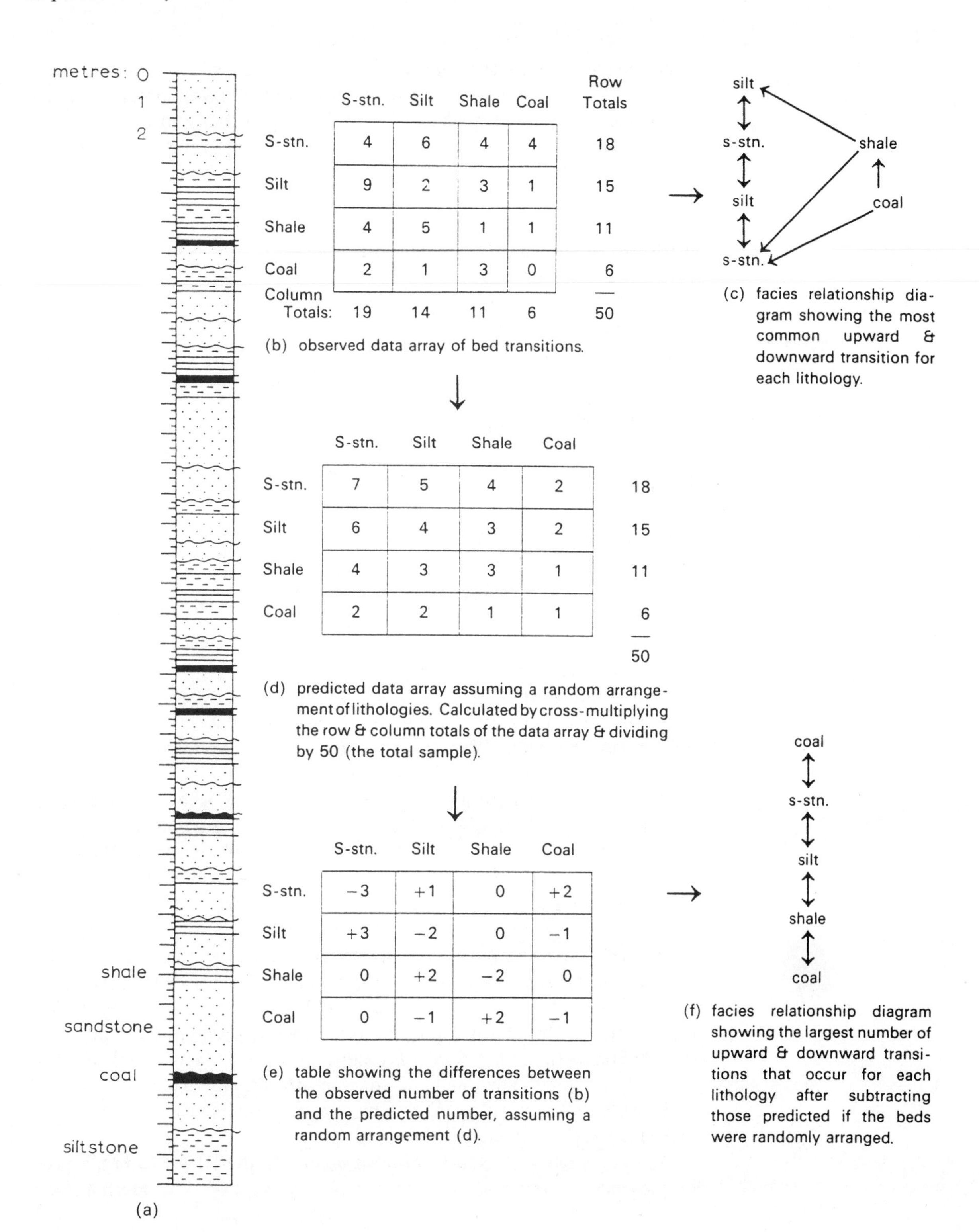

	S-stn.	Silt	Shale	Coal	Row Totals
S-stn.	4	6	4	4	18
Silt	9	2	3	1	15
Shale	4	5	1	1	11
Coal	2	1	3	0	6
Column Totals:	19	14	11	6	50

(b) observed data array of bed transitions.

(c) facies relationship diagram showing the most common upward & downward transition for each lithology.

	S-stn.	Silt	Shale	Coal	
S-stn.	7	5	4	2	18
Silt	6	4	3	2	15
Shale	4	3	3	1	11
Coal	2	2	1	1	6
					50

(d) predicted data array assuming a random arrangement of lithologies. Calculated by cross-multiplying the row & column totals of the data array & dividing by 50 (the total sample).

	S-stn.	Silt	Shale	Coal
S-stn.	−3	+1	0	+2
Silt	+3	−2	0	−1
Shale	0	+2	−2	0
Coal	0	−1	+2	−1

(e) table showing the differences between the observed number of transitions (b) and the predicted number, assuming a random arrangement (d).

(f) facies relationship diagram showing the largest number of upward & downward transitions that occur for each lithology after subtracting those predicted if the beds were randomly arranged.

identified cycles in an artificial rock sequence generated from random numbers in a telephone directory (Zeller, 1964). There is now, however, a range of statistical gymnastic methods for confirming cyclicity in a sedimentary sequence. These range from simple methods that even a field geologist can use (e.g. Selley, 1970), illustrated in Figure 1.11, to techniques at the apogee of the statistician's art (Schwarzacher, 1975).

Studies of modern sedimentary processes show not only that genetic increments are inbuilt into the sedimentary model, but so also sometimes are genetic sequences. The upward-coarsening cyclicity generated by the repeated crevassing and switching of delta distributaries is an example (see p. 134). Cycles due to the sedimentary process are termed 'autocyclic'. Cycles due to processes external to the sedimentary model are termed 'allocyclic' (Beerbower, 1964). Climatic, tectonic and eustatic changes are examples of allocyclic generating mechanisms.

SEISMIC SEQUENCE STRATIGRAPHY AND THE ART OF ZEN

This book is principally concerned with the diagnosis of the depositional environments of ancient sedimentary rocks. Normally such a study is carried out simultaneously with a stratigraphic analysis of the formations studied. It is necessary, therefore, to consider the relationship between facies analysis and stratigraphy.

For over a century fossils provided a vital link between these two pursuits, palaeoecology aiding the environmental interpretation, and biostratigraphy establishing the chronology. Now, however, palaeontology is joined by seismic surveying as a linking tool.

The advent of modern high resolution seismic has lead to a new discipline, known by its advocates as seismic sequence stratigraphy (see Payton, 1977; Schlee, 1984; Berg and Wolverton, 1985; and Wilgus *et al.*, 1988).

Before considering the impact of these geophysical studies on stratigraphy and facies analysis it may be useful to review the story so far (Figure 1.12). Sequences of sedimentary rocks may be delineated by facies, by lithostratigraphy, and by chronostratigraphy. Facies are defined by their geometry, lithology, sedimentary structures, palaeocurrent pattern and palaeontology. Facies may be subdivided in to sub-facies and genetic increments and sequences may be recognized. The lithostratigraphic framework will define mappable rock units arranged in a hierarchy of groups, formations and members. The chronostratigraphy is commonly based on biostratigraphy, and may be used to attribute the rocks to geological systems, stages or series. These units may be divided in to upper, middle and lower, as appropriate.

The foregoing are essentially observational activities. They form the basis for subsequent interpretations. The facies may be studied to interpret their

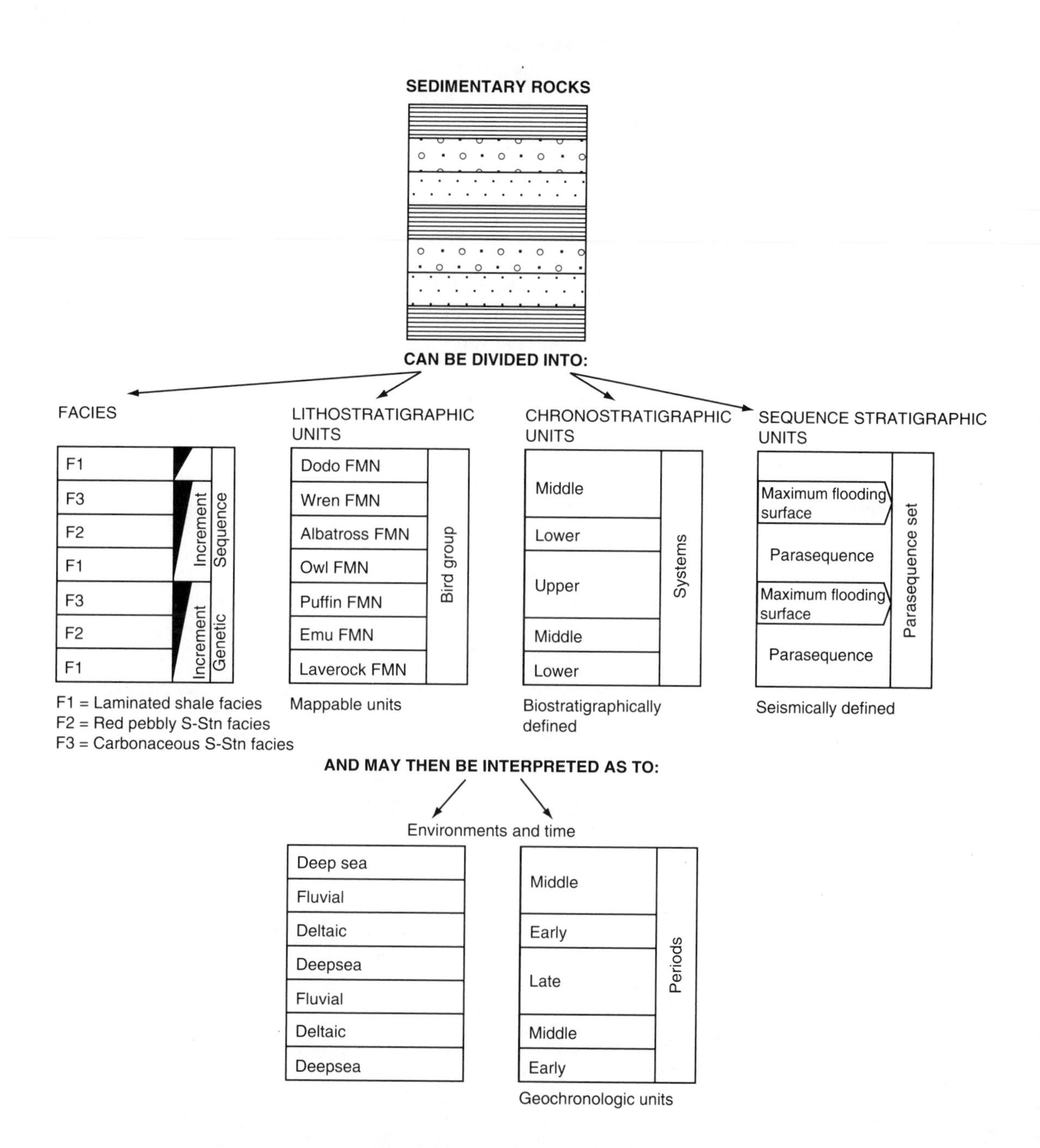

Figure 1.12 Diagram to show how sedimentary rocks may be differentiated by facies, by lithostratigraphy, by chronostratigraphy or by sequence stratigraphy, and may then be interpreted as to environment and time.

environments of deposition. The chronostratigraphy may be interpreted to establish geochronology. Geochronologic units are the intervals of time during which a particular chronostratigraphic unit was deposited. They are arranged in the hierarchy of periods, epochs or ages. Geochronologic units may be subdivided in to early, middle and late, as appropriate. A useful way of distinguishing between chronostratigraphic and geochronologic units is to consider an hourglass, or its miniaturized version, the eggtimer. The sand that flows through the hourglass is the chronostratigraphic unit. The time taken is the geochronologic unit.

Seismic sequence stratigraphy is 'basically a geologic approach to the stratigraphic interpretation of seismic data' (Vail *et al.*, 1977, p. 51). It is based largely on two assumptions: that seismic reflectors are time lines, and that regressions and transgressions are caused by global eustatic changes (Vail *et al.*, 1977). Of fundamental importance is the seismic delineation of depositional sequences. A depositional sequence is defined as a 'stratigraphic unit composed of a relatively conformable succession of genetically related strata, bounded at top and bottom by unconformities or their correlative conformities' (Vail, *ibid.* p. 53). Depositional sequences are bounded by two types of unconformity:

- Type 1 unconformity: subaerial and submarine;
- Type 2 unconformity: subaerial only.

The term Parasequence is now proposed for Busch's Genetic Increment, and Parasequence Set for Genetic Sequence (Van Wagoner *et al.*, 1988; 1990). Table 1.3 shows the hierarchies and terminology of the old established lithostratigraphy and chronostratigraphy, and of the new sequence stratigraphy.

Table 1.3 Comparative hierarchies of sequence stratigraphy, lithostratigraphy and chronostratigraphy, modified from Weimer (1992). Sequence stratigraphic terms defined by Van Wagoner, *et al.* (1998), genetic sequences and increments by Busch (1971), and lithostratigraphic and chronostratigraphic hierarchies by the American Commission on Stratigraphic Nomenclature (1961). Cross hierarchy correlations are only valid where indicated by the = sign.

Sequence stratigraphy	*Lithostratigraphy*		*Chronostratigraphy*
	Informal terms	*Formal terms*	
Sequence	Facies and sub-facies		System
System track		Supergroup	Series
High stand (HST)	= Regressive deposits	Group	Stage
Transgressive (TST)	= Transgressive deposits		
Lowstand (LST)	= Lowstand deposits		
Parasequence set	= Genetic sequence	Formation	Lithochronozone
Parasequence	= Genetic increment	Member	

Many geologists find it hard to accept the two fundamental assumptions of sequence stratigraphy, namely that seismic reflectors are time lines, and that sea-level changes are global and eustatic. These two assumptions deserve scrutiny.

Assumption 1 Seismic reflectors are time lines

The assumption that seismic reflectors are time lines may be questioned both on the large scale of regional unconformities, and on the small scale of stratal surfaces. Hubbard *et al.* (1985) have shown how, in a rift-drift continental margin, sediments of diverse age may be truncated by, and may onlap a single unconformity (Figure 1.13). Several wells penetrate the unconformity, but in every instance the ages of the truncated and onlapping sediments is different. The unconformity cannot, therefore, be considered to be a time line in any meaningful sense.

Moving from regional unconformities to the small scale of stratal surfaces, several authors have shown that lithologic changes below the limits of present seismic resolution also invalidate the assumption that seismic reflectors are time lines. Two particular cases illustrate the problem, onlaps and *en echelon* sand bodies. Thorne (1992) and Cartwright *et al.* (1993) have both pointed out that where beds onlap or downlap onto older beds it is normal for the onlapping or downlapping beds to thin out laterally in the form of condensed sequences that may extend for many kilometres. Yet because the

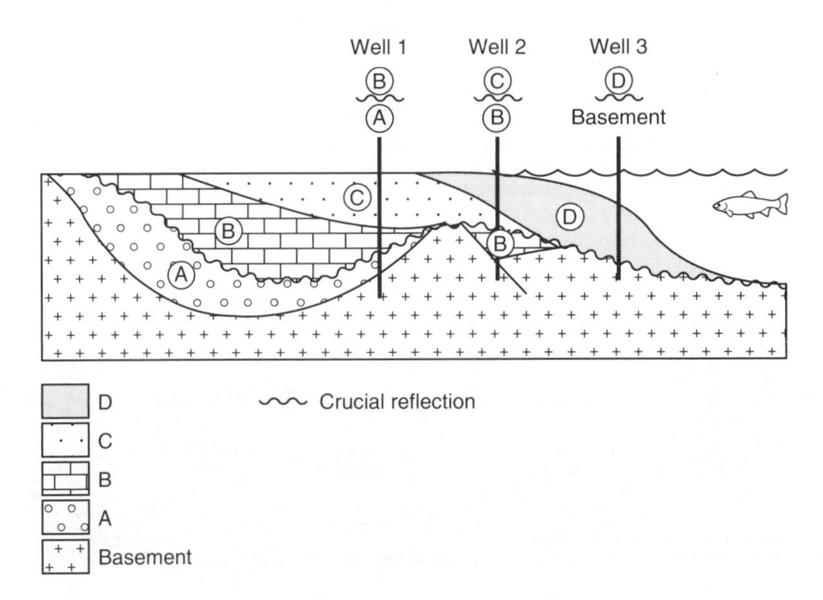

Figure 1.13 Illustration to show how, on a regional scale, a single seismic reflector need not be a time line. Note how, at the well locations, the crucial reflector separates different rock units in every case. The reflector cannot therefore be considered to be a time line (from Hubbard *et al.*, 1985)

seismic method is presently unable to image beds less than a few metres thick, it will be unable to delineate the true lateral extent of the rock units and their concomitant time lines (Figure 1.14).

Thorne (1992) and Tipper (1993) have both also drawn attention to the problem of *en echelon* arranged laterally stacked coastal sands. Individual coastal sand units are seldom more than 20 metres in thickness (see Chapter 7).

(a)

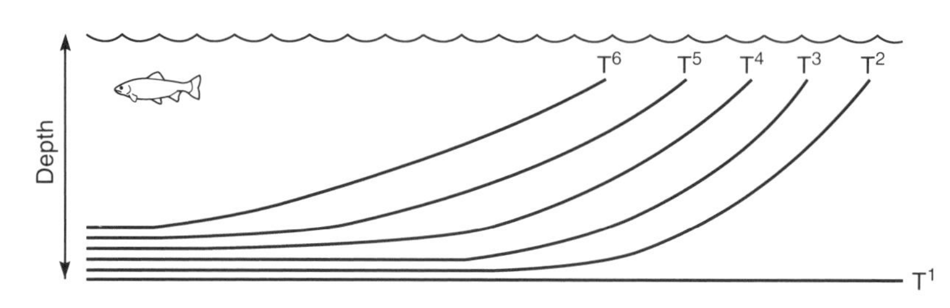

1. Geology: The foresets of a prograding delta extend out over the basin floor as a condensed sequence

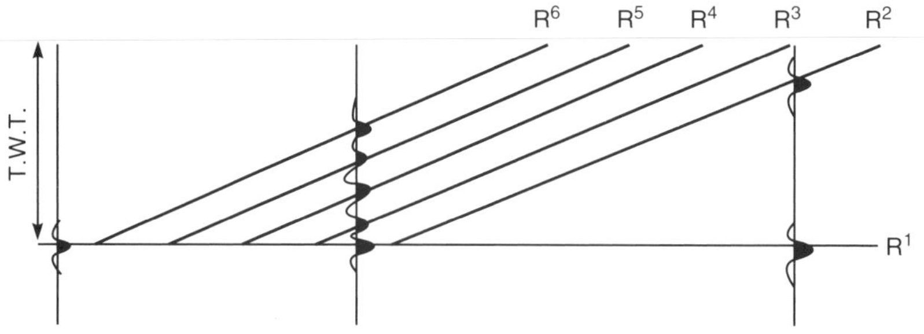

2. Seismic: The foresets appear subsumed on to a single reflector, R^1, and are interpreted as downlap

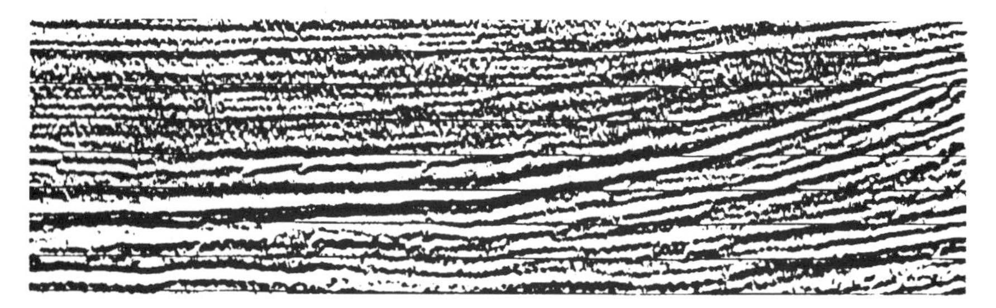

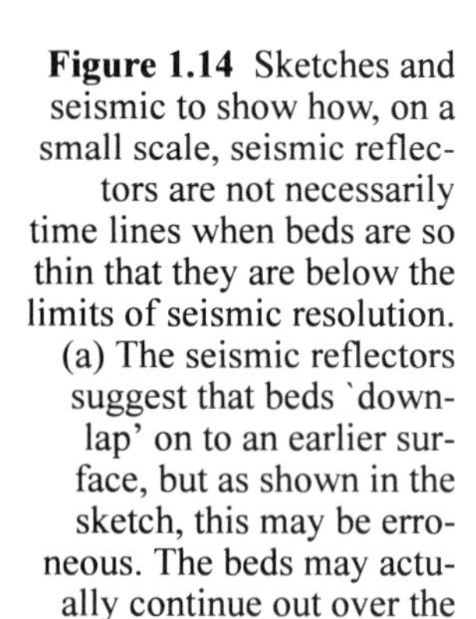

Figure 1.14 Sketches and seismic to show how, on a small scale, seismic reflectors are not necessarily time lines when beds are so thin that they are below the limits of seismic resolution. (a) The seismic reflectors suggest that beds 'downlap' on to an earlier surface, but as shown in the sketch, this may be erroneous. The beds may actually continue out over the basin floor as a condensed sequence.

(b)

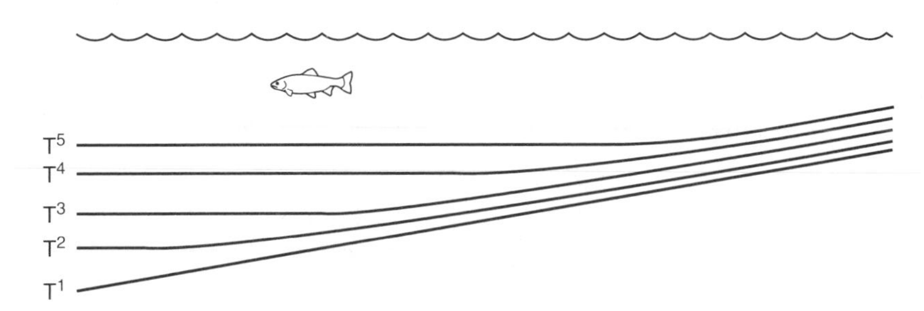

Geology: Basin fill sediments thin up onto a slope in a condensed sequence

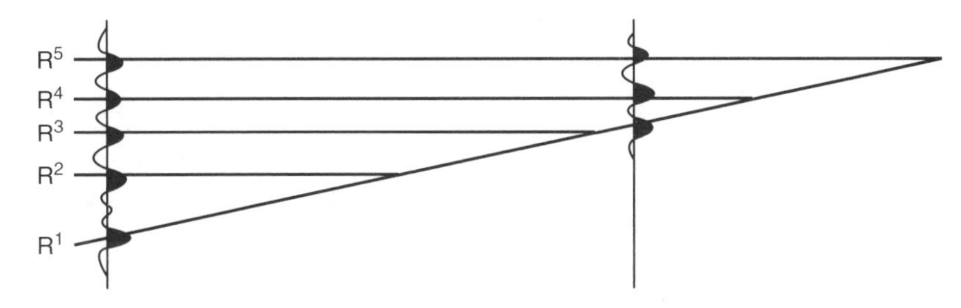

Geophysics: Seismic reflectors R^2 - R^5 apparently onlap reflector R^1

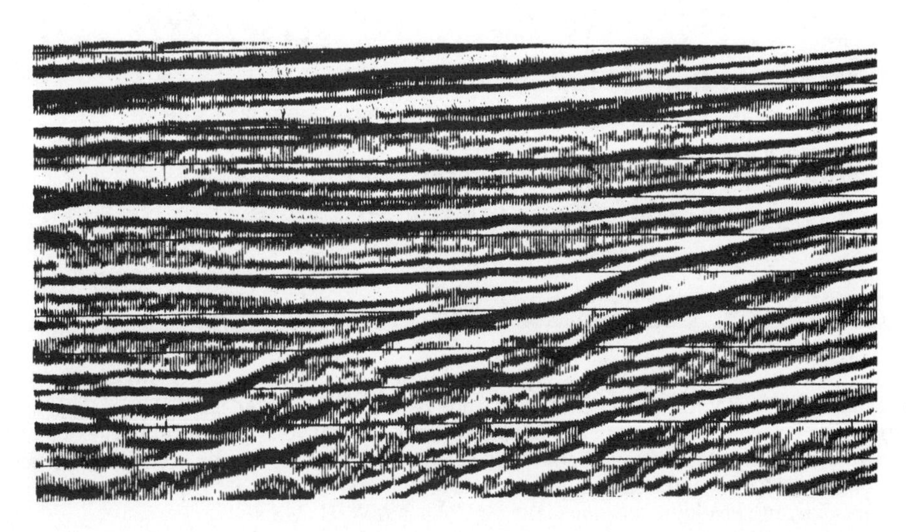

Figure 1.14 (b) The seismic reflectors suggest that beds 'onlap' on to an earlier surface. But, as shown in the sketch, beds may be too thin to be imaged seismically, and may actually extend laterally in a condensed sequence.

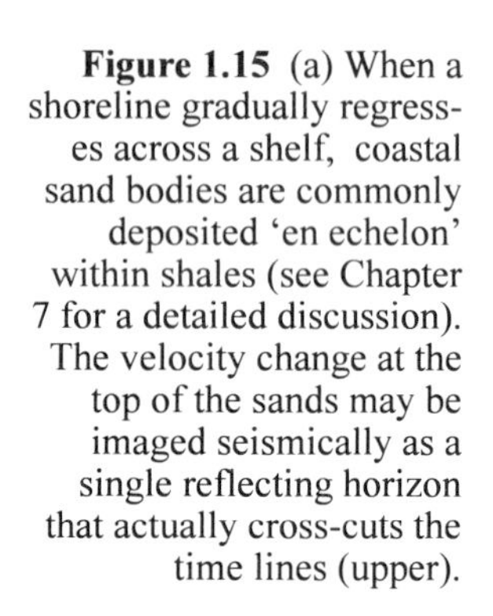

Figure 1.15 (a) When a shoreline gradually regresses across a shelf, coastal sand bodies are commonly deposited 'en echelon' within shales (see Chapter 7 for a detailed discussion). The velocity change at the top of the sands may be imaged seismically as a single reflecting horizon that actually cross-cuts the time lines (upper).

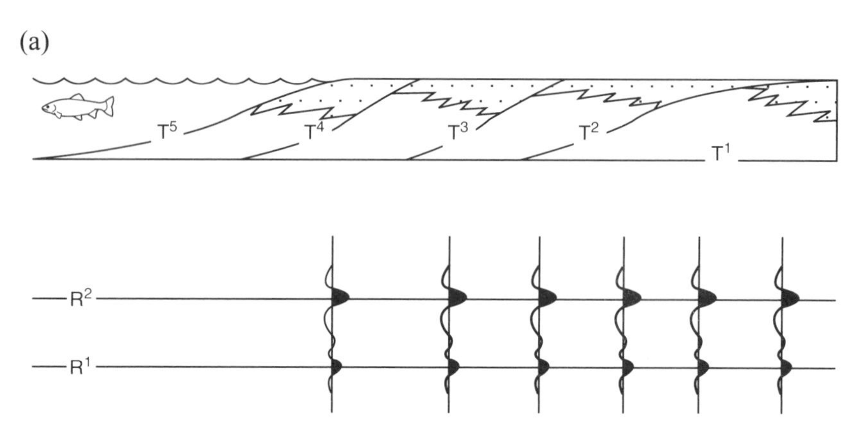

(b)

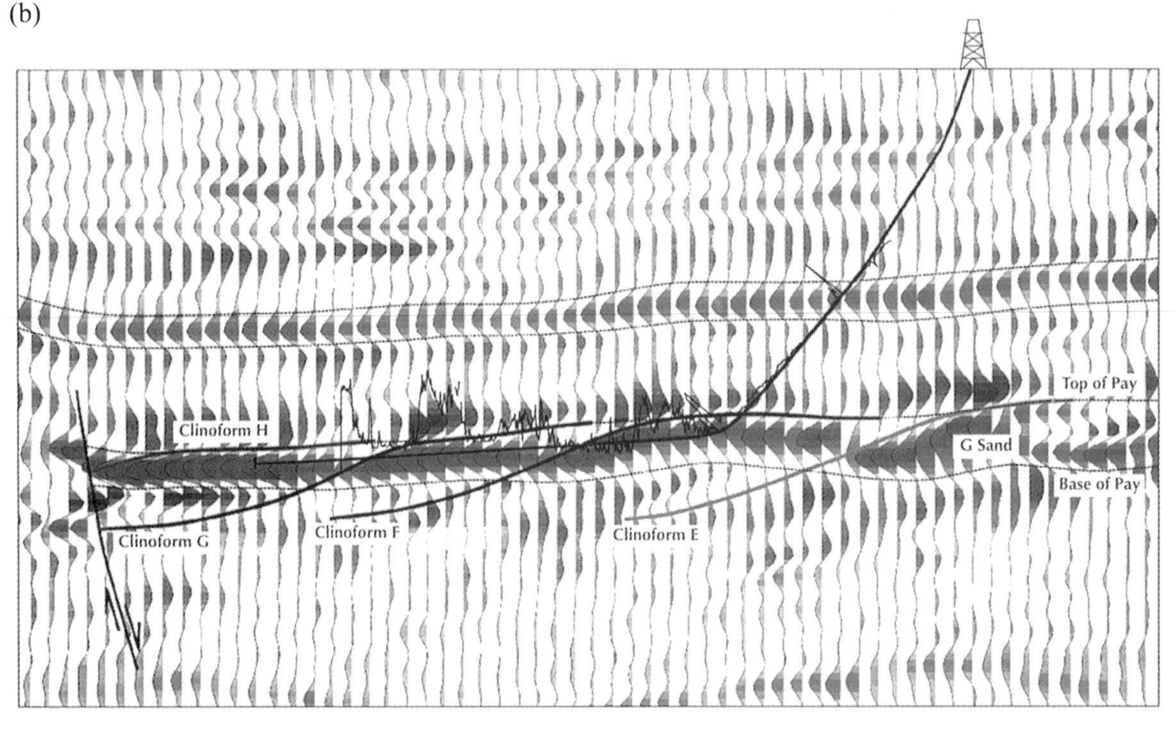

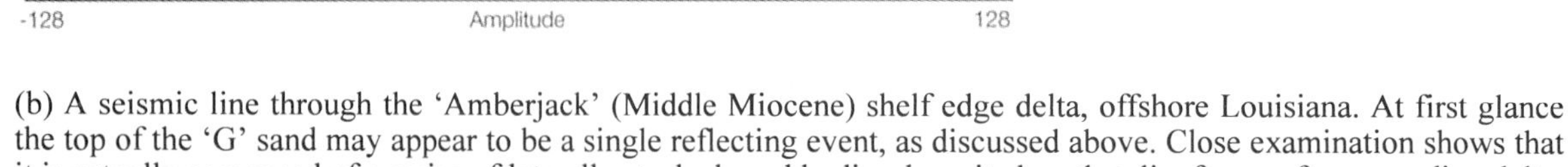

(b) A seismic line through the 'Amberjack' (Middle Miocene) shelf edge delta, offshore Louisiana. At first glance the top of the 'G' sand may appear to be a single reflecting event, as discussed above. Close examination shows that it is actually composed of a series of laterally stacked sand bodies deposited on the clinoforms of a prograding delta. Note the track of a well, carefully drilled to penetrate the various reservoirs of the 'G' sand horizontally. Courtesy of BP Gulf of Mexico Mississippi Canyon Development, especially B. Cohn and W. H. Mills.

This may well be below the limit of seismic resolution. The sand-bearing zone may be delineated, and this may be economically useful information to have, but it is clear from Figure 1.15 that the seismic reflector cross-cuts time lines.

There are therefore, numerous instances, on scales large and small, to show that seismic reflectors need not be time lines. It is, perhaps, only the horizontal reflectors of pelagic basin floor sediments, and the sigmoidal reflectors of prograding coastal wedges that can be assumed to be time lines with any degree of confidence. But this assumption cannot be extended to the sequence boundaries above and below (Figure 1.16).

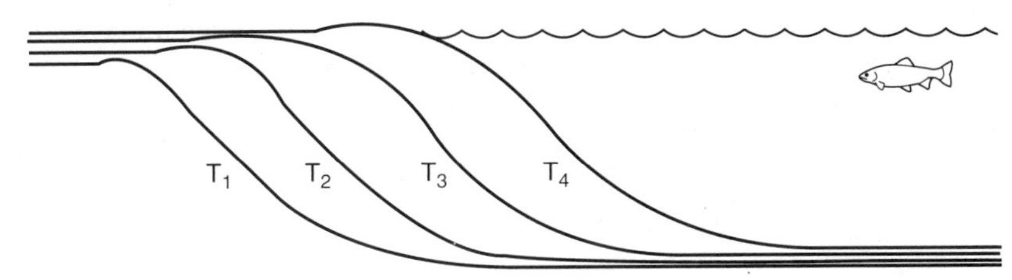

1. Geology: Prograding delta coastline. The topsets illustrated here are sensitive to sea level changes and thus the top of the progrades may sometimes offlap or be truncated.

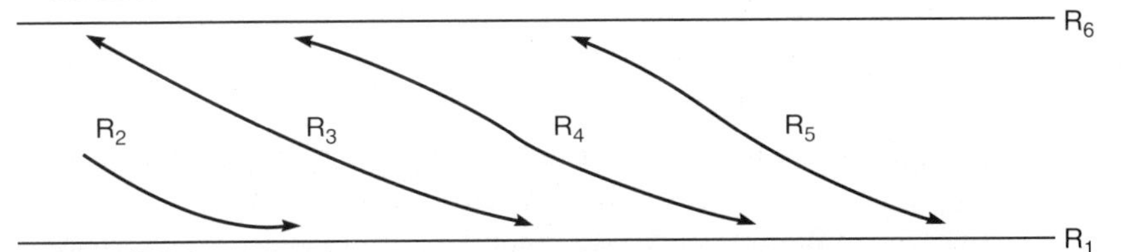

2. Geophysics: R_2, R_3, R_4 and R_5 may well be time lines. R_1 probably subsumes condensed toesets of T_1–T_4. R_6 subsumes top set beds, offlap and truncation events.

Figure 1.16 There are some instances where seismic reflectors may be synchronous, notably where a delta progrades, but note the problems of down lap and onlap illustrated in Figure 1.14.

Assumption 2 Sea-level changes are global and eustatic climatic

The second crucial assumption of seismic sequence stratigraphy is that shifts in the positions of shorelines reflect global changes in sea level.

The stratigraphic record contains abundant evidence of shifting shorelines. Once upon a time these were termed transgressions, if the sea advanced over the land, and regressions, if the land advanced over the sea. Now these are referred to as 'high stands' and 'low stands' respectively.

The second important tenet of seismic sequence stratigraphy is that advances and retreats of shorelines can be correlated globally, and that they are caused by eustatic changes in sea level. Sea level curves have been published by Vail *et al.* (1977) and Haq *et al.* (1988) (Figure 1.17).

These purport to show, not only how sea level has varied globally with time, but also that there is a periodicity to the fluctuations. Indeed it is argued that there are at least three orders of cycles, with periodicities of 200–300 my, 10–80 my and 1–10 my. Enthusiasts also argue for fourth- and fifth-order cycles, with frequencies down to 0.5–0.45 my. Driving mechanisms for these cycles may include changes in the rate of sea-floor spreading, the waxing and waning of polar ice in response to global warming and cooling, and so forth. It would be comfortable to identify driving mechanisms for cycles. This is quite separate from establishing whether such cycles exist. As observed earlier in this chapter, when discussing cyclic sedimentation, we are conditioned to find patterns in random data. Thus Miall (1992) found a 77% correlation between randomly generated stratigraphic sequences and the Exxon global sea-level cycle chart.

There is plenty of evidence for global climatic change, and evidence for global changes in sea level. Every geologist knows, however, that sea level varies at a point in response to:

1. global changes in ocean water volume – probably driven by climatic change;
2. local tectonic uplift and subsidence of the crust – probably driven by plate movements;
3. local changes in the availability of sediment supply – probably driven by climatic change, changing rates of sediment runoff, and the initiation and extinction of drainage systems.

Sadly it is normally impossible to determine which of these mechanisms, singly or together, may be responsible for a particular sea-level change. There is obviously no universal bench mark or Plimsoll Line against which global changes in sea level may be calibrated, and which may be used therefore to differentiate the various processes of sea-level change (Figure 1.18).

Thus the assumptions that seismic reflectors are time lines, and that sea-level change is global and cyclic can be challenged. This does not negate the stimulating effect that seismic sequence stratigraphic analysis has had on sedimentary geology. In their enthusiasm seismic stratigraphers have often ignored well-established geological terminology and principles. Their extremists happily draw sequence boundaries across palaeontologically defined biostratigraphic zones, and through sedimentologically defined genetic increments. Whole oil company exploration groups forget that they are employed to find drillable prospects, to debate emotively whether a particular seismic reflector is a Type 1 or Type 2 unconformity.

None the less seismic sequence analysis has stimulated geologists to look much more closely at seismic data, and to re-evaluate old established stratigraphic and sedimentological principles. Perhaps one of the most important contributions of sequence stratigraphy has been to focus attention on how a drop in sea level exposes a continental shelf, favouring the deposition of shelf-derived deep sea sands at the foot of the shelf slope (p. 290).

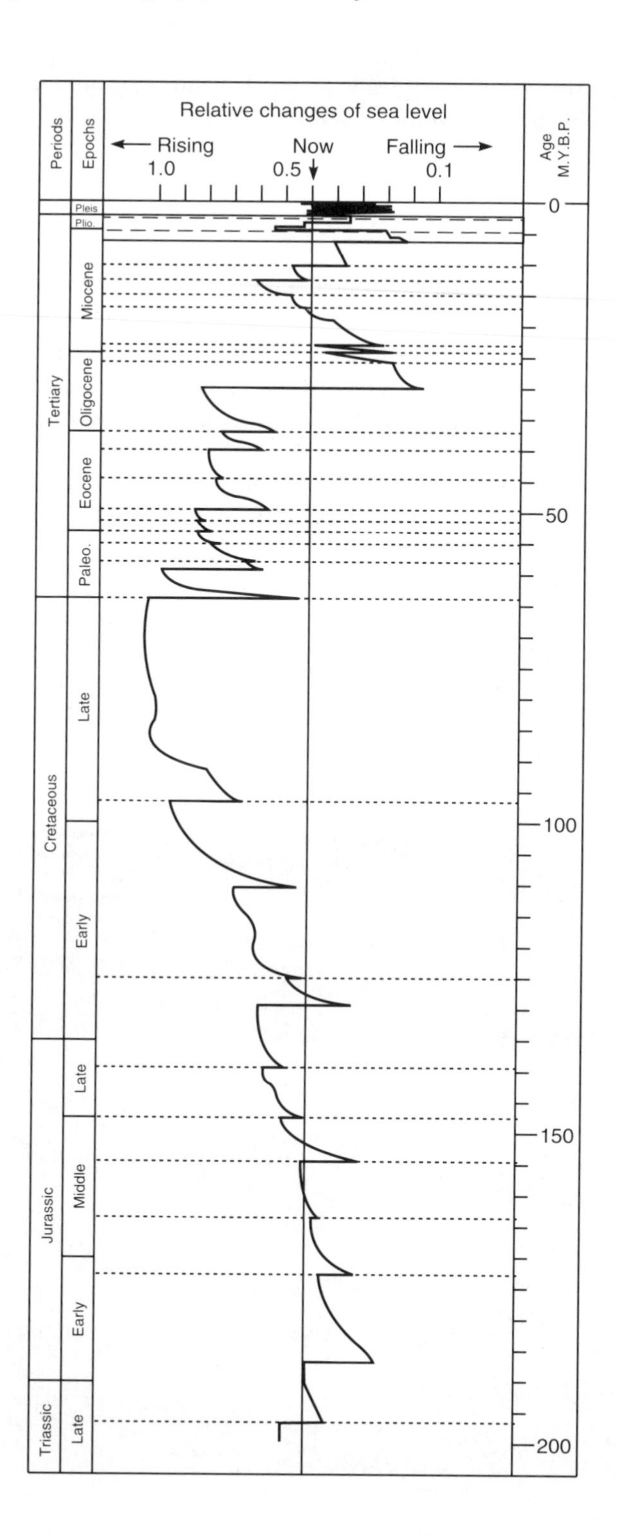

Figure 1.17 A global sea level curve for the last 200 million years, or so. For sources see Vail *et al.* (1977) and Haq *et al.* (1988).

Conversely a rise in sea level leads to the deposition of organic rich shales over shelf-wide unconformities (p. 290).

SUMMARY AND CONCLUSIONS

This chapter has discussed the fundamental geological concepts on which the interpretation of the depositional environments of sedimentary rocks are based, so it may be appropriate to close it a with a summary and synthesis.
A sedimentary environment is a part of the earth's surface that is physically, chemically and biologically distinct from adjacent areas. Environments may be erosional, equilibrial or depositional. Depositional sedimentary environments generate sedimentary facies. A sedimentary facies is a mass of rock defined and differentiated from others by its geometry, lithology, sedimentary structures, palaeocurrent pattern and palaeontology.

Each depositional sedimentary environment can be characterized by a sedimentary model. Most depositional environments migrate across the face of the earth generating a characteristic sedimentary sequence, often with a characteristic, but seldom unique, vertical profile of grainsize. This sequence is termed a genetic increment. Sedimentary sections tend to be made up of genetic sequences composed of genetic increments generated by the same sedimentary model.

Figure 1.18 Geologist L. Baron checking for evidence of global sea level change, courtesy of A. Gale.

It follows on from the concept of laterally prograding environments that conformable rock-unit boundaries are commonly diachronous. Since the seismic method images tangible contrasts in velocity between different rock types, rather than intangible time, the assumption that seismic reflectors are time lines should be treated with care.

Stratigraphic evidence for transgressions and regressions may be due to local changes in tectonics or sediment supply, as well as to global eustatic sea-level changes. It may be dangerous to assume that only the last of these is responsible and ubiquitous. From Galileo to Ernest Hemingway it has been known that the earth moves – '*Eppur si muove*'.

REFERENCES

Ager, D. V. (1993) *The Nature of the Stratigraphical Record*, 3rd edn, Wiley, Chichester.

Anderton, R. (1985) Clastic facies models and sedimentary facies analysis. In P. J. Brenchley and B. P. J. Williams (eds), *Sedimentology Recent Developments and Applied Aspects*, London Geological Society, 31–48.

Beerbower, J. R. (1964) Cyclothems and cyclic depositional mechanisms in alluvial plain sedimentation, *Bulletin of Kansas University Geological Survey*, **169**, 35–42.

Berg, R. E. and Wolverton, D. G. (1985) *Seismic Stratigraphy II: An Integrated Approach*, American Assoc. Petrol. Geol., **39**.

Blatt, H., Middleton, G. and Murray, R. (1980) *Origin of Sedimentary Rocks*, 2nd edn, Prentice-Hall, New Jersey.

Bolviken, E., Storvik, G., Nilsen, D. E., Siring, E. and Well, D. V. D. (1992) Automated prediction of sedimentary facies from wireline logs. In A. Hurst, C. M. Griffiths and P. F. Worthington (eds), *Geological Aspects of Wireline Logs*, Vol. II, Geological Society Special Publication, **65**, 123–40.

Brown, L. F. and Fisher, W. L. (1977) Seismic-stratigraphic interpretation of depositional systems: examples from Brazilian rift and pull-apart basins. In C. E. Payton (ed.), *Seismic Stratigraphy Applications to Hydrocarbon Exploration*, American Assoc. Petrol. Geol., **26**, 213–48.

Busch, D. A. (1971) Genetic units in delta prospecting, *Bulletin of American Assoc. Petrol. Geol.* **55**, 1137–54.

Cartwright, J. C., Haddock, R. C. and Pinheiro, L. M. (1993) The lateral extent of sequence boundaries. In *Tectonics and Seismic Sequence Stratigraphy*, London Geological Society Special Publication, **71**, 15–34.

Goldring, R. (1991) *Fossils in the Field*, Longman, Singapore.

Gressly, A. (1838) Observations géologiques sur le Jura Soleuroi, *Neue Denskschr. allg. schweiz. Ges. ges. Naturw.* **2**, 1–112.

Haq., B. U., Hardenbohl, J. and Vail, P. (1988) Mesozoic and Cenozoic chronostratigraphy and eustatic cycles. In C. K. Wilgus, H. Posamentier, C. A. Ross and C. G. St C. Kendall (eds), *Sea-level Changes: An Integrated Approach*, Soc. Econ. Pal. Min. Sp. Pub. **42**, 71–108.

Hubbard, R. J., Pape, J. and Roberts, D. G. (1985) Depositional sequence mapping as a technique to establish tectonic and stratigraphic framework and evaluate hydrocarbon potential on passive continental margin. In R. E. Berg and D. G. Wolverton (eds), *Seismic Stratigraphy II: An Integrated Approach*. American

Assoc. Petrol. Geol., **39**, 79–92.
Khayyam, O. (*c.* 1100) *The Rubayat*, English translation by E. Fitzgerald, W. Brendon, London.
Lyell, Sir C. (1865) *Elements of Geology*, John Murray, London.
Miall, A. D. (1992) Exxon global cycle chart: an event for every occasion?, *Geology*, **20**, 787–90.
Middleton, G. V. (1973) Johannes Walther's Law of the Correlation of Facies, *Bull. Geol. Soc. Amer.*, **84**, 979–88.
Mojsisovics, M. E. von (1879) *Die Dolomit-Riffe Von Sud Tirol und Venetien*, A. Holder, Vienna.
Moore, R. C. (1949) Meaning of facies, *Bull. Geol. Soc. Amer.*, **39**, 1–34.
North American Code of Stratigraphic Usage, 1982 (1983) *Bull. Amer. Assoc. Petrol. Geol.*, **67**, 841–75.
Payton, C. E. (1977) *Seismic Stratigraphy – Application to Hydrocarbon Exploration*, Amer. Assoc. Petrol. Geol., Mem. No. **2**.
Potter, P. E. and Pettijohn, F. J. (1977) *Paleocurrents and Basin Analysis*, 2nd edn, Springer-Verlag, Berlin.
Prevost, C. (1838) *Bull. Soc. Geol. France*, **9**, 90–5.
Reading, H. G. (ed.) (1986) *Sedimentary Environments and Facies*, 2nd edn, Blackwell, Oxford.
Rider, M. H. and Laurier, D. (1979) Sedimentology using a computer treatment of well logs, *Trans. Soc. Prof. Well Log Analysts*, 6 European Symposium, London, Paper J.
Schlee, J. S. (1984) *Interregional Unconformities and Hydrocarbon Accumulation*, Amer. Assoc. Petrol. Geol., **36**.
Scholle, P. A., Bebout, D. G. and Moore, C. H. (eds) (1983) *Carbonate Depositional Environments*, Amer. Assoc. Petrol. Geol. Bull., Mem. No. **33**, Tulsa.
Scholle, P. A. and Spearing, D. R. (eds) (1982) *Sandstone Depositional Environments*, Amer. Assoc. Petrol. Geol., Mem. No. **31**, Tulsa.
Schwarzacher, W. (1975) *Sedimentation Models and Quantitative Stratigraphy*, Elsevier, Amsterdam.
Selley, R. C. (1970) Studies of sequence in sediments using a simple mathematical device, *Quart. Jnl. Geol. Soc. London*, **125**, 557–8.
—— (1988) *Applied Sedimentology*, Academic Press, London.
Serra, O. and Abbott, H. T. (1980) *The Contribution of Logging Data to Sedimentology and Stratigraphy*, Soc. Pet. Eng. Paper 9270.
Shakespeare, W. (1623) *The Merchant of Venice*.
Teichert, C. (1958) Concept of facies, *Bull. Am. Ass. Petrol. Geol.*, **42**, 2718–44.
Thorne, J. A. (1992) An analysis of the implicit assumptions of the methodology of seismic sequence stratigraphy. In J. S. Watkins, F. Zhiqiang and K. J. McMillan (eds), *Geophysics of Continental Margins*, Amer. Assoc. Petrol. Geol., Mem. No. **53**, 375–394.
Tipper, J. C. (1993) Do seismic reflectors necessarily have chronostratigraphic significance?, *Geol. Mag.*, **130**, 47–55.
Vail, R., Mitchum, R. M., Todd, R. G., Widmier, J. M., Thompson, S., Sangree, J. B., Bubb, J. N. and Hatledid, W. G. (1977) Seismic stratigraphy and global changes in sea level. In C. E. Payton (ed.), *Seismic Stratigraphy – Application to Hydrocarbon Exploration*, Amer. Assoc. Petrol. Geol., Mem. No. **26**, 49–212.
Van Wagoner, J. C., Mitchum, R. M., Campion, K. M. and Rahmanian, V. D. (1990) *Siliciclastic Sequence Stratigraphy in Well Logs, Cores and Outcrops*, Amer. Assoc. Petrol. Geol. Tulsa, Methods in Exploration Series 7.
Van Wagoner, J. C., Posamentier, H. W, Mitchum, R. M., Vail, P. R., Sarg, J. F., Loutit, T. S. and Hardenbol, J. (1988) An overview of the fundamentals of

sequence stratigraphy and key definitions. In C. K. Wilgus, B. S. Hastings, H. Posamentier, C. A. Ross and G. St C. Kendall (eds), *Sea-level Changes: An Integrated Approach*, Soc. Econ. Min. Pal. Sp. Pub. **42**, 39–46.

Walker, R. G. (ed.) (1984) *Facies Models*, Geoscience Canada Reprint Series 1, Geol. Ass. Canada, Toronto.

Walther, J. (1893) *Eileitung in die Geologie als Historische Wissenschaft band 1. Beobachtun uber die Bildung der gesteine ind ihter organischen Einschlusse*, G. Fischer, Jena.

Weimer, R. J. (1992) Developments in sequence stratigraphy: foreland and cratonic basins, *Amer. Assoc. Petrol. Geol. Bull.* **76**, 965–82.

Wilgus, C. K., Posamentier, H., Ross, C. A. and Kendall, C. G. St C. (eds) (1988) Sea-level changes: an integrated approach, *Soc. Econ. Pal. and Min. Sp. Pub.*, **42**.

Zeller, E. J. (1964) Cycles and psychology, *Geol. Surv. Kansas Bull.*, **169**, 631–6.

2 Methods of environmental interpretation

INTRODUCTION

Chapter 1 reviewed fundamental concepts of stratigraphy, facies and environments. The object of Chapter 2 is to describe the various techniques available to interpret the depositional environment of sedimentary rocks. It will quickly be realized that many diagnostic techniques are available, some are only suitable for studies where sediments crop out at the surface, others are only suitable for subsurface studies, while a few techniques are applicable to both. In the 25 years since the first edition of this book appeared methods of surface studies have changed little, but subsurface methods have undergone major changes due to a revolution in geophysics, both in seismic surveying and in wireline logging. The technicalities of the various methods will not be described here, but it is assumed that readers will be attending, or have attended, courses in sedimentology, seismic geophysics and wireline logging. If not, do not panic, the following text is presented in such a way as to make it accessible to all readers, even engineers.

Chapter I showed how facies may be usefully defined according to the five parameters of geometry, lithology, palaeontology, sedimentary structures and palaeocurrents. This provides a convenient and logical way of arranging the material in this chapter.

GEOMETRY

The external shape of a sedimentary facies is a function of the depositional topography, the syndepositional geomorphology of the environment, and post-depositional history. For example, the geometry of a blanket sand burying an old land surface may be a function of the erosive process and its geomorphic stage which previously sculpted the land surface. This may be true whether the sand is continental or marine in origin. Likewise post depositional compaction, erosion and tectonism may all, singly or together, so modify the geometry of a facies as to render it non-diagnostic.

Figure 2.1 illustrates some of the more important facies geometries. This shows that some shapes are produced by several different environments. For example, channels may be fluvial, deltaic, tidal, or submarine. Similarly fans may be alluvial, deltaic or submarine. All too often facies appear to have a blanket geometry. This may appear, at first, to have little diagnostic value. Detailed study may show, however, that the blanket is composed of genetic increments of laterally stacked sand bodies in the forms of channels, shoestrings or other smaller bodies.

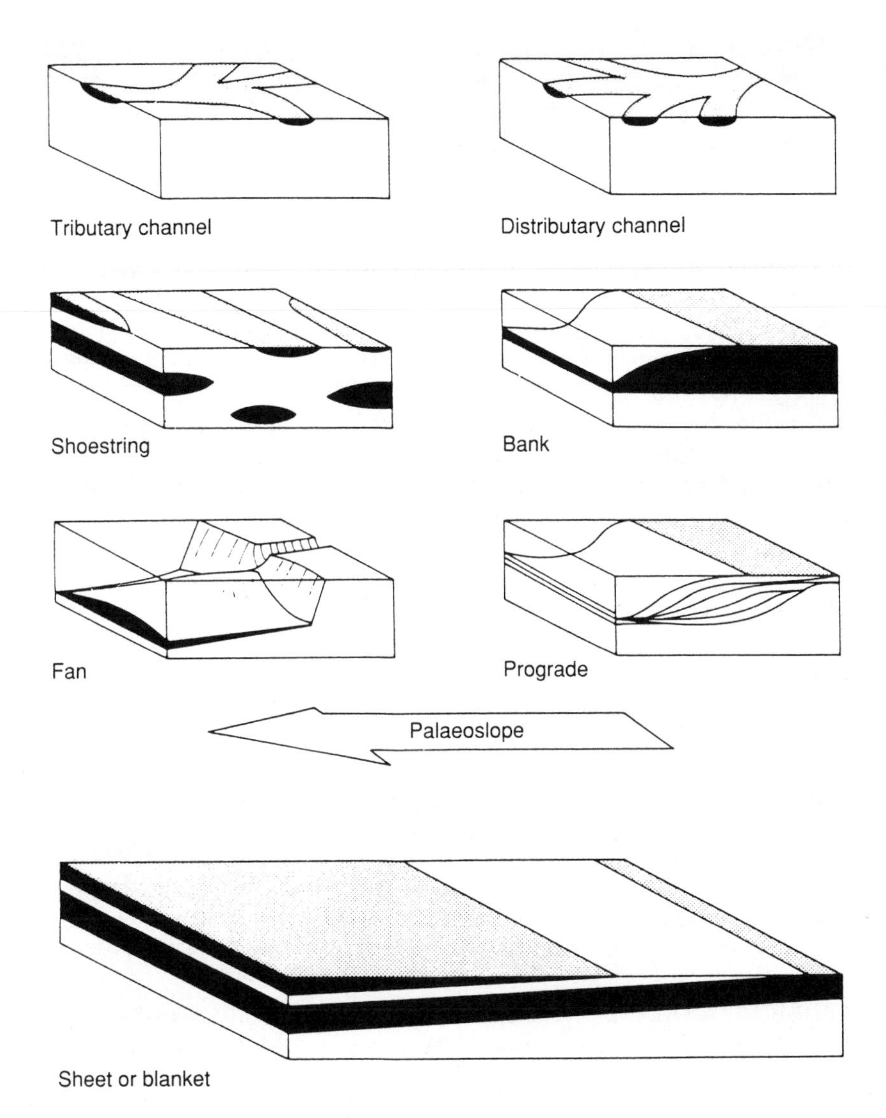

Figure 2.1 Cartoons of various facies shapes. Note that shape alone may not be diagnostic. Channels and fans occur in a range of environments from fluvial to deep sea. Shoestring sands may be beaches, barrier islands or tidal sand waves.

In surface studies the geometry of a facies is exhumed, and can be mapped on the ground or from the air. Examples of exhumed reefs and channels are illustrated elsewhere in the text. In the subsurface different methods must be employed. In the olden days the geometry of a facies was mapped out using borehole data. In the oil industry subsurface mapping was, and remains, a crucial exercise, both as a means of predicting where to drill additional wells, and to calculate petroleum reserves. Often the geometry of a reservoir facies was not resolved, and could not therefore be used to interpret the depositional environment and to predict future well locations, until the field had been almost completely delineated.

Nowadays all that has changed, thanks to geophysics. At very shallow depths ground probing radar (GPR) can be used to delineate the geometry of facies to depths of several metres. This is very useful for shallow site investigations (Stephens, 1994). In the deeper subsurface 2D, and more especially 3D, seismic surveys can delineate the geometry of facies, and lead to the interpretation of its depositional environment, before a single well has been drilled. It is important to note that when seismic is that good, it is unnecessary actually to bother to interprete the environment of deposition. It then becomes a matter of purely academic interest, since the main reason why petroleum geologists are interested in environmental intepretation is as a means of predicting reservoir geometry. If seismic can do that instantaneously, who needs sedimentology?

Detailed discussion of the seismic method is outside the scope of this book, and the reader is referred to the standard texts. An account of how the seismic method is used in subsurface facies analysis is relevant, however, and will be discussed in some detail.

Seismic stratigraphy has developed as three discrete, though interrelated, disciplines: seismic sequence analysis, seismic facies analysis, and seismic attribute analysis (Pacht *et al.*, 1993) (Table 2.1).

Table 2.1 Summary of the parameters examined and interpretations made for the three-fold hierarchy of seismic stratigraphy.

SEISMIC ANALYSIS	PARAMETER STUDIED	INTERPRETATION
SEQUENCE ANALYSIS	Sequences defined and mapped. Nature of sequence contacts recorded.	Sequences correlated with global sea level changes (if possible), and time calibrated.
FACIES ANALYSIS	Description of sequence-defined seismic character	Recognition of channeled, mounded and pelagic sediments.
ATTRIBUTE ANALYSIS	Analysis of wave shape, amplitude polarity and continuity	Identification of vertical changes in rock properties, including grain-size profiles, and direct hydrocarbon indicators.

Seismic sequence analysis is concerned with the large scale interpretation of seismic data and the mapping of sequence boundaries (Figure 2.2). Attempts are then made to correlate sequence boundaries with geological time and global changes in sea level. Seismic sequence analysis has largely been dealt with in Chapter 1. It also includes, of course, the mapping of facies boundaries, and thus is the source of information that allows the geometry of a facies to be used in environmental interpretation. For example the facies geometry delineated in Figure 2.3 may be used to interpret the depositional environment as a submarine fan. Other examples, such as channels, deltas and reefs, will be illustrated in subsequent chapters.

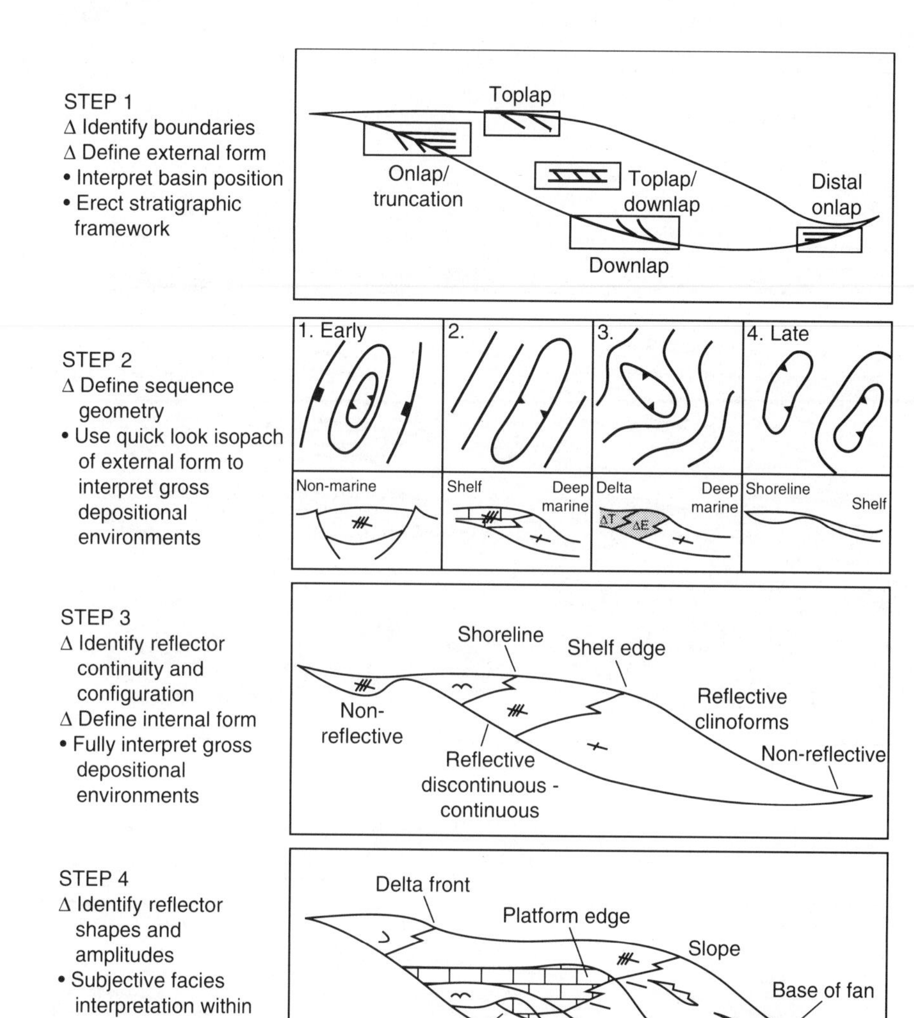

Figure 2.2 Illustration to show some of the terminology applied to sequence boundaries in seismic stratigraphy, and the steps involved in seismic sequence analysis. Based on Hubbard *et al.* (1985).

Seismic facies analysis consists of the delineation and interpretation of packages of seismic reflections (Mitchum *et al.*, 1977). The internal parameters include the continuity and amplitude of reflectors, and their phase, frequency and interval velocity. Figure 2.4 shows examples of some of the different internal reflection types that can be recognised, and their possible interpretations.

Seismic attribute analysis also may be used to infer vertical profiles of grainsize in clastic sequences (Figure 2.5). As discussed in Chapter 1, vertical grainsize profiles are important aids in environmental interpretation.

Thus the seismic method is a very important tool for diagnosing the depositional environment of a sedimentary facies. In many cases, it may be all

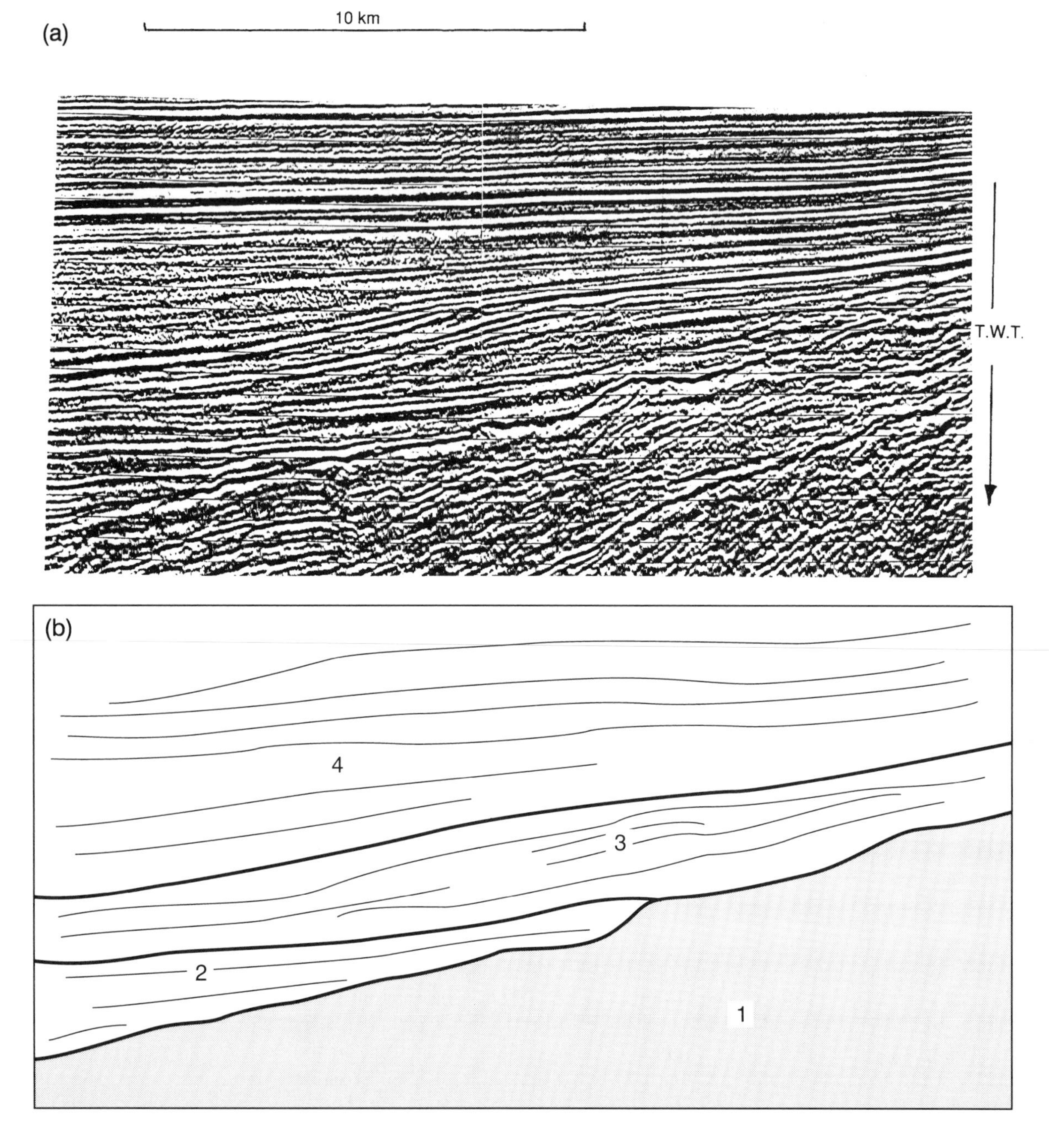

Figure 2.3 An example of seismic sequence analysis. (a) Seismic line and (b) geological interpretation. Basinal shales (2.) unconformably overlie basement (1.). A submarine fan (3.) prograded out over the basinal shales from right to left, and is in turn overlain by later basinal shales (4.). (Enthusiasts may spot apparent 'onlaps' and 'downlaps', but remember the warnings about identifying such phenomena given in Chapter 1.)

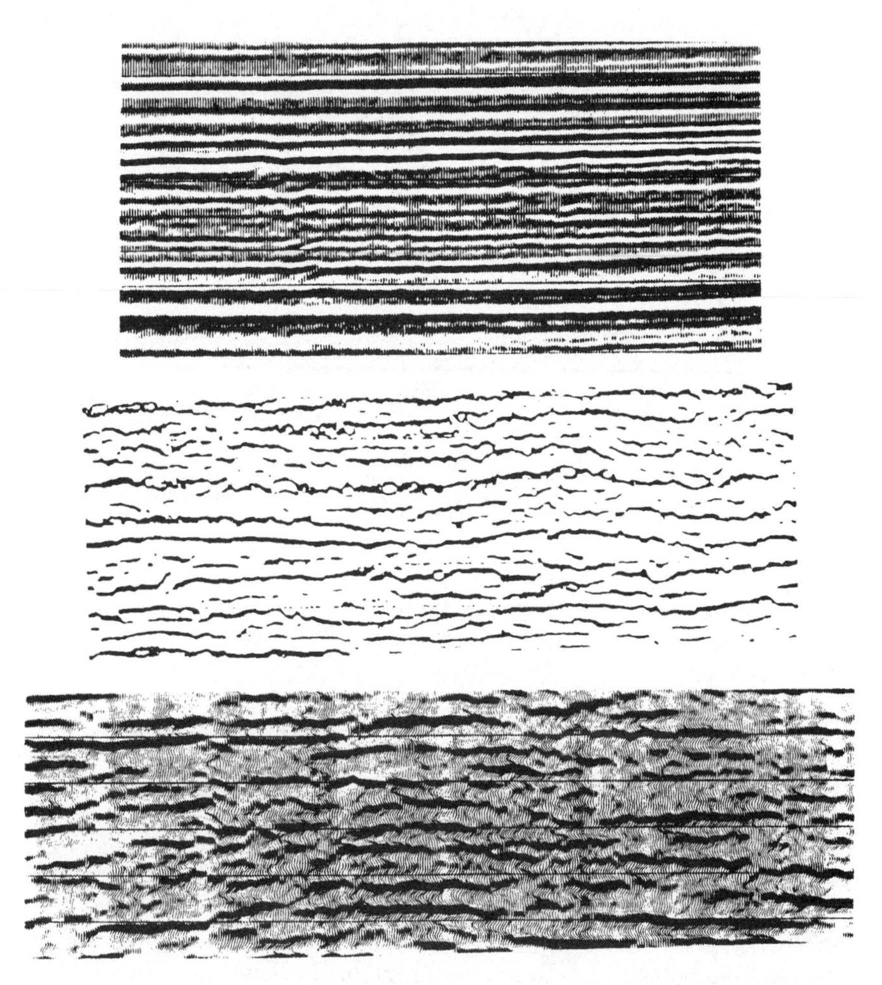

Figure 2.4 Fragments of seismic lines to illustrate seismic facies analysis. Upper: continuous high amplitude reflectors suggest basinal shales with interbedded cherts or pelagic limestones. Middle: hummocky mounded reflectors suggest submarine fans. (Bizarre mode of display is because the horizontal scale has been exaggerated to enhance the hummocky nature of the reflectors.) Lower: short high amplitude reflectors suggest channels in alluvium.

that is required. The seismically mapped shape may make the environmental interpretation so obvious that no well need be drilled to provide lithological or other data to complete the diagnosis.

LITHOLOGY

Lithology is the second of the five facies-defining parameters whose environmental significance should be considered. In surface studies the amount of lithological information will, of course, be contingent upon the extent of rock outcrop. Lithology and diverse other sedimentological data will be recorded on measured sections logged at suitable localities, such as stream sections, along escarpments and outlying jebels. There are as many methods of recording sedimentological logs as there are sedimentologists, and the reader is referred to Johnson (1992) for a thoughtful discussion of the philosophic

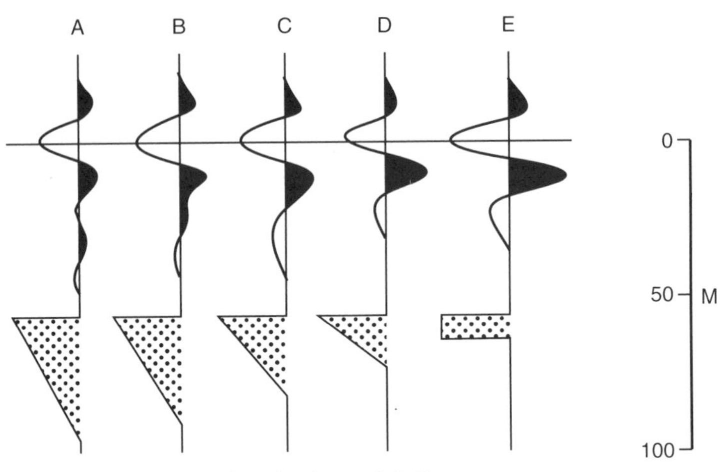

Figure 2.5 Seismic attribute analysis. Modelled wave traces for sand bodies with different scales and grain size profiles. Compare with Figure 1.10 to see how environmental interpretations may be made. (Developed from Meckel and Nath, in Payton, 1977.)

problems of devising appropriate sedimentological logs, and Lewis and McConchie (1994) for the practicalities of their construction. Most schemes are a variants of Figure 0.1 (see frontispiece to this book), which is the scheme that is used throughout the text. This method contains sufficient sedimentological data to enable an environmental interpretation to be made, but is not so complicated that a training course is required by composer or reader.

Subsurface lithological information is acquired with greater difficulty, and a correspondingly wider range of methods. Lithological information may be gathered from boreholes, both directly or indirectly. Samples of the rock may come to the surface, either in the form of cores, or brought up in the drilling mud as small fragments termed drill cuttings, or ditch samples. Coring is common in mineral exploration, but in petroleum exploration it is restricted to the crucial petroleum producing zones. When a core is pulled to the surface it contains a wealth of information. In petroleum exploration it is, of course, the porosity, permeability and petroleum saturation that are most important, more important even than the lithology itself. But the core can be treated as a narrow surface outcrop and can be sedimentologically logged just like a cliff or stream section.

When coring is not carried out the well is drilled by a cable tool or rotary drill system. The resultant rock fragments are brought to the surface by baling, during cable tool drilling, and by the circulation of the mud system, during rotary drilling. When washed clean and examined under the microscope the lithology of the strata penetrated is carefully logged.

Lithological information can also be gathered from wireline logging. After

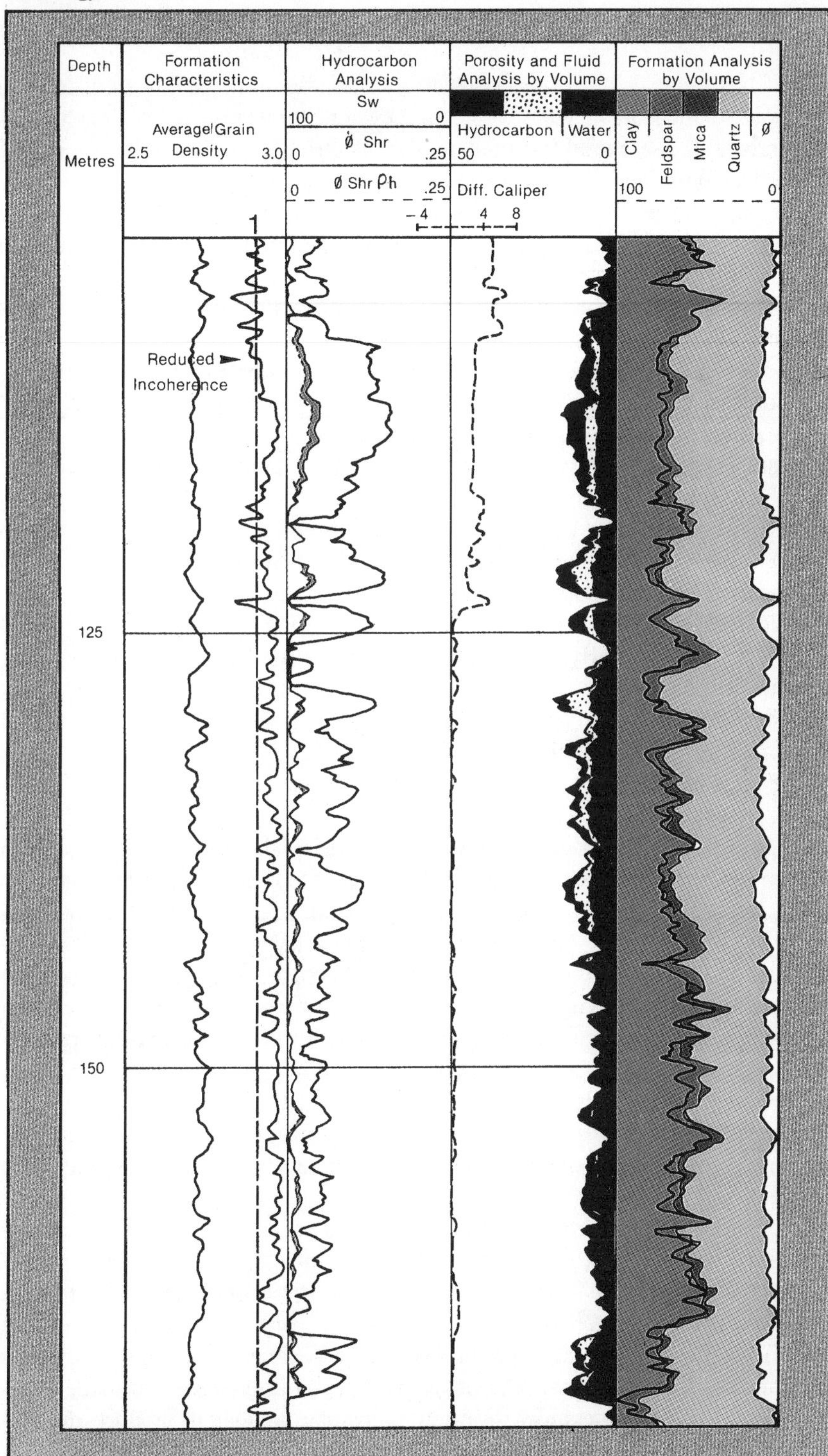

Figure 2.6 Though the lithology of a facies may best be studied from the rocks themselves, their chemical composition can be remotely sensed by mineralogical borehole logs. These are computed by integrating the data from as many as a dozen separate logging tools, measuring electric, radiometric, acoustic and other physical and chemical properties of the rock formations. Courtesy of Schlumberger. (Also reproduced as Plate 1.)

a well has been drilled various electronic sondes may be lowered down a borehole to measure a wide range of physical properties of the formations penetrated. These variables include the electric, acoustic and radiometric properties. Wireline logs were originally developed by the petroleum industry to measure the porosity and petroleum saturation of rock formations. Now they can also be used to interpret the lithology, sedimentary structures and dip of the formations penetrated.

This vast topic variously termed formation evaluation or borehole geophysics is outside the scope of this book. The technicalities of the various tools will not be discussed here. For further information see the sales brochures of the various wireline service companies. When a range of physical parameters has been measured for a borehole it is possible to compute the percentages of the different minerals present, and thus to generate a vertical profile of lithology (Figure 2.6).

Geophysics can also be used to determine lithology on a basinwide scale. Gravity and magnetics can be used separately, or better still together, to delineate basement from sedimentary cover. These techniques, supplemented by seismic data, can also be used to identify and map lithology on a regional scale. This is not nearly as reliable as borehole evidence (yet) because the acoustic velocity of a particular lithology has a wide range depending on its porosity, fluid saturation and purity. Mixed lithologies such as shaley sand and dolomitic limestone will give particular problems.

The most reliable way of interpreting depositional environment from lithology is of course when real rock samples are available from outcrop core or cuttings. An important distinction must be made at the outset about what can be learnt from sandstone lithology, and what can be learnt from carbonates.

Terrigenous sands are composed largely of grains of quartz and other detrital minerals. For many years geologists studied the grainsize and sorting of sands, in the hope that these parameters could be used to indicate depositional process, and hence environment. It was thought that the hydrodynamics of a given environment would deposit sediments with distinctive textural properties. Studies did indeed suggest that, in modern environments, it may be possible to differentiate sands deposited by windblown, fluvial, beach and turbidity flow processes on the basis of their texture, using complex statistical gymnastics. Attempts to extrapolate this method to ancient sandstones has met with failure.

There are two problems. The size, shape and sorting of a sediment sample is a function not only of its ultimate depositional process. A sand newly weathered and eroded from igneous or metamorphic basement will possess granulometric parameters inherited from the size, shape and abundance of the minerals in the parent rock. Similarly, recycled sediments will inevitably bear the imprint of earlier transportational episodes. Furthermore it is hard, if not impossible, to disaggregate a lithified sandstone so that its texture directly replicates that which it had when first deposited. Clays may have infiltrated, or formed from the breakdown of unstable minerals, or they may have been precipitated as cement. Chemically unstable sand grains may have leached out. Overgrowths may have increased particle size. The small diameter of a borehole is unlikely to collect statistically valid samples in coarse, poorly sorted sediment, particularly grain flows and debris flows.

For all of these reasons the statistical analysis of the texture of a lithified sediment is an unreliable method of diagnosing environment. The mineralogy of a terrigenous sand might also be sought for clues. Sadly the detrital mineralogy is, as with texture, to a large extent inherited from the rocks, be they igneous, metamorphic or sedimentary, from which the sediment was derived. There are, however, some constituents which may provide clues to the environment of deposition. The presence of glauconite, shell debris, mica and carbonaceous detritus may all be environmental indicators, as may the presence of red coloration (Selley, 1976). These five constituents will be considered one by one.

The term glauconite is variously applied to: pretty green grains within marine sediments attributed to glaucony facies (defined shortly), or to: the minerals that compose the green grains, or to: the micaceous end members of the glauconite group of minerals (see Odin, 1988). Briefly, green clays form on continental shelves in cool reducing conditions, with slow sedimentation rates. This material infills pores, both within the sediment and within microfossil cavities. It replaces faecal pellets and detrital clay. The resultant green clays are sometimes referred to as glaucony facies. Fluctuations of sea level generate alternating episodes of glaucony formation and reworking, giving rise to the sand sized green clay pellets colloquially referred to as glauconite.

Once formed, glauconite is stable in sea water. The grains may be deposited in continental shelf sands, carried up on to beaches, or transported off the shelf edge and down in to deep water. Glauconite does not form in continental conditions, except in one or two rare instances. It is rapidly oxidized during weathering, so second cycle glauconite is unknown, almost. Though where glauconitic sandstones crop out on continental shelves, glauconite may be redeposited into younger shallow marine sands. This may be seen on modern continental shelves, where Cretaceous greensands contribute to modern continental shelf sands.

Thus one may conclude that the presence of glauconite grains in a sand is an indicator of a marine environment. The reverse is, of course, not necessarily true. An absence of glauconite is not an indicator of non-marine conditions. The presence of shell debris may be a similar indicator of a marine environment. Obviously the presence of identifiable fossil fragments is of inestimable value as an environmental indicator. But the presence of sand sized shell fragments of indeterminate origin may be similarly used. It is true, of course, that marine invertebrates secrete lime skeletons in non-marine environments. But the preservation of calcareous material in the continental realm is much lower than in marine environments. Sea water is neutral, and most connate waters are alkaline, but there is nothing new in acid rain. Meteoric waters have always tended to be acidic, and thus leach out calcareous fossils and lime fragments in continental sediments. Shell debris can therefore, like glauconite, be used as a pointer towards marine conditions.

Carbonaceous detritus is another constituent of sandstones that may be of environmental usefulness. Carbonaceous detritus is a term that embraces material from fossilized plant remains down to disseminated kerogen. Carbonaceous detritus may be land-derived humic material, or it may be of marine algal origin. In high-energy environments, be they marine or non-marine, oxidation and abrasion will tend to destroy all organic matter. Carbonaceous detritus can only be deposited and preserved in environments where sediment is rapidly dumped and subjected to little reworking. Such environments include deltas in particular, especially such locales as crevasse splays and delta slopes, submarine fans, lakes and alluvial backswamps. As with glauconite, an absence of carbonacous detritus is not necessarily an indicator of a high-energy environment, though it may be circumstantial evidence. Mica may be used in the same way as carbonaceous detritus. It is, of course, a detrital mineral, but like carbonaceous detritus it will only settle out in those environments where sedimentation rates are high, and turbulent reworking minimal.

The fifth environmentally diagnostic attribute of sandstone lithology is its colour. Colour is largely determined during penecontemporaneous diagenesis. Sands deposited below the water table within the phreatic zone have pores that are water saturated. They tend to undergo early diagenesis in somewhat reducing conditions. Organic matter may be preserved, and iron compounds tend to be stabilized in the grey-green ferrous state. Thus most subaqueous sands, marine and non-marine, tend to a drab grey-green hue. By contrast, sediments that are deposited by wind, or by ephemeral floods in piedmont fan and braided channel systems undergo early diagenesis above the water table in the vadose zone. They tend to be red coloured because penecontemporaneous diagenesis takes place in a pore system open to the atmosphere. In these oxidizing conditions organic matter tends to be destroyed and iron compounds age to red ferric oxides (Figure 2.7).

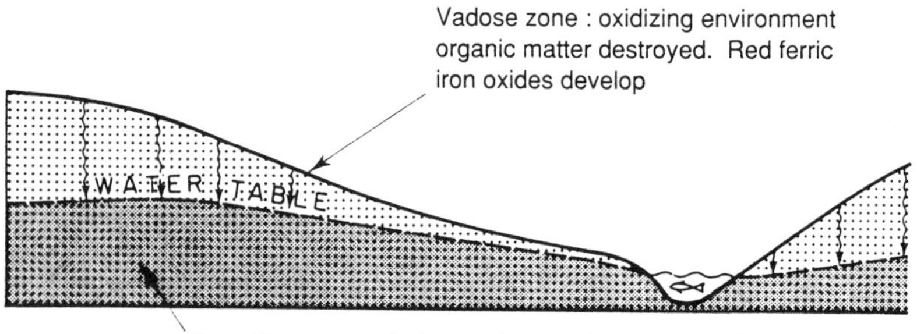

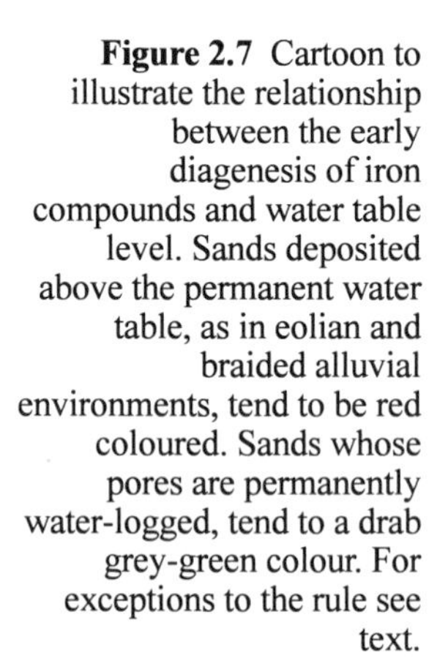

Figure 2.7 Cartoon to illustrate the relationship between the early diagenesis of iron compounds and water table level. Sands deposited above the permanent water table, as in eolian and braided alluvial environments, tend to be red coloured. Sands whose pores are permanently water-logged, tend to a drab grey-green colour. For exceptions to the rule see text.

All generalizations are dangerous, including this one. Notable exceptions include deep water red clays and limestones, such as the 'Ammonitico Rosso' of the palaeo-Tethys (see p. 285). Reddening is common beneath unconformities, especially when the overlying sediments are of continental origin. The reddening of Carboniferous Coal Measures beneath the Permian desert sands of north-west Europe is a case in point. Finally, in oil fields the red colour of continental sands is often masked, both by the presence of petroleum within the reservoir, and by a reduced halo several metres deep in the water zone beneath and adjacent to the petroleum:water contact. Figure 2.8 shows the genesis and distribution of these five constituents of sands that may be environmentally significant.

Moving on from petrography, there is one other technique that can be applied to sandstone lithology to diagnose depositional environment. In Chapter I the discussion on sedimentary models pointed out that each depositional system generates a characteristic, though seldom unique, vertical profile of grainsize (refer back to Figure 1.10). Now geophysical logs not only elucidate lithology, but necessarily record the nature of lithological

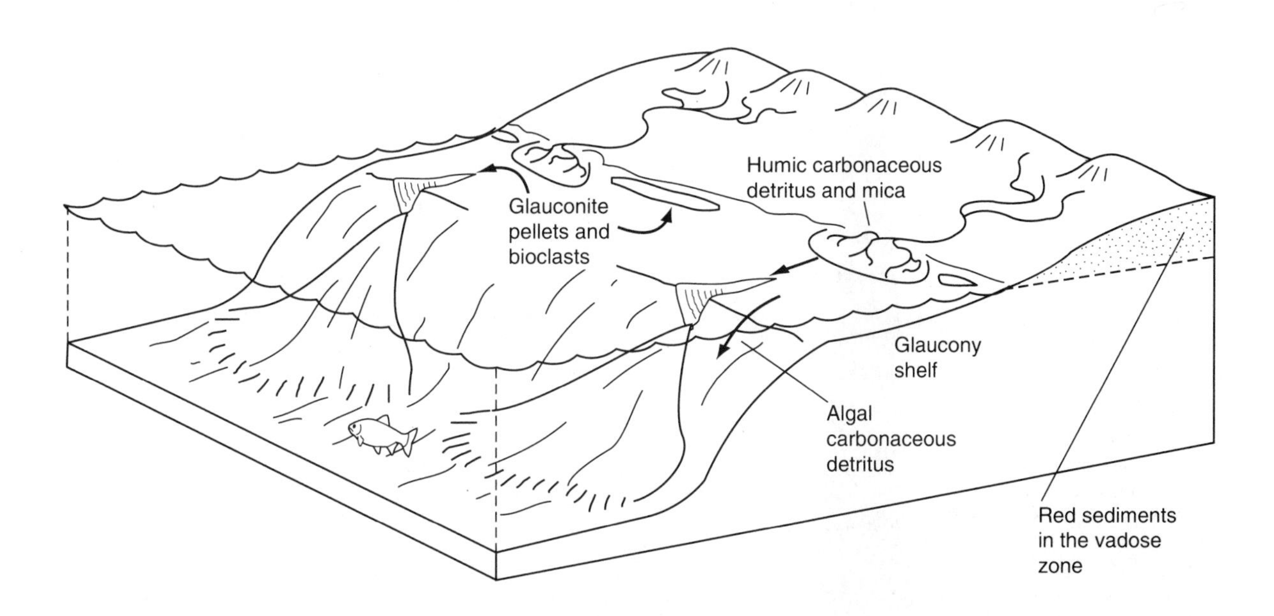

Figure 2.8 Cartoon to show the correlation between the distribution of glauconite, shell debris, carbonaceous detritus and red coloration in different environments. This combination of parameters is one of the most effective ways of using lithology as an indicator of the depositional environment of sandstones. Note, of course, that all available data are integrated in environmental diagnosis. This is just one particular technique, developed by Selley (1976).

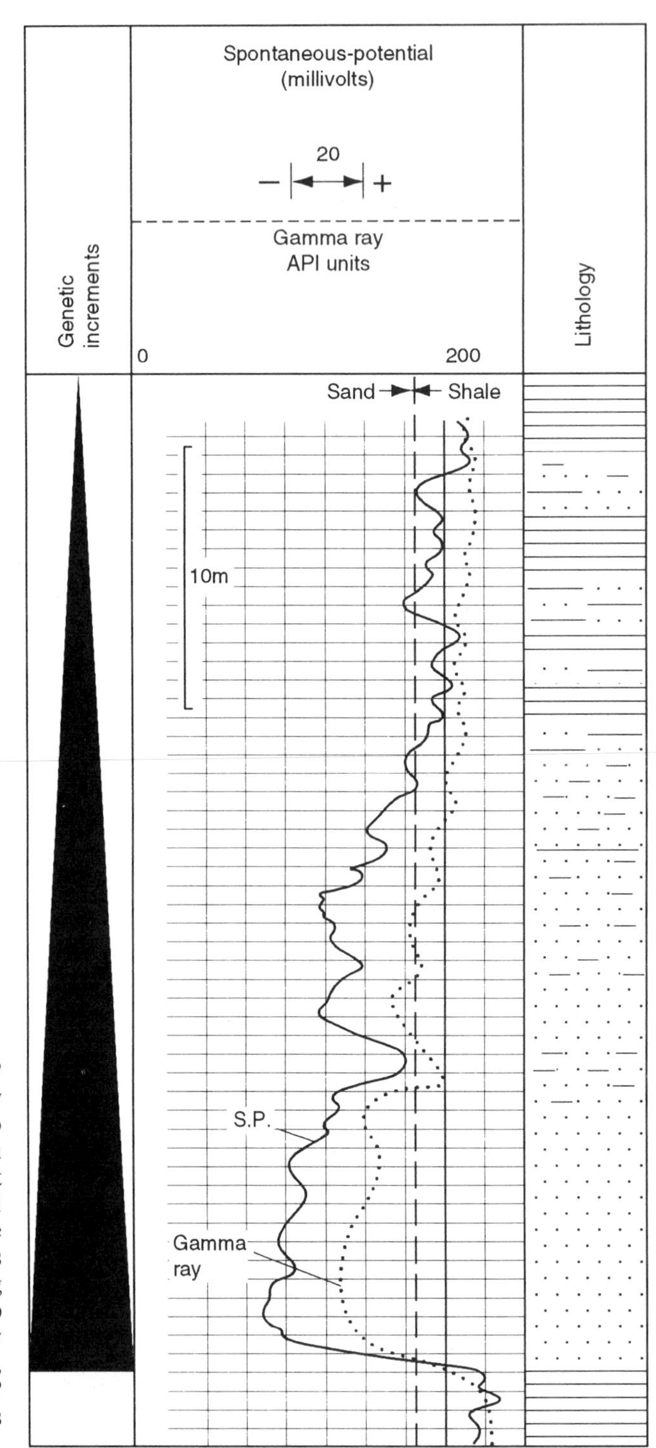

Figure 2.9 Gamma and SP borehole logs to show how they may be used to differentiate sand from shale, to delineate abrupt and gradational lithological boundaries, and to define genetic increment. In this example an upward-fining sand with a sharp (erosional?) base is clearly seen. A channel, perhaps, but in what environment? Log motif alone is non-diagnostic.

transitions. Two tools are particularly useful for this purpose namely the Gamma log and the SP (Self-Potential or Spontaneous Potential) log. This is a well-established technique that has been used for many years.

The gamma log records the natural radioactivity of rock formations as a sonde with a scintillometer is drawn up the borehole. Most radioactivity in sediments is found in potassium and most potassium occurs in clays. Thus high radioactivity indicates shale, low radioactivity indicates sand or carbonates. As a sand fines up into shale, so the gamma reading will gradually increase, and vice versa. Sharp lithological boundaries will be marked by correspondingly sharp deflections in the gamma curve. Hence it is possible to use the gamma to draw the bed boundaries and define the genetic increments (Figure 2.9).

The foregoing is, of course, a gross simplification. For a gruesome account of all the exceptions see Rider (1990). Some of the exceptions to the general rule include the fact that not all clays contain potassium. Kaolinitic clays, typically continental in origin, are low in potassium and consequently low in radioactivity. Potassium occurs in minerals other than clays, notably feldspar, so arkosic sands may appear 'dirty'. Micas and glauconite are also potassium-bearing. Potassium is not the only radioactive element in sediments. Many of the heavy minerals, such as zircon, are highly radioactive. Thus placer streaks in sands may simulate shale beds in a gamma log. Most of these mineral identification problems may be overcome by looking at other logs over the same internal. One however is caused by intraformational shale conglomerates on channel floors (Figure 2.10). Despite these complications, surface studies, in which portable scintillometers are carefully lowered down cliffs, demonstrate the general reliability of the method (e.g. Slatt *et al.*, 1992).

The SP log can be used in much the same way as the gamma log. The SP records the self-potential or spontaneous potential current that occurs between an electrode mounted in a sonde, and one at the surface. As the sonde is moved up the borehole it measures the current in millivolts. The precise origin of this charge is a matter for debate, even today, but there is agreement that it is a measure of permeabilitv, in a purely qualitative way. Deflection of the log to the right indicates impermeability, deflection to the left indicates permeability. Now the coarser a sediment, the higher its permeability. Because of this the SP curve can be used just like the gamma, as a way of defining genetic increments in sand:shale sequences, and indicating grainsize profiles, and the sharpness of lithological boundaries (Figure 2.9). As with the gamma log, all sorts of things can inhibit the use of the SP as a grainsize profile. For example, it cannot differentiate cemented, and thus impermeable, sands from shale. It only works well when there is a good contrast between the salinity of the formation water, and of the drilling mud filtrate.

Though both the SP and the gamma ray tools have problems, experience has shown over many years that they can be used as vertical profiles of grainsize in sand:shale sections. Generally they are seen to track together (Figure 2.11). If they do not, one should ponder why. This figure also shows that the

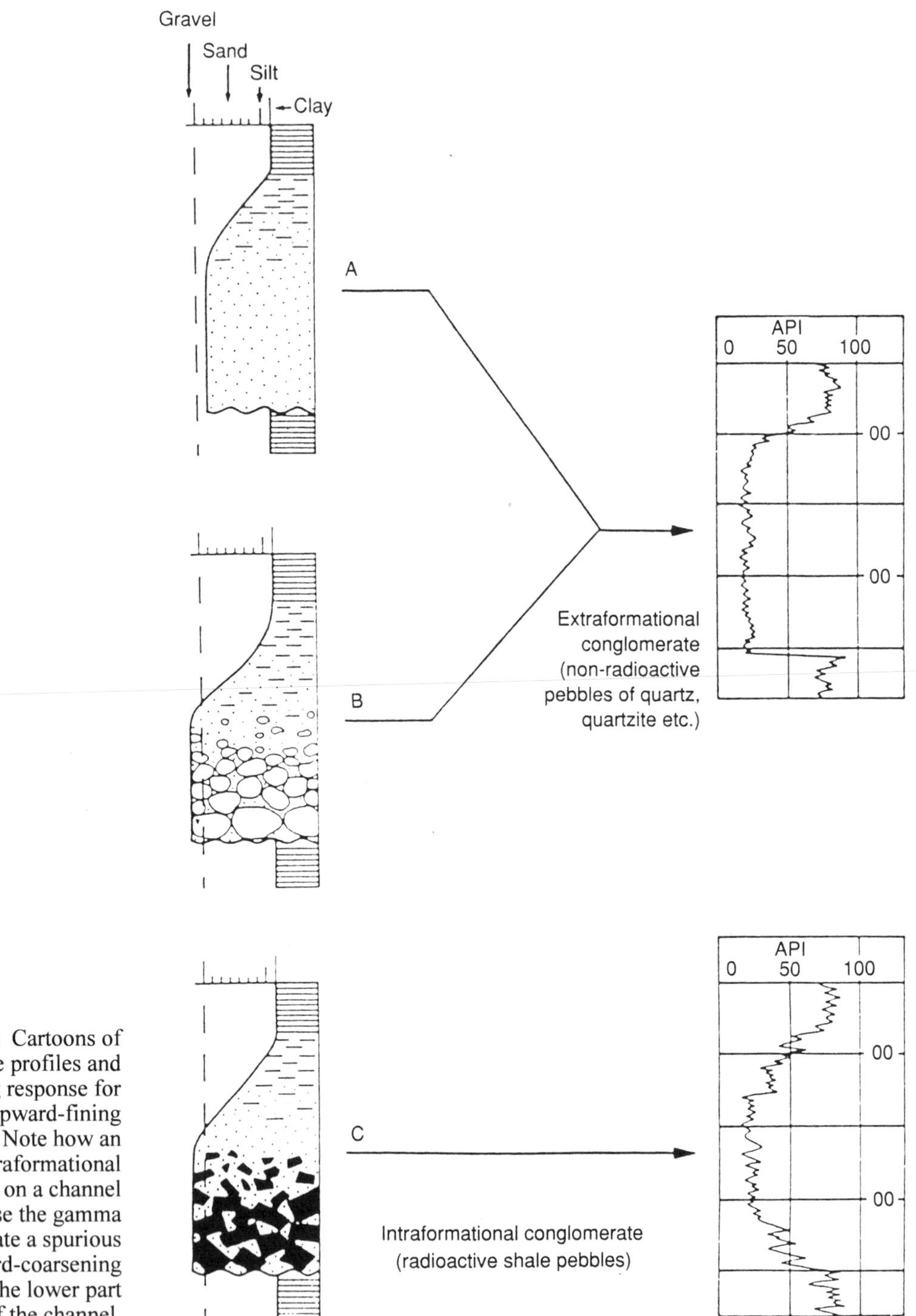

Figure 2.10 Cartoons of grain-size profiles and gamma log response for some upward-fining sequences. Note how an intraformational conglomerate on a channel floor may cause the gamma log to indicate a spurious upward-coarsening sequence in the lower part of the channel.

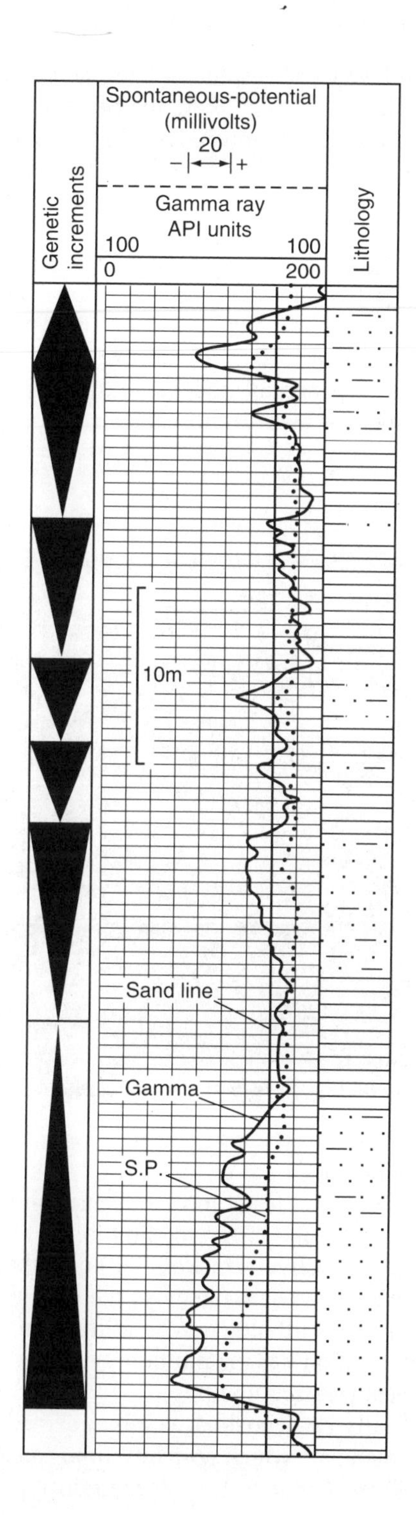

Figure 2.11 Gamma and SP borehole logs to show how they may be used to differentiate sand from shale, to delineate abrupt and gradational lithological boundaries, and to define genetic increments. Note how the SP curve is much less sensitive to lithological changes than the gamma log, lacking the same degree of bed resolution. In Figure 2.9 the single upward-fining genetic increment was obvious, but in this log the recognition of increments is by no means as clear. There are no 'right' answers, and when lithological boundaries are gradational, then so are the limits of the increments.

gamma log gives much better bed resolution than the SP. The bed resolution of the gamma tool is in the order of 0.25 m, the SP is not nearly so sensitive, and can seldom differentiate beds much less than one metre in thickness. Despite these limitations Figure 2.11 shows how the gamma and SP curves can be used to identify the lithological boundaries and genetic increments of sand:shale sequences.

Finally it must be noted that neither of these logs can be used as vertical profiles of grainsize in carbonate sequences. Limestones are normally of low radioactivity, irrespective of whether they are mudstone or grainstone. Similarly the SP can seldom be used as a grainsize profile in carbonates. This is because limestones have generally undergone much more diagenesis than sandstones. Their porosity and permeability is generally secondary, and unrelated to the primary porosity and permeability of the sediment when first deposited.

Figure 2.12 Photomicrograph of a sandstone. It is not possible to give a reasoned interpretation of the depositional environment of this lithology, without additional information. Contrast this with Figure 2.13.

This point usefully concludes the discussion of what can be learnt from the lithology of sandstones, and leads in to a discussion of the environmental information that can be gleaned from the lithology of limestones. Look at Figure 2.12. This figure effectively makes the point that, saving the little tricks just discussed, one can seldom diagnose the depositional environment of a terrigenous sand from its lithology. With a limestone, however, the key to environmental interpretation lies in its lithology (Figure 2.13).

This is because limestones are composed of various types of grain that tend to form, and normally be deposited within, a narrow range of environments. The fossil fragments of which so many limestones are made, are an obvious key to environment, but so are other grain types such as peloids,

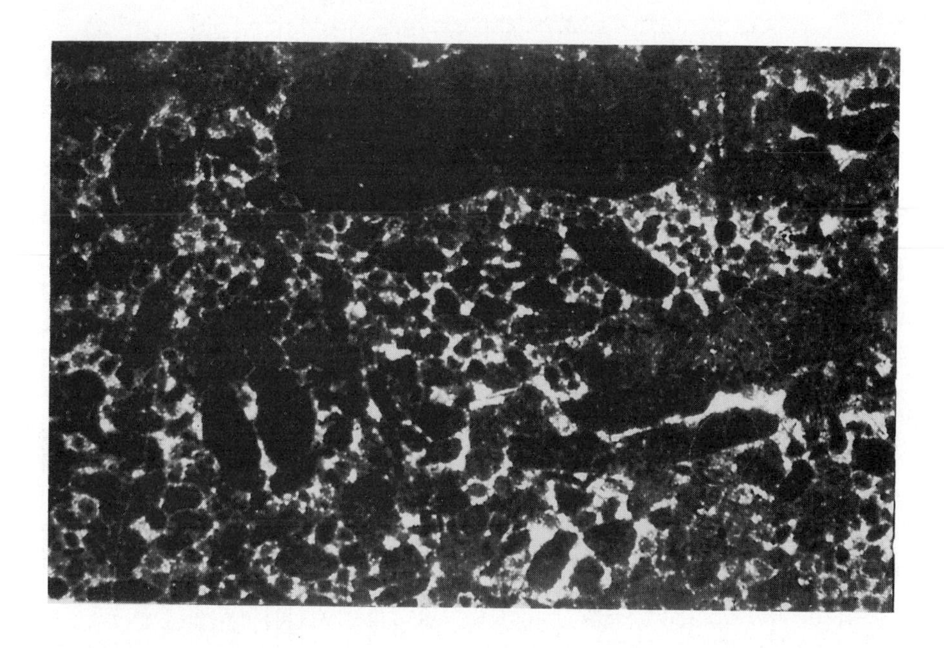

Figure 2.13 Photomicrograph of a limestone. Describable as packstone in Dunham's classification, or as an intrapelmicrite according to Folklore. The intraclasts suggest reworking of penecontemporaneously cemented, eroded and redeposited lime mud. The peloids could be attributed to a faecal origin, excreted by marine invertebrates of small rectal calibre. The overall texture of a grain supported sand with minor mud matrix suggests neither a quiet, nor a too vigorous energy level. These features are consistent with deposition in a shallow nearshore shelf environment, perhaps a lagoon?

pellets, grapestone, ooids, oncolites and many more. These will be discussed in greater detail in the relevant chapters on carbonate environments. The point is made here, and will be embellished and justified later. It is only when limestones have undergone extensive diagenesis, particularly if dolomitized, that petrography ceases to be an environmental indicator.

PALAEONTOLOGY

Having considered the uses of facies, geometry and lithology in environmental interpretation, it is now appropriate to consider palaeontology. Fossils are perhaps one of the oldest means of diagnosing the depositional environment of a sediment.

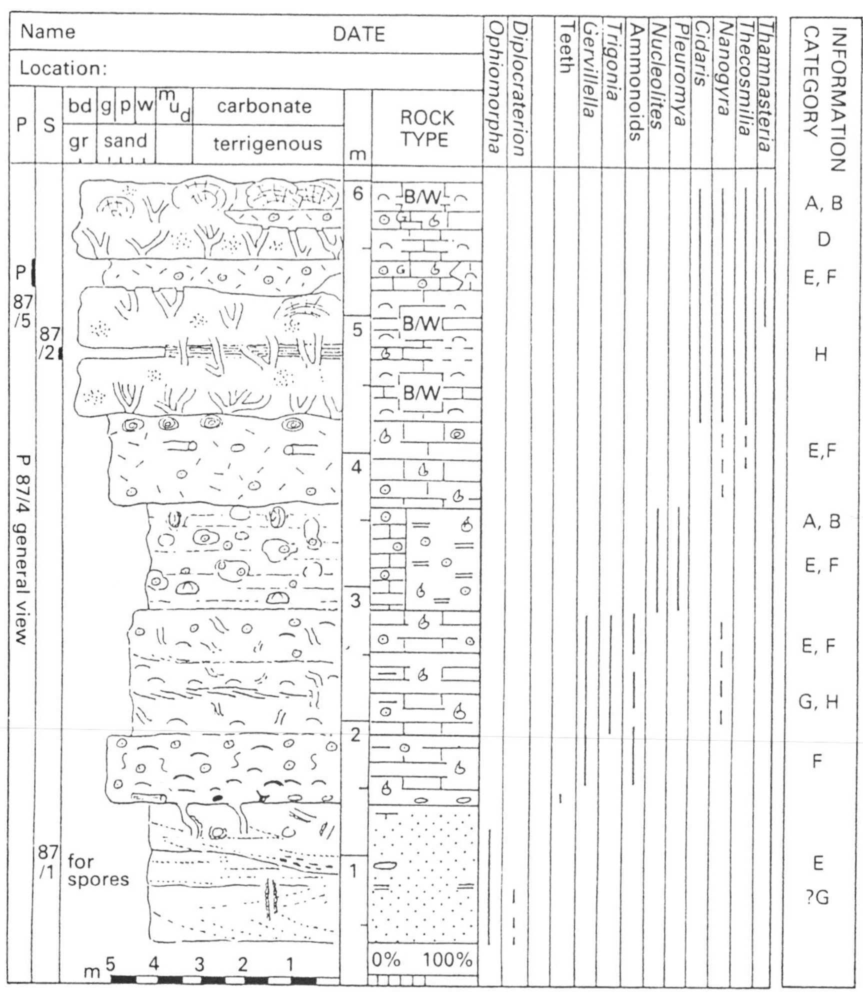

Figure 2.14 Measured section of a limestone formation, showing details of palaeontological data recorded (from Goldring, 1991, Figure 1.2 © Longman Technical and Scientific).

Palaeontology is a vast topic. All geologists are exposed to courses on the subject, and will have studied fossils during field work, but few will have specialized in palaeontology sufficiently to use them to their best advantage. Palaeontology has two uses, to define biostratigraphy, and to deduce environment. It is of course only the last of these that is relevant within the terms of reference of this book. Dodd and Stanton (1990) provide an excellent introduction to palaeoecology, and Goldring (1991) describes the practicalities of collecting fossils and their palaeoecological interpretation.

As with geometry and lithology, the methods of collecting palaeontological information differ at outcrop and subcrop. For surface studies palaeontological data are collected and logged together with sedimentology on measured sections, as described earlier in this chapter. Figure 2.14 shows how to record palaeontological information over a particularly interesting section.

In subsurface geology data can only be collected from cores and cuttings. Cores can, of course, be treated as outcrop sections, and logged accordingly as earlier described. Cuttings present an obvious problem, but are still very useful. Macrofossils are smashed up by the drill bit, but real experts can still identify fossils, down sometimes to generic level, from their skeletal fabric. It is however the microfossils that are so important in well cuttings, principally for biostratigraphic information, but also for clues as to depositional environment. Foraminifera, radiolaria, conodonts, palynomorphs and many other curious little bioclasts can survive the trauma of the drill bit, to be collected safely at the surface, logged, identified and studied for biostratigraphy and palaeoecology.

Once the palaeontological data have been collected, be it from outcrop or subcrop, the interpretation begins. Two questions must be answered:

1. Did the fossils live in the environment in which they were buried, or were they carried into it dead or alive, if alive killed and buried, or buried alive, and then killed?
2. Is the habitat of the fossil really well understood?

Consider the first question. Fossilization is generally considered to be a very rare and chancy business. Sometimes fossils are found obviously in growth position, as with plants on palaeosols, and colonial animals such as corals. Fossils are sometimes preserved however, in a particular environment, not because they liked it, but because it was so hostile that it killed them. Think of all the drowned cats washed out in to the North Sea by the River Thames. Or think of the Archaeopteryx fossilized in the Solenhofen Limestone. Did some early birds try to fly under water? Surely not. Thus palaeontologists have always tried to distinguish between life assemblages and death assemblages. Whereas this may be fairly easy at outcrop (refer back to Figure 2.13 again), it may be very difficult, if not impossible, when trying to interpret fossil assemblages from drill cuttings.

The second problem is that the natural habitat of extinct fossil groups may be poorly known. It has been observed that today bears inhabit every climatic zone, but deserts. Were only polar bears to be extant, every fossil bear bone would lead to thoughts of past glaciations. More relevant examples to consider might include extinct fossil groups such as the Stromatoporoids, the Archaeocyathids and the Conodonts.

There is however, one particular group of fossils that even ordinary geologists can use, with little specialist training. These are the trace fossils, the tracks, trails and burrows that are common in so many sedimentary rocks. Trace fossils are, of course, obviously in place, thus overcoming the first problem outlined above. Second, it has been noted that, in both modern and ancient sediments, there are characteristic trace fossil assemblages, or ichnofacies. Each ichnofacies consists of a suite of characteristic trace fossils which occur in characteristic sedimentary facies, and whose environment

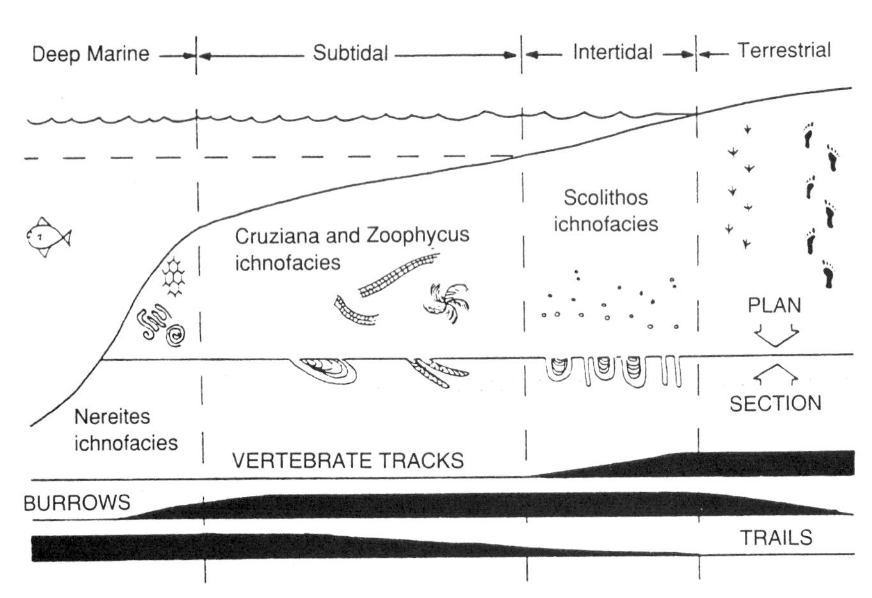

Figure 2.15 Characteristic ichnofacies for various environments (from Selley, 1988) by courtesy of Academic Press Inc.

may be determined by independent criteria. These trace fossil assemblages are independent of time. As different groups of organisms evolve and die out they inhabit a particular ecological niche.

Thus, in very simple terms, every little beast that lives on a tidal flat responds to the ebbing tide by making a vertical burrow. The facts of life are the same for worms, crustaceans or bivalves: burrow down or get eaten by pterodactyls or sea gulls (Larsen,1993). Detailed accounts of ichnology and of ichnofacies will be found in Bromley (1990). Figure 2.15 illustrates the concept of ichnofacies.

This is just a brief review of a very important diagnostic tool of sedimentary environments, where the geological 'general practitioner' is too often at the mercy of the specialist.

SEDIMENTARY STRUCTURES

Sedimentary structures are the fourth of the five parameters that define a sedimentary facies. They are very important in helping to interpret the depositional environment of a facies (Collinson and Thompson, 1988). As with the facies parameters previously discussed, there are differences in the ways in which structures may be studied at outcrop and in the subsurface. At outcrop sedimentary structures are recorded as part of section measuring. Where cliffs or bedding surfaces are extensive, however, it is possible to produce sections and plans of great lateral and horizontal continuity (e.g. Miall and Tyler, 1991).

Access to sedimentary structures in the subsurface is much more difficult.

By definition sedimentary structures are below the (present) limits of seismic resolution. They can be studied, with some degree of discomfort, in mine shafts and adits. The restricted circumference of boreholes necessarily inhibits observations of sedimentary structures, unless cores are available. As previously remarked, cores can be logged in the same way as for a surface sections. There are several techniques for imaging sedimentary structures by using geophysical well logs. Two main methods have been developed, acoustic and resistivity logging.

Acoustic imaging tools were first developed in the 1960s and are currently marketed by several service companies. An early version was the Borehole Televiewer* (BHT). Current versions include the Ultrasonic Borehole Imager* (UBI) and the Circumferential Acoustic Scanning Tool (CAST⁺). These acoustic imaging tools all work on a similar principle. A rotating sonic transducer, that emits acoustic pulses, is drawn up the borehole on a sonde. The pulses reflected from the side of the well bore are sensed by the transducer. Thus the two-way travel time of the acoustic pulse may be measured. This will vary according to the acoustic velocity of the strata adjacent to the borehole. Resolution down to beds 2.5 cm thick can now be achieved. The returning data can be used to produce a monochrome image of the borehole.

The images are clearest where there are the largest velocity contrasts adjacent to the well bore. Thus the best results tend to be produced in carbonates, because they are acoustically much faster than sands and shales. Acoustic images of limestone and dolomite often reveal solution vugs and fractures. Sedimentary structures are less easily imaged in either carbonates or clastics, because the acoustic contrast between sediment layers is less marked than for large pores. The acoustic method has the advantage over the resistivity logging tools to be discussed shortly, in that it can be used with oil-based drilling mud systems.

Resistivity imaging tools rely on the fact that sands tend to have a higher resistivity (lower conductivity) than shales. The closer together the pair of electrodes that are used to measure resistivity, then the thinner the bed that can be delineated. Today it is possible to delineate beds down to 5 mm. The present resistivity tools evolved originally from the single microresistivity log. In the 1950s tools were built that ran 3 micro-logs together, and which could thus be used to define the direction and amount of dip around the borehole. This tool, and its successors, is described more fully in the next section on palaeocurrent analysis. Over the years more and more electrodes were added to the sonde to increase the accuracy of bed dip delineation. The first imaging tool, the Formation MicroScanner* (FMS), consisted of four electrode bearing arms. Two arms carried 27 button electrodes, producing a total of 54 micrologs (Figure 2.16).

These could be processed to produce a continuous vertical image of some 20% of the well bore. With repeated runs coverage could be increased to some 50%. For details of the tool, and some results see Harker *et al.* (1990)

* Trademark of Schlumberger
⁺ Trademark of Haliburton

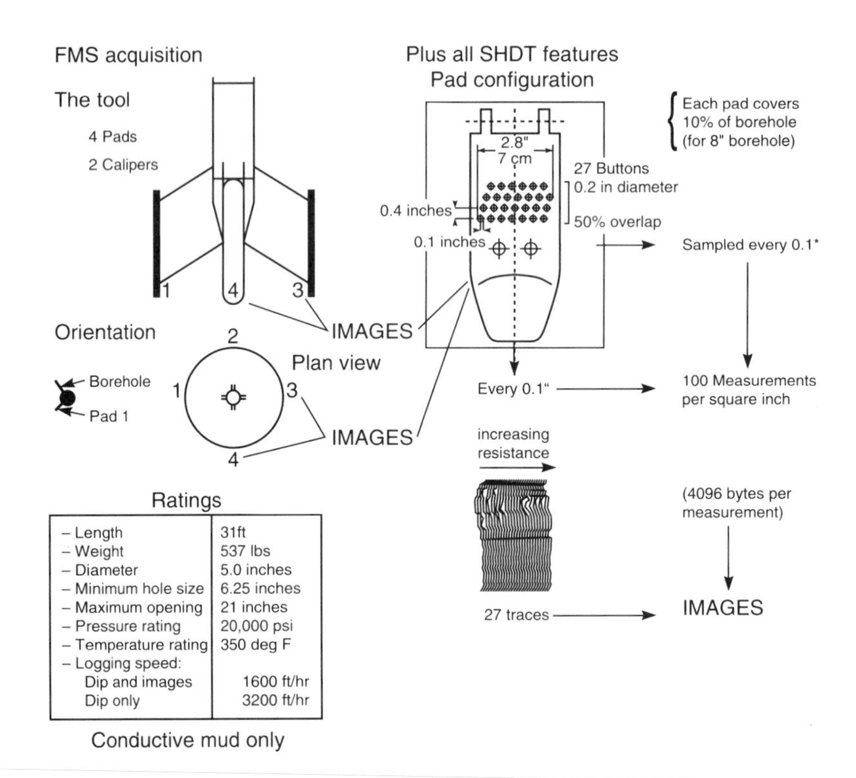

– Length	31ft
– Weight	537 lbs
– Diameter	5.0 inches
– Minimum hole size	6.25 inches
– Maximum opening	21 inches
– Pressure rating	20,000 psi
– Temperature rating	350 deg F
– Logging speed:	
Dip and images	1600 ft/hr
Dip only	3200 ft/hr

Figure 2.16 Diagrams to show the workings of the Schlumberger Formation MicroScanner. This was the first resistivity borehole imaging device now superseded by the Formation MicroImager FMI.

and Bourke (1992) respectively. The FMS has now been succeeded by the Fullbore Formation MicroImager* (FMI). This tool has 48 sensors on each arm, producing a total of 192 micrologs, and producing an image of 80% of the borehole, with bed resolution down to less than 0.5 cm.

The images produced are fantastic (Figure 2.17), and are the next best thing to lowering an anorexic geologist down the hole. The newest device to aid the study of sedimentary structures in boreholes is termed **Diamage***. When a core is brought to the surface it is photographed on a rotating stage, using the **Autocar** system, to produce a flat image of the core surface. This photograph is scanned, and the image loaded on to a workstation. The FMI log for the same borehole interval may then be loaded and displayed on the screen beside the core photo (Figure 2.18).

This powerful technique, by comparing real rock with remotely sensed images, gives the geologist great confidence when interpreting sedimentary structures from borehole data.

Thus today it is now possible not only to record sedimentary structures at outcrop, but also from borehole images in the subsurface. Sedimentary structures give very valuable information (Figure 2.19).

Sedimentary structures give clues as to the depositional process. No individual process is restricted to any one environment, however, so all sedimentary structures are to be found in a wide range of environments. For

* Trademark of Schlumberger

A

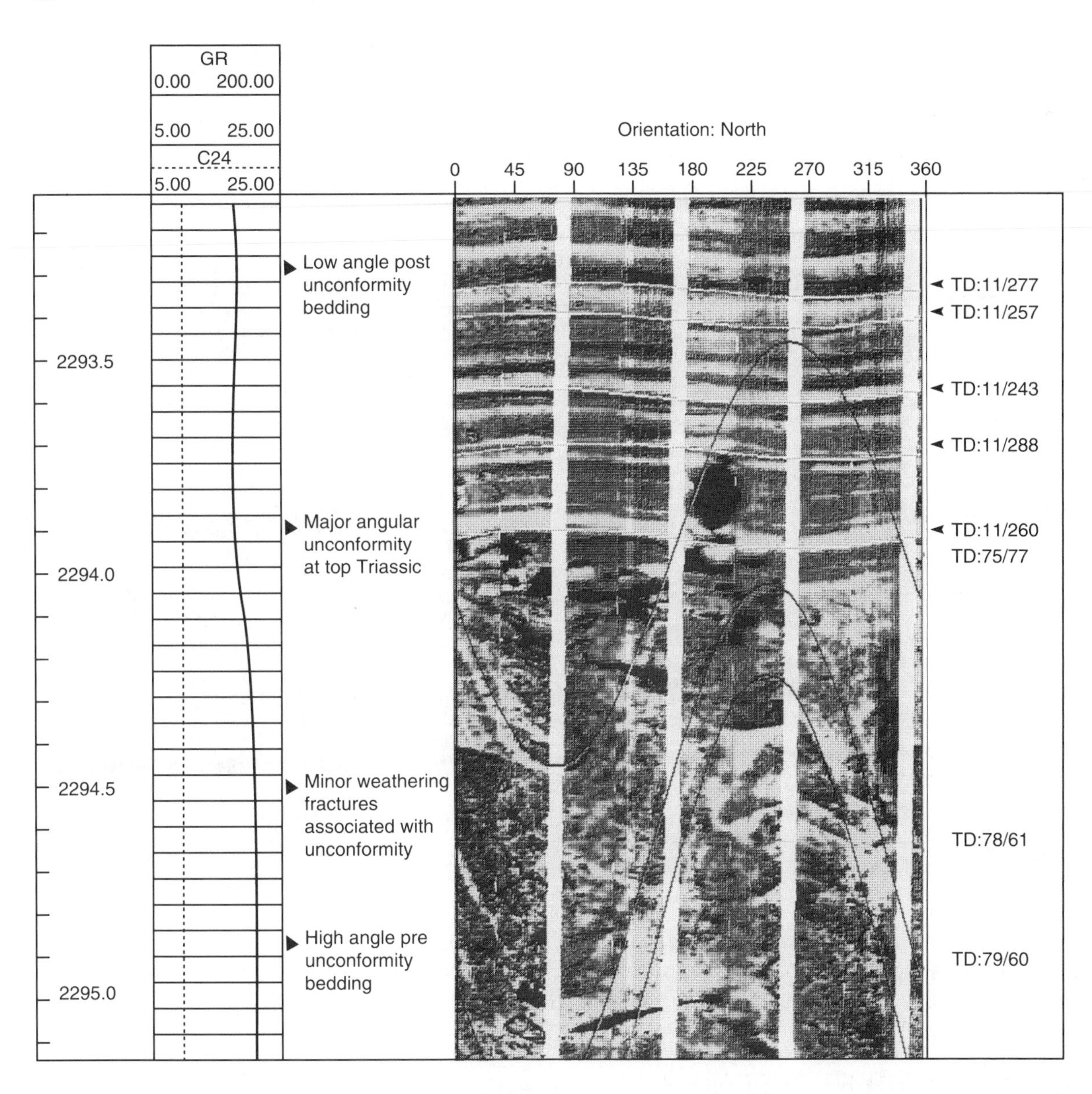

Figure 2.17 Images produced by Schlumberger's Formation MicroImager (FMI). A: An unconformity, courtesy J Roestenburg (also reproduced as Plate 2(a)).

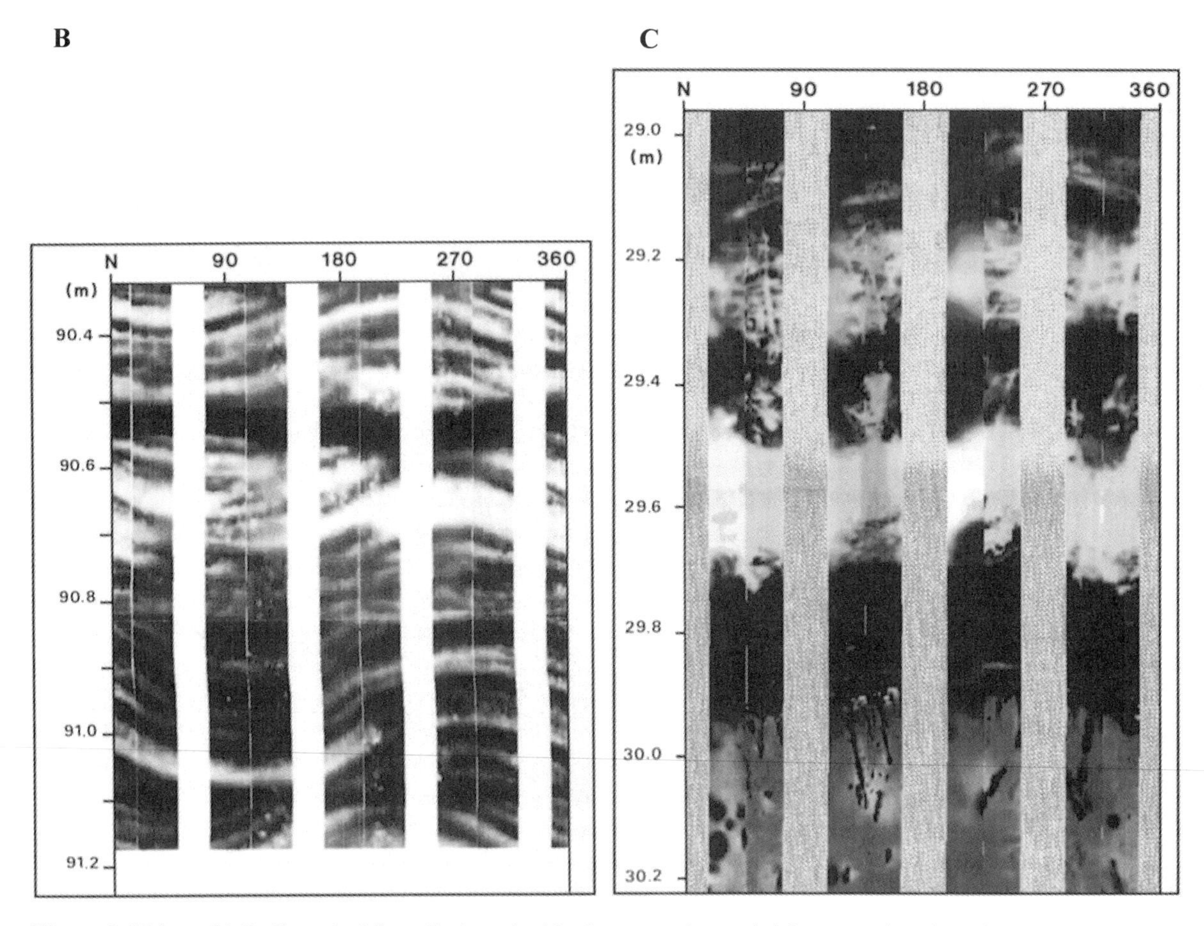

Figure 2.17 (contd.) B: Cross bedding. C: A coal with plant roots beneath (also reproduced as Plates 2(b) and (c)).

instance cross-bedding occurs in eolian sands, in river channels and on continental shelves. Desiccation cracks may indicate emergence, but this could be in a flood plain, in a playa lake, a tidal flat, or a lagoon. But suites and sequences of structures can be very compelling indicators of the depositional environment, as will be demonstrated throughout this book.

PALAEOCURRENT PATTERNS

Many sedimentary structures can be used to determine the sense, and sometimes the flow direction of the current from which they were deposited. Palaeocurrent analysis, as this technique is termed, is very important for mapping the palaeogeography of sedimentary basins, and as an aid to interpreting depositional environment. Furthermore it is a very powerful tool in

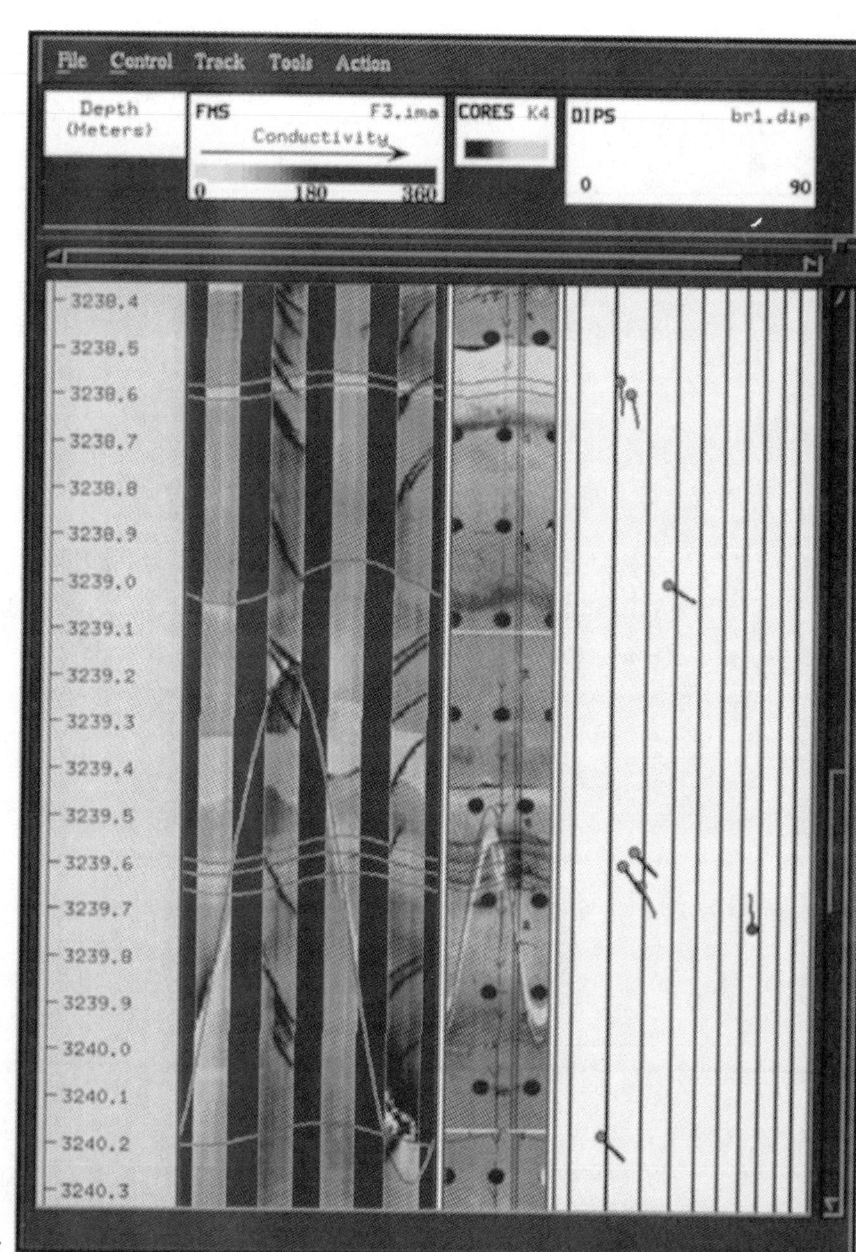

Figure 2.18 Monitor image produced by DIAMAGE*. This system enables cores to be photographed, the photos loaded and displayed on a screen beside corresponding FMI images and compared with dips calculated from the cored interval. Courtesy of Schlumberger (also reproduced as Plate 3).

* Trademark of Schlumberger

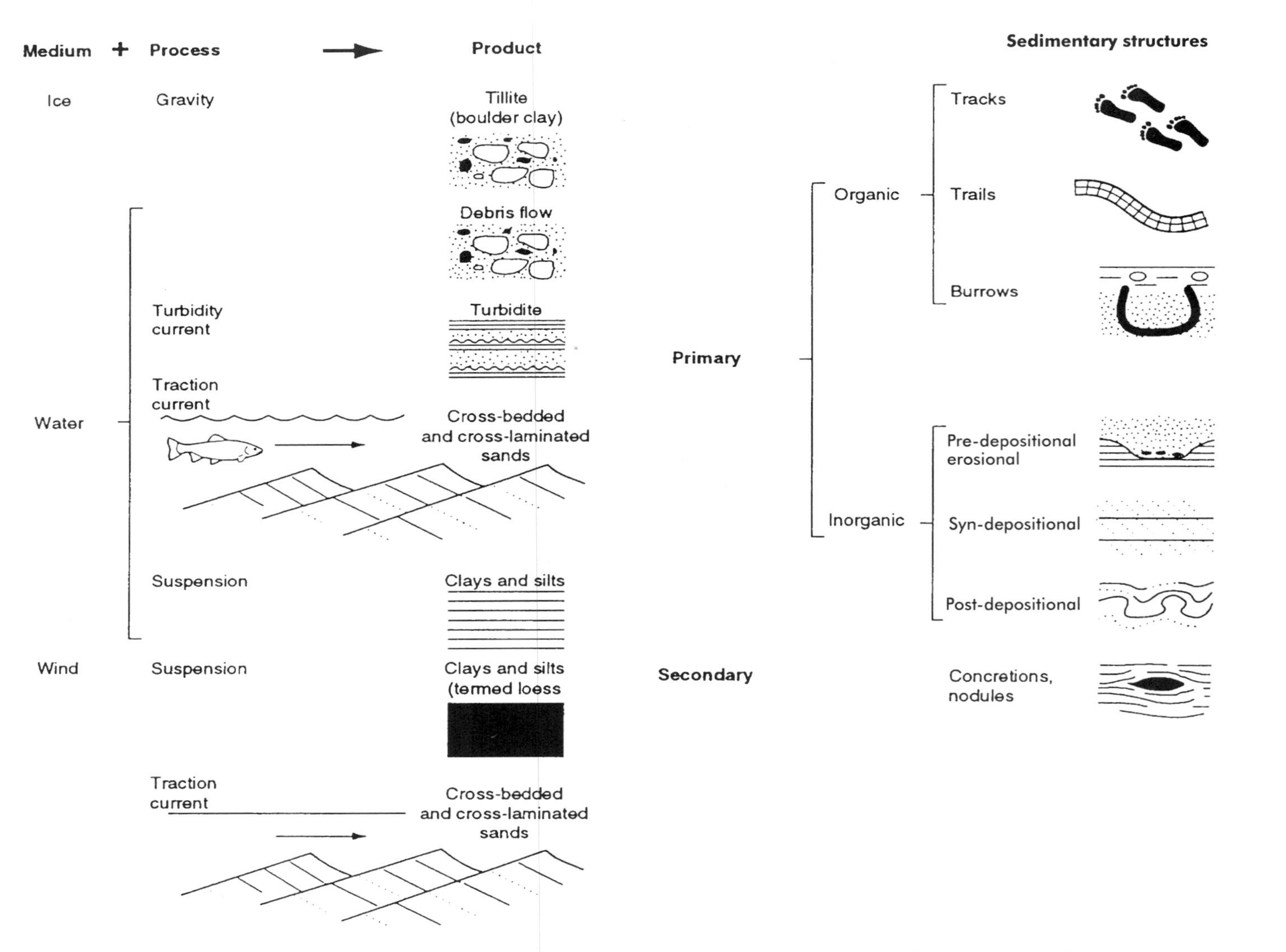

Figure 2.19 Left: a summary of the sedimentary processes and products. Right: a summary of the various types of sedimentary structure. Note the difference between bed forms and sedimentary structures. Ripples give rise to cross-lamination (normally < 3 cm high). Megaripples, or dunes, give rise to cross-bedding (normally > 0.5 cm high). Note the similarity between the bed forms and structures produced by traction currents in air and under water.

predicting the geometry and trend of mineral deposits and petroleum reservoirs in sedimentary rocks. The structures that may be used to interpret palaeocurrents range in scale from channels, down to the orientation of fossils and sand grains. By far the most widely used structure, however, is cross-bedding.

The orientation of sedimentary structures can be measured at outcrop, and recorded on measured sections, or on plans and sections, where outcrop permits. The technique for collecting palaeocurrent data at outcrop, for removing tectonic tilt, and for further statistical gymnastics are given in Lewis and McConchie (1994). In the subsurface cross-bedding and channel axes can be observed in mine workings. They have been used in coal measures to predict the distribution of channels. These may not only cut out coal beds, but may also permit catastrophic water flooding. Cross-bedding has been used to map the trend of roll-front uranium ore bodies in fluvial sediments, and in gold mines, such as those of the Witwatersrand of South Africa, to predict concentrations of ore (see Chapter 3 for details).

In boreholes palaeocurrents can be interpreted from sedimentary structures. As alluded to in the previous section, microresistivity logs form the basis for this method. This was initiated in the 1950s with the invention of a tool termed the dipmeter. The dipmeter consisted of three pairs of electrodes mounted on spring-loaded arms 120° apart. The tool might, of course, rotate as it was drawn up the borehole, so a magnetic compass was included in the sonde, providing a reference from which the effects of rotation could be removed. The amount and orientation of any deviation of the borehole from true vertical depth were also recorded.

In the early days the micrologs were examined visually, events were correlated from track to track, and dips laboriously calculated by sliding little sticks of wood to and fro (a quaint biodegradable device probably unfamiliar to younger readers, called a 'slide rule'). The early dipmeter was used principally to identify structural dip and strike (Figure 2.20).

The resultant data are plotted on a log known colloquially as a 'tadpole plot'. This shows the amount and direction of dip at each depth for which a correlateable event has been picked and measured. Regular patterns are often readily discernible on tadpole logs and colour coded (Figure 2.21).

The steady low angle dips of the 'green motif', may normally be confidentally attributed to structural dip in shale. There is never a unique explanation for the upward-declining 'red' and for the upward increasing 'blue' motifs. They can both be produced by structural as well as sedimentological phenomena. Random 'bag o-nails' motifs may be an artefact of borehole conditions as well as due to geological phenomena, such as fractures, slump bedding, debris flows, and so forth. In the old days sedimentary dips could only be recognized and interpreted with considerable imagination.

Over the years the 3-arm dipmeter has been replaced by 4-arm, and 6-arm types. The slide rule was replaced by the computer, which can not only do

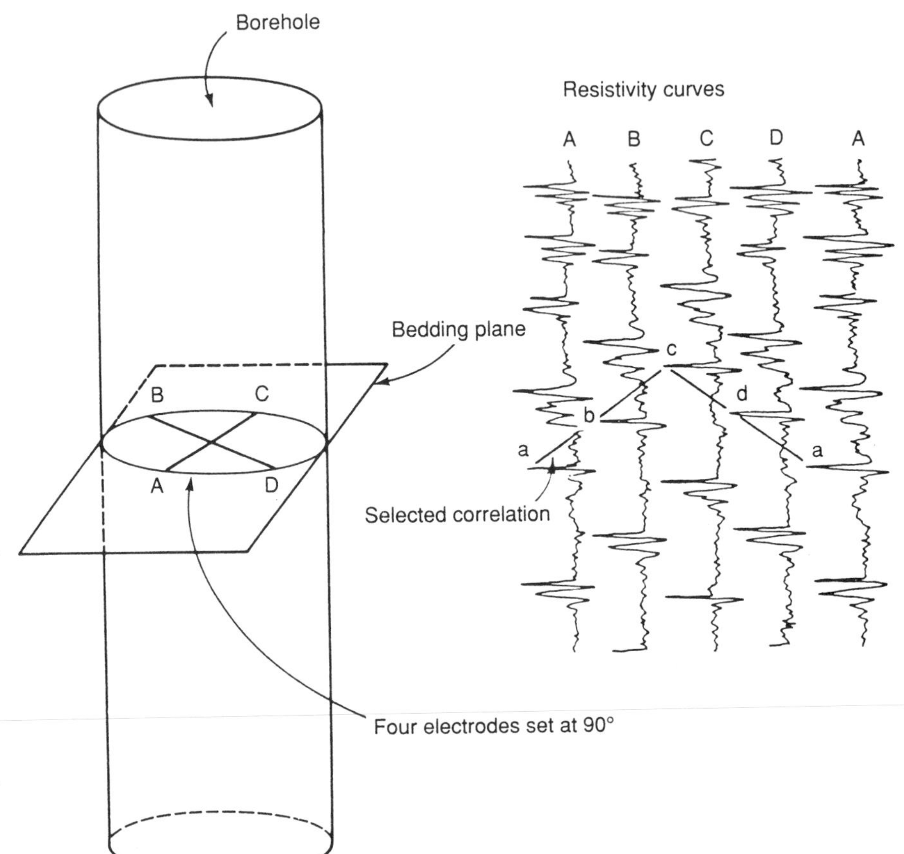

Figure 2.20 Sketch to show the mode of operation of a dipmeter log, and the way in which dip directions are calculated from selected correlations. This is a 4 arm example, mid way between the early 3 arm types and modern 4 arm multiple track imaging tools.

the calculations, but also identify correlatable events on the micrologs. For many years geologists were nervous about the validity of the dips calculated from dipmeter logs, because they were so hard to verify. Serra (1985) described the art of dipmeter interpretation at its zenith, before the advent of imaging logs.

Some ten years ago the dipmeter rapidly metamorphosed into the borehole imaging logs described in the previous section. Now, therefore, it is possible to have a borehole image displayed on a monitor, with the log of your choice to one side, together with a tadpole plot. In the same way that a plane can be flown either by autopilot or manually, so is it possible for either the operator or the computer to select the correlative events on the image.The computer will then calculate their amount and direction, remove tectonic tilt (if necessary), display the dips singly on a tadpole plot, and group them at selected intervals into rose diagrams. Now geologists can really see the sedimentary structures from which the dips are calculated (Figure 2.22).

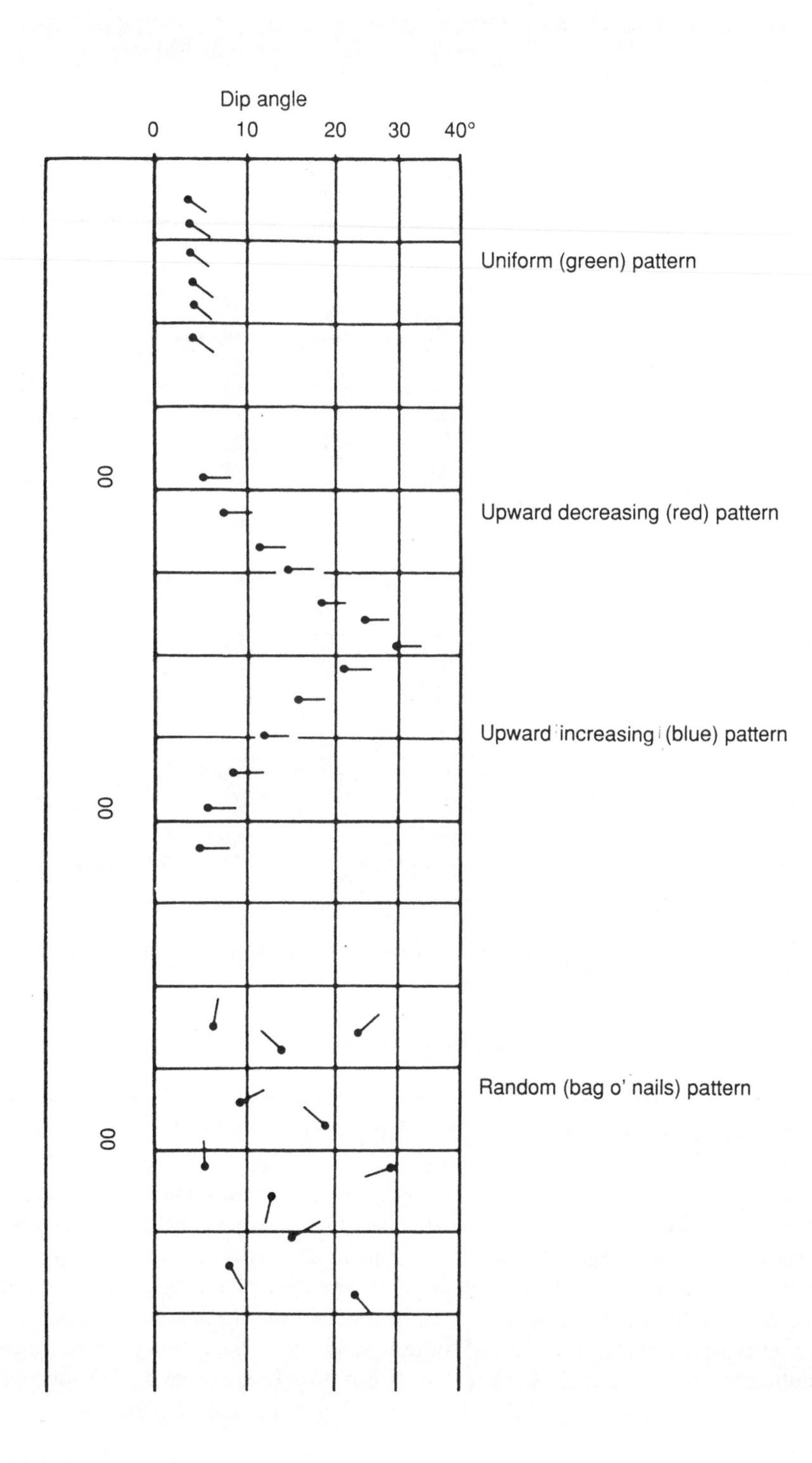

Figure 2.21 Sketch of a tadpole log, the conventional way of displaying dipmeter data. The head of the tadpole indicates the depth at which the calculation has been made, and the amount of dip is read off the horizontal scale. The tadpole tail points in the direction of dip. Green patterns are normally indicative of tectonic tilt in shales. Blue and red patterns can be produced by many different structural and sedimentological phenomena. Random dips may be an artefact of the tool, as well as geologically significant.

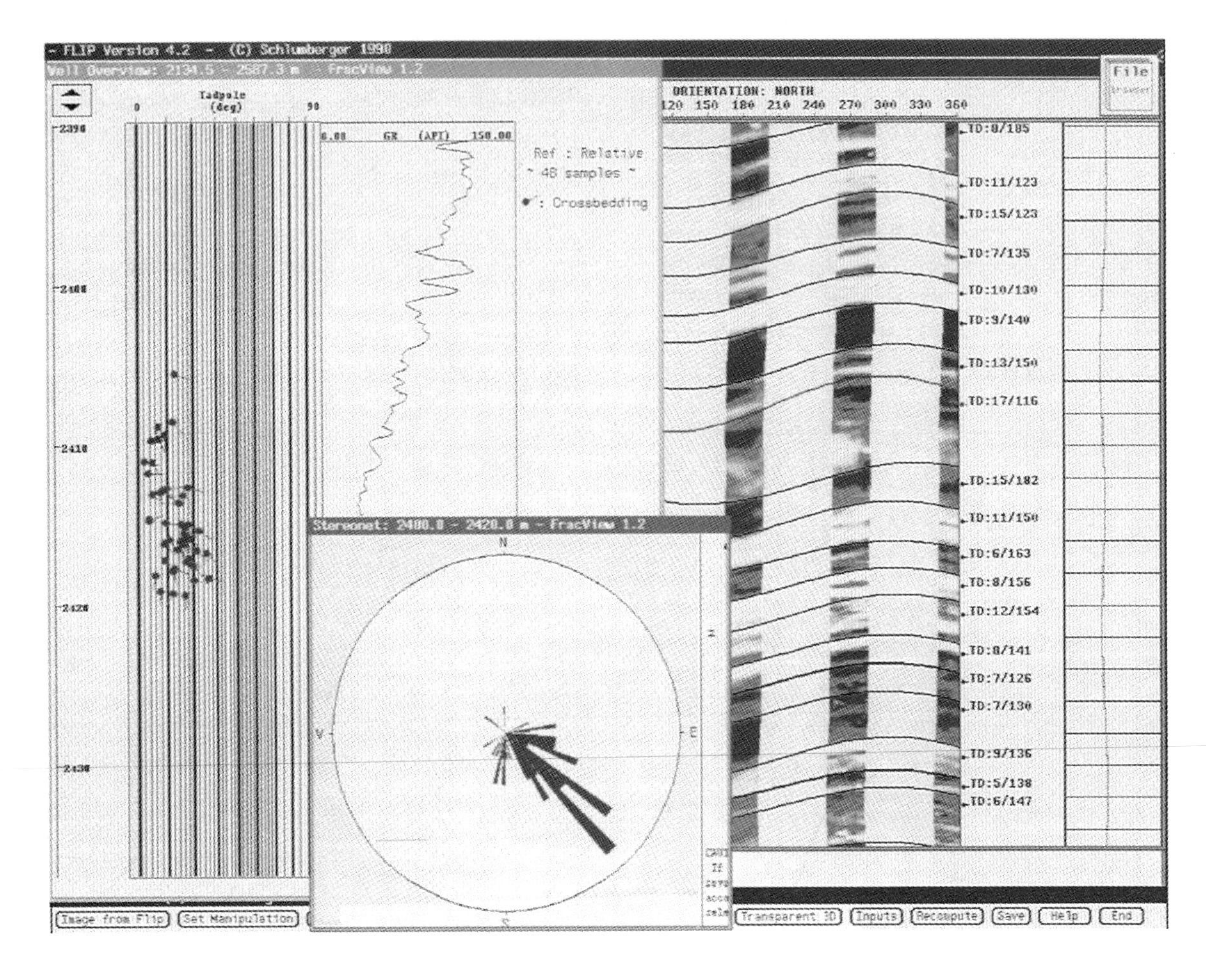

Figure 2.22 Work station display of borehole images, gamma calibration log, tadpole plot and dip azimuth. Courtesy of J. Roestenburg and Schlumberger (also reproduced as Plate 4).

With the advent of **Diamage*** one can take one step more. It is possible, not only to match directly core photographs with the borehole image, but also to see if the directional measurements calculated by the computer actually exist in the rock (refer back to Figure 2.18).

SUMMARY AND CONCLUSION

This chapter has covered a great deal of ground, so a summary is appropriate. Each parameter that defines a sedimentary facies can be used to diagnose depositional environment. The approach for each one will now be summarized.

Geometry is of limited value at outcrop because of the blanket shape of most facies. Reefs and channels may be exceptions. In the subsurface, the geometry of a facies used to be of limited value, because it was the last parameter to be known during the drilling of a field. Now, however, 3D seismic may delineate facies geometry before a single well is drilled, and may

* Trademark of Schlumberger

instantly indicate its depositional environment. The main impetus for the diagnosis of ancient depositional environments was to predict reservoir geometry and trend. 3D seismic is now reducing environmental interpretation to a topic of only academic interest.

Lithology is of limited use for interpreting the depositional environment of clastics, though vertical grainsize profiles are very helpful, and these can often be identified in geophysical well logs. With limestones, however, grain type and texture are of great diagnostic value.

Palaeontology is a very important indicator of depositional environment, but it presents pitfalls. To use fossils to interpret environment requires knowledge of the ecology of extinct fossil groups, and the ability to differentiate life assemblages from death assemblages. Trace fossils are of great use in environmental interpretation, because they circumvent both of these problems.

Sedimentary structures indicate depositional process. Since most processes operate in several environments, no single structure indicates any one environment. It is suites and sequences of structures that are environmentally significant.

Sedimentary structures can be used to infer palaeocurrent direction. Some palaeocurrents indicate palaeoslope, others do not. Palaeocurrent pattern is of importance in environmental diagnosis, since unipolar patterns indicate unidirectonal flow, as in rivers and deltas. Bimodal patterns may be an indicator of tidal flow in tidal channels, and on shallow marine shelves. Regional palaeocurrent studies help to delineate palaeogeography. Vertical changes in palaeocurrents in slope-controlled environments help to unravel the interrelationship between structural movement and sedimentation.

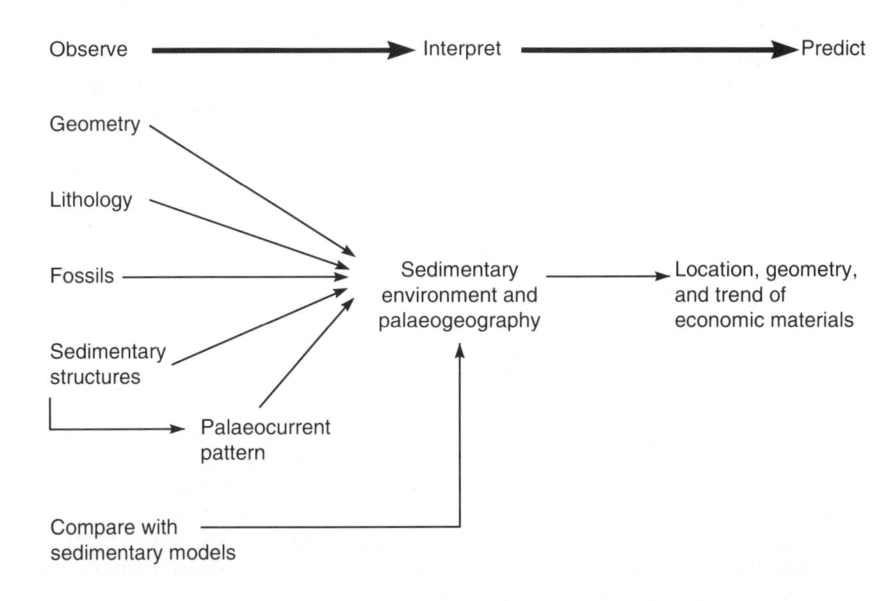

Figure 2.23 Flow diagram to illustrate the methodology of environmental diagnosis. Note that four of the five parameters that define a sedimentary facies are observational. The fifth, palaeocurrent pattern, is actually interpreted from sedimentary structures.

Figure 2.23 illustrates the multimethod approach to finding out how a sedimentary facies was deposited. It is generally unwise to diagnose the depositional environment of a sedimentary facies using any one parameter or technique. Geology is at best an imprecise science in which it is seldom possible to make deterministic statements. Geology deals in probabilities, not certainties. The depositional environment of a sedimentary facies cannot be 'proved' in the same way that a mathematical proposition may be proved, nor may an environmental interpretation be repeated and verified, like a laboratory experiment. The student reader must learn to beware of the kind of dogmatic statements with which this chapter abounds. Studious readers may care to immure themselves in the nearest geological library for the next year checking references. Out and out sceptics may prefer to spend the next ten years in the field. Let the rest read on remembering that all generalizations are dangerous including this one.

REFERENCES

Bourke, L. T. (1992) Sedimentological borehole image analysis in clastic rocks: a systematic approach to interpretation. In A. Hurst, C. M. Griffiths and P. F. Worthington (eds), *Geological Applications of Well Logs*, Vol. II, Geol. Soc. Lond. Sp. Pub. **65**, 31–42.

Bromley, R. G. (1990) *Trace Fossils: Biology and Taphonomy*, Unwin Hyman.

Collinson, J. D. and Thompson, D. B. (1988) *Sedimentary Structures*, 2nd edn, Allen & Unwin, London.

Dodd, J. R. and Stanton, R. J. (1990) *Palaeoecology, Concepts and Applications*, 2nd edn, Wiley, Chichester.

Goldring, R. (1991) *Fossils in the Field*, Longman, Singapore.

Harker, S. D., McCann, G. J. and Adams, J. T. (1990) Methodology of formation microscanner tool image interpretation in Claymore and Scapa Fields (North Sea). In A. Hurst, M. A. Lovell and A. C. Morton (eds), *Geological Applications of Wireline Logs*, Geological Society of London Sp. Pub., **48**, 11–25.

Hubbard, R. J., Pape, J. and Roberts, D. G. (1985) Depositional sequence mapping as a technique to establish tectonic and stratigraphic framework and evaluate hydrocarbon potential on a passive continental margin. In O. R. Berg and D. G. Wolverton (eds), *Seismic Stratigraphy*, Vol. II, Amer. Assoc. Petrol. Geol., **39**, 79–92.

Johnson, M. R. (1992) A proposed format for general purpose comprehensive graphic logs, *Sed. Geol.*, **81**, 289–98.

Larsen, G. (1993) *Far Side Calendar*, Farworks Inc., Hong Kong.

Lewis, D. W. and McConchie, D. (1994) *Analytical Sedimentology*, Chapman & Hall, London.

Miall, A. D. and Tyler, N. (eds) (1991) The three-dimensional facies architecture of terrigenous clastic sediments and its implications for hydrocarbon discovery and recovery, *Soc. Econ. Pal. and Min., Concepts on Sedimentology and Paleontology*, **3**.

Mitchum, R. M. (1977) Glossary of terms used in seismic stratigraphy. In C. E. Payton (ed.), *Seismic Stratigraphy – Applications to Petroleum Exploration*, Amer. Assoc. Petrol. Geol., Mem. No. **26**, 205–12.

Odin, G. S. (ed.) (1988) Green marine clays, *Developments in Sedimentology*, **45**, Elsevier, Amsterdam.

Pacht, J. A., Bowen, B., Schaffer, B. L. and Pottorf, W. R. (1993) Systems tracts, seismic facies, and attribute analysis within a sequence-stratigraphic framework – example from the offshore Gulf coast. In E. G. Rhodes and T. F. Moslow (eds), *Marine Clastic Reservoirs*, Springer-Verlag, Berlin, 3–21.

Rider, M. H. (1990) Gamma-ray log shape used as a facies indicator: critical analysis of an oversimplified methodology. In A. Hurst, M. A. Lovell and A. C. Morton (eds), *Geological Applications of Well Logs*, Geol. Soc. Lond. Sp. Pub. **48**, 27–37.

Sclatt, R. M., Jordan, D. W., D'Agostino, A. E. and Gillespie, R. H. (1992) Outcrop gamma-ray logging to improve understanding of subsurface well correlations. In A. Hurst, C. M. Griffiths and P. F. Worthington, *Geological Applications of Well Logs*, Vol. II, Geol. Soc. Lond. Sp. Pub. **65**, 3–20.

Selley, R. C. (1976) Subsurface environmental analysis of North Sea sediments, *Bull. Amer. Assoc. Petrol. Geol.*, **60**, 184–95.

Serra, O. (1985) *Fundamentals of Well Log Interpretation*, 2 vols, Elsevier, Amsterdam.

Stephens, M. (1994) Architectural element analysis within the Kayenta Formation (Lower Jurassic) using ground-probing radar and sedimentological profiling, southwestern Colorado, *Sed. Geol.*, **90**, 179–211.

3 River deposits

INTRODUCTION: RECENT ALLUVIUM

Most of the Earth's land surfaces are areas of erosion or equilibrium. Continental depositional environments are relatively rare, so ancient continental deposits are volumetrically subordinate to marine ones in the stratigraphic record. Significant continental depositional environments include:

- fanglomerate
- fluvial – braided channel systems
  – meander channel systems
- lacustrine
- eolian.

This chapter deals with fanglomerate and fluvial sediments. Chapters 4 and 5 deal with eolian and lacustrine deposits respectively. Fluvial sediments are very important, not only because they are volumetrically significant in the stratigraphic record, but also because they are important aquifers, they contain placer deposits of diverse heavy minerals. They contain deposits of coal, and ancient alluvium also makes outstanding petroleum reservoirs. The outline used in this, and subsequent chapters, is first to describe the modern environment and resultant sedimentary model, then to present a case history, then discuss environmental diagnosis, and finally to consider the economic importance of the deposits of the environment in question.

MODERN FLUVIAL ENVIRONMENTS

Fluvial deposits form part of an interrelated series of depositional systems that may extend down the depositional slope for hundreds of kilometres from mountain front to coast. The dominance of any one system depends on climate, gradient and grainsize. A spectrum of processes and products operate across these environments (Callow and Petts, 1992; Fielding, 1993). The deposits of fanglomerate, braided and meander channel systems will be dealt with in turn.

Fanglomerates

Fanglomerate is a term that aptly describes alluvial fan conglomerates.

These deposits commonly occur at basin margins, especially along active fault scarps, and adjacent to mountain fronts. When sediment is in short supply fanglomerates are separated from braided alluvial outwash plain by a scoured pediment. With abundant sediment supply, however, fanglomerates pass down gradient into braided alluvial outwash fans without a break.

Figure 3.1 Modern fanglomerates. Wastwater screes, the Lake District, England. Photograph courtesy of M. Bremner.

Recent fanglomerates (Figure 3.1) occur in periglacial climates such as those of the northern Rocky Mountains of North America and Greenland, and in the arid terrains of modern mountainous deserts such as the Tibesti mountains of the Sahara and the Oman mountains of southern Arabia. The depositional processes of fanglomerates include gravity collapse and slide, debris flows, especially after heavy rain, and broader sheet floods that episodically transport sands and gravel. The episodic nature of fanglomerate transportational processes is due to ephemeral flash floods in arid terrains, and to the seasonal melt waters that flow off ice caps and glaciers in periglacial climates.

The resultant sediments of these processes are, of course, very coarse-grained, because they have just been eroded off a steep slope. They range in particle size from boulders downwards and are very poorly sorted. Sedimentary structures are restricted to poorly developed bedding. This commonly has a depositional dip that can be used as a palaeocurrent indicator in ancient fanglomerates. When plotted out the ancient depositional dips may exhibit a radial pattern that can be used to delineate individual fans, and identify the palaeowadis from which they radiated (Figure 3.2). It is, of course, commonplace for fans to debouch from many wadis along a mountain front, and to coalesce to form a composite wedge of fanglomerates.

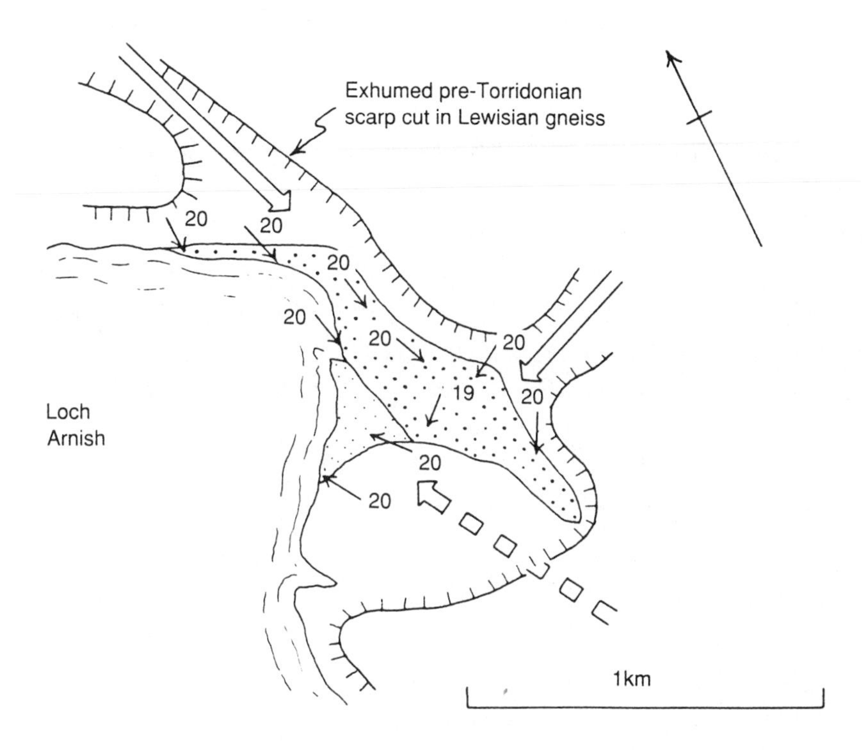

Figure 3.2 Map of PreCambrian fanglomerates at Torran, on the Isle of Raasay, Scottish Hebrides. Depositional dips (small arrows) radiate from exhumed palaeowadis cut into Lewisian Gneiss basement. Thus may sediment flow directions be deduced (large arrows). From Selley, 1965, Journal of Sedimentary Petrology; by courtesy of the Society of Economic Paleontologists and Mineralogists.

Sometimes fanglomerate builds out not over land, but into lakes or seas. These deposits thus merge imperceptibly in to those of the fan delta environment (see pp. 150–151). For further accounts of modern fanglomerates, and their ancient counterparts see Nilsen (1982).

Alluvium of braided channel systems

Fanglomerates commonly pass down slope in to finer grained alluvium. In arid environments these two facies may be separated by a wind-scoured pediment surface. Geomorphologists have identified several different types of channel system, and have attempted to identify the physical processes that control their formation (e.g. Bevan and Kirkby, 1993; Stansitreet and McCarthy, 1993). The two main types that are commonly recognized are braided and meandering channel systems (Figure 3.3).

Braided channel systems tend to occur on steeper gradients and in coarser sediments than meander channels. Flow tends to be ephemeral rather than permanent. Recent braided fluvial systems thus occur at the edges of mountains

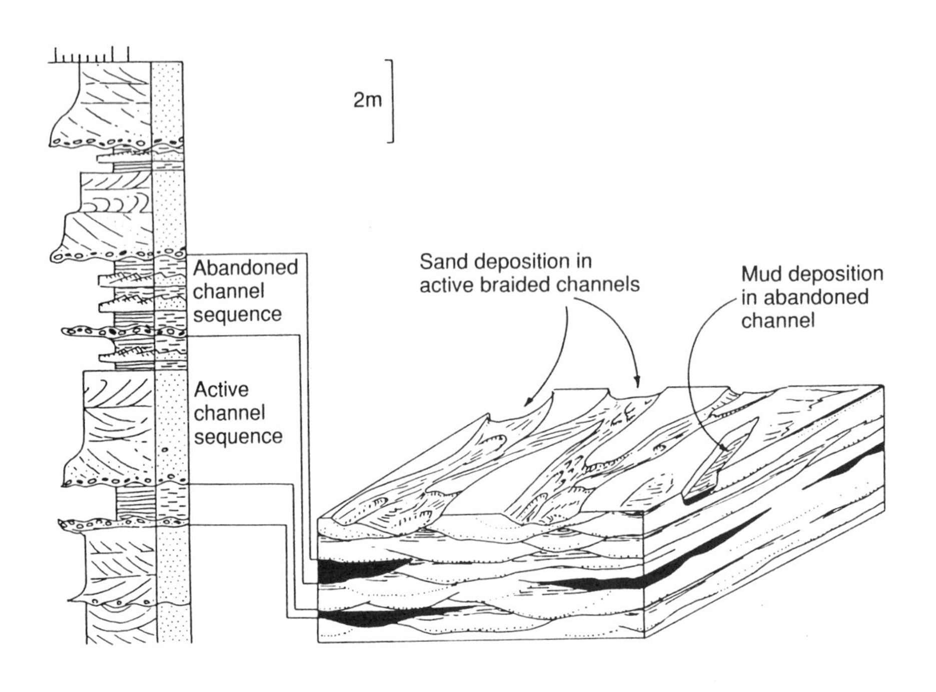

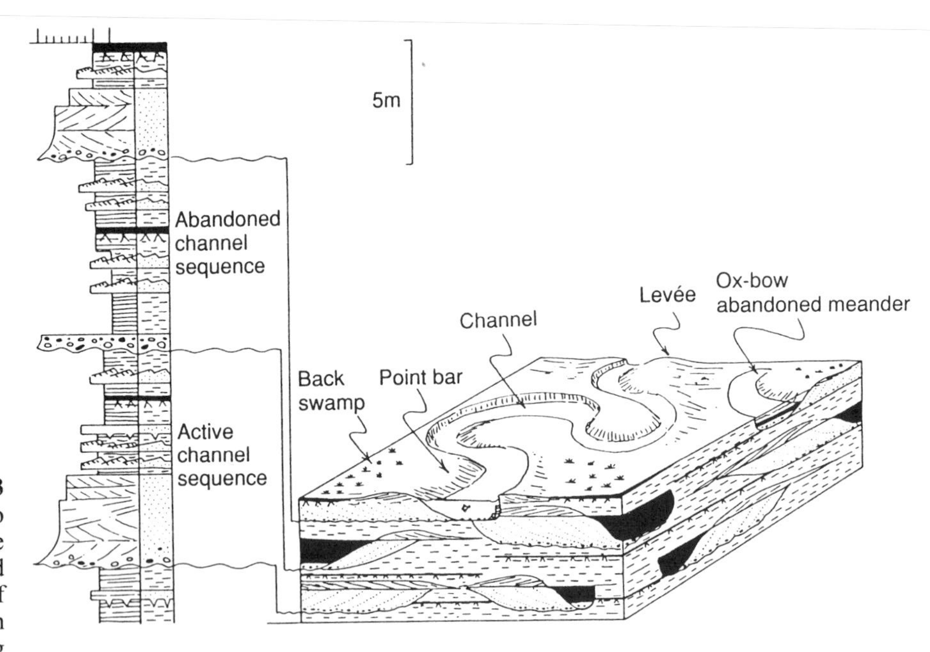

Figure 3.3 Geophantasmogram to illustrate the subenvironments and sedimentary sequence of braided alluvial outwash (upper) and meandering fluvial channels (lower) from Selley (1988).

and ice sheets, in deserts, and less commonly in tropical rain forest. In these regions erosion is rapid, discharge sporadic and high, and there is little vegetation to hinder run off. Because of these factors rivers are generally overloaded with sediment. A channel is no sooner cut than it is choked with its own detritus. This is dumped as bars in the centre of the channel, around which two new channels are diverted. Repeated bar formation and channel branching generates a network of braided channels over the whole depositional area.

Channels tend to be broad and relatively shallow, in contrast to the narrower deeper channels of meandering rivers. Within the channels the traction currents transport the sediment in the form of longitudinal and transverse bar systems. These deposit cross-bedded sands, with occasional shooting flow flat bedding.

Recumbent foresets and other penecontemporaneous quicksand deformational structures are common. Unlike the alluvium of meandering rivers systems, there is no clearly defined flood plain on which fine sediment may settle out of suspension. The only environment in which clay and silt are deposited and preserved is in rare abandoned channels. Channel abandonment is caused by channel choking and switching, and by stream piracy due to rapid headward erosion in the soft unconsolidated sediments. Lag conglomerates on the floors of ancient shale-infilled channels point to abandonment being a catastrophic event (Figure 3.4).

Thus the alluvium of ancient braided channel systems often tend to lack a regular organization of upward-fining or upward-coarsening sequences (Figure 3.3, upper). There are some exceptions. Coarsening-upward sequences have been described from what are interpreted as the alluvium of ancient braided channels. This can occur from the episodic advance and abandonment of fanglomerate lobes over braided outwash (Nilsen, 1982), or from the in-filling of periglacial braided channels at times of rising flood stages (Costello and Walker, 1972). Miall (1977) has proposed four subtypes of the braided model; some lack vertical organization, but one, the Donjek type, exhibits upward-fining increments. For further detailed accounts of modern braided channel systems and alluvial fans see Billi *et al.* (1992) and Newsom (1994).

Alluvium of meandering channel systems

The second main type of fluvial channel is that of the meander system. As previously noted meandering channels normally occur where gradients are low, current velocities slower, and sediment finer-grained, than is found in braided channel systems (Friend, 1993). Furthermore, it is not only lower current velocities that are noted, but also the fact that current is less episodic, and more uniform. Indeed it has been noted that alluvium attributable to meander channel systems is absent in pre-Devonian strata. This may be because the

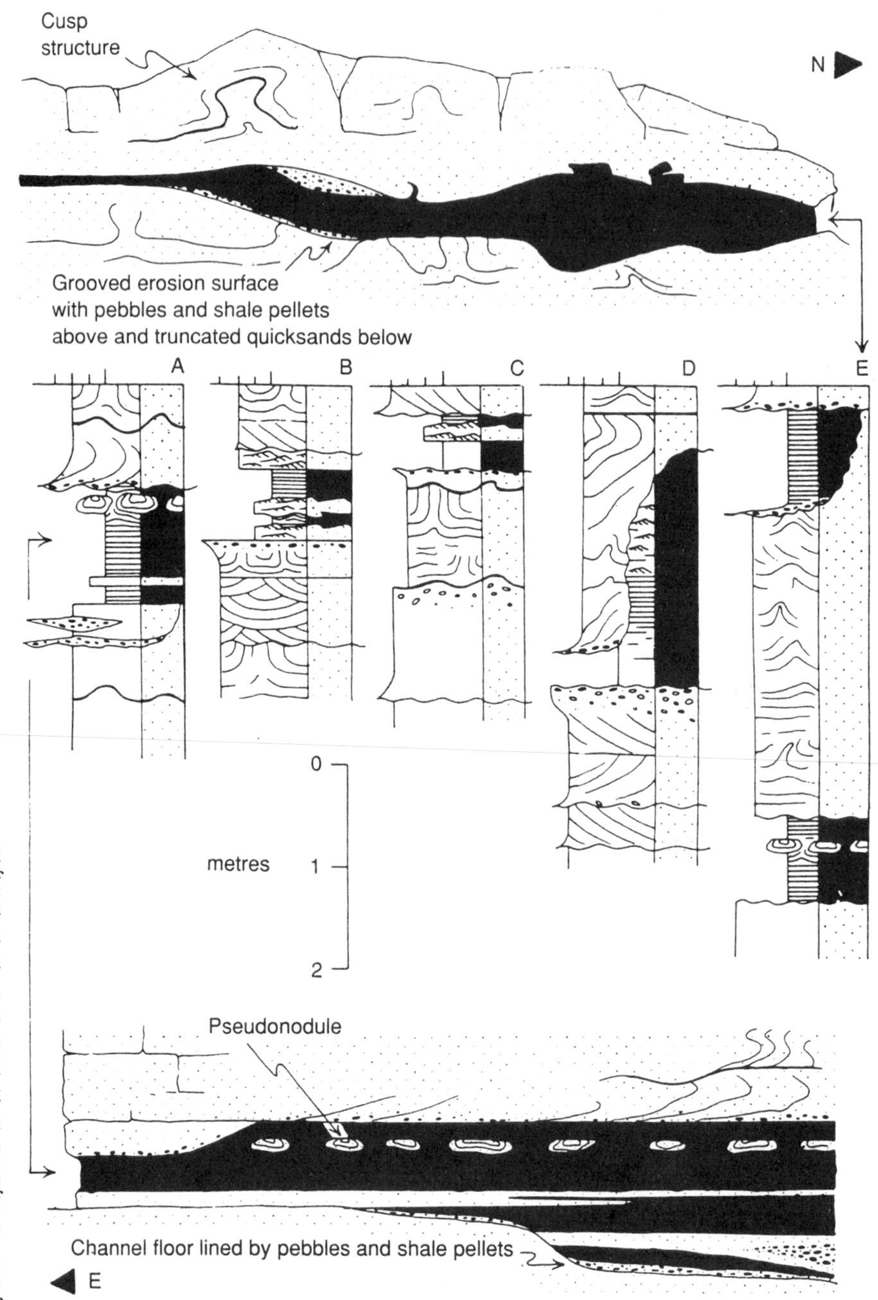

Figure 3.4 Measured sections and sketches of shale-infilled abandoned channels in Torridonian (PreCambrian) braided alluvium, Isle of Raasay, Scottish Hebrides. This is one of the diagnostic features of braided alluvium. Note how easy it may be to recognize such channels from borehole cores in the subsurface, and thus to infer the lack of continuity of such shale beds. From Selley, 1969, Fig. 4; by courtesy of the Scottish Journal of Geology.

land was not colonized by plants until that time. Thus there was no vegetation to slow down run off. Rainfall would have tended to have discharged rapidly and episodically. Hence the dominance of braiding in pre-Devonian alluvium. The Mississippi is a good example of a Recent meandering river. The deposits of a meandering river can be sub-divided into three main sub-facies due to deposition in three different sub-environments.

Floodplain sub-environment

Sheets of very fine sand, silt and clay are deposited on the over bank areas of the river's floodplain. These are laminated, ripple-marked, and often contain horizons of sand-filled shrinkage cracks, suggesting sub-aerial exposure. The presence of soils may be indicated by carbonate caliches, ferruginous laterites, and rootlet horizons. Peat may form and detrital plant debris be preserved on bedding surfaces. This sub-facies is deposited largely out of suspension during floods when the river breaks its banks.

Channel sub-environment

The lateral migration of a meandering channel erodes the outer concave bank, scours the river bed, and deposits sediment on the inner bank (point bar). This produces a characteristic sequence of grainsize and sedimentary structures. At the base an erosion surface is overlain by extraformational pebbles, intraformational mud pellets, fragmented bones, and waterlogged drift wood. These originated as a lag deposit on the channel floor, and are overlain by a sequence of sands with a general vertical decrease in grainsize. Massive, flat bedded and trough cross-bedded sands grade up into tabular planar cross-bedded sands of diminishing set height. These in turn pass up into micro-crosslaminated and flat-bedded fine sands which grade into silts of the floodplain sub-facies (Figure 3.3, lower).

Abandoned channel sub-environment

When a river channel meanders back on itself short-circuiting the flow, it causes the formation of abandoned channels, sometimes called ox-bow lakes. These become infilled with fine-grained sediment, often organic-rich, and sometimes hosting peat-forming marsh. The resultant facies are similar to those of the floodplain but are distinguishable from them by their channel geometry and because they abruptly overlie channel lag conglomerates with no intervening point bar sand sequence.

Now that the sedimentary models of fluvial systems have been outlined, a case history of an ancient fluvial deposit will be described, and its depositional environment diagnosed.

THE CAMBRO-ORDOVICIAN SANDSTONES OF THE SAHARA AND ARABIA

From Doughty (1888) to Husseini (1990, 1991) geologists have remarked on the uniformity of Lower Palaeozoic stratigraphy from the Atlantic coast of North Africa east to the shores of the large gulf between Arabia and Iran. Laterally extensive blankets of sandstone and shale crop out around the PreCambrian shields of the Sahara and Arabia, and can be traced intermittently northwards and eastwards towards the Mediterranean and Gulf coasts respectively (Figure 3.5).

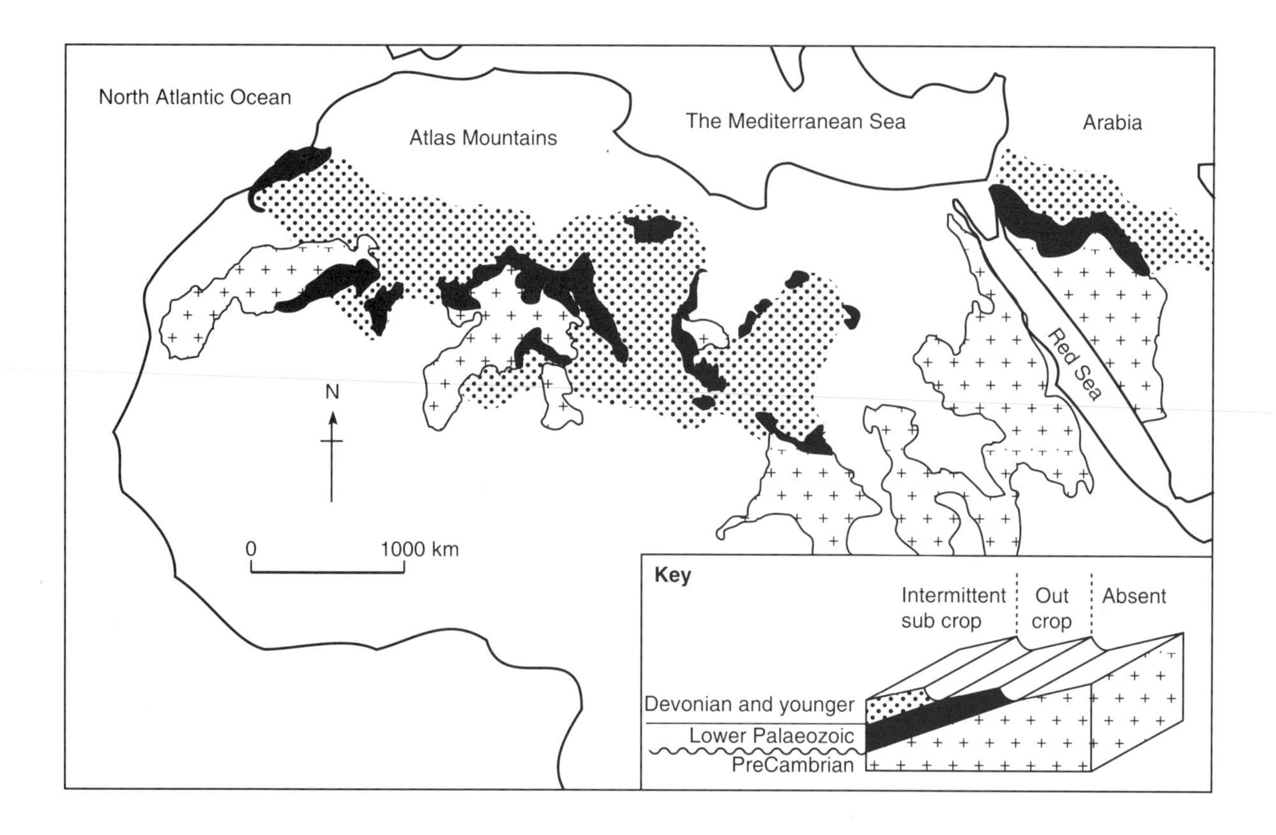

Figure 3.5 Distribution of Lower Palaeozoic sediments in North Africa and Arabia.

These formations show a remarkably uniform vertical sequence of facies. The sequence normally commences with coarse pebbly channeled sands (attributable to fluvial environments). These pass up into better sorted finer sands, commonly bioturbated (attributable to shallow marine environments). This facies passes up in to graptolitic shales, which in turn pass up, often via turbidite sands, to prograding deltaic, and finally fluvial sands (Figure 3.6).

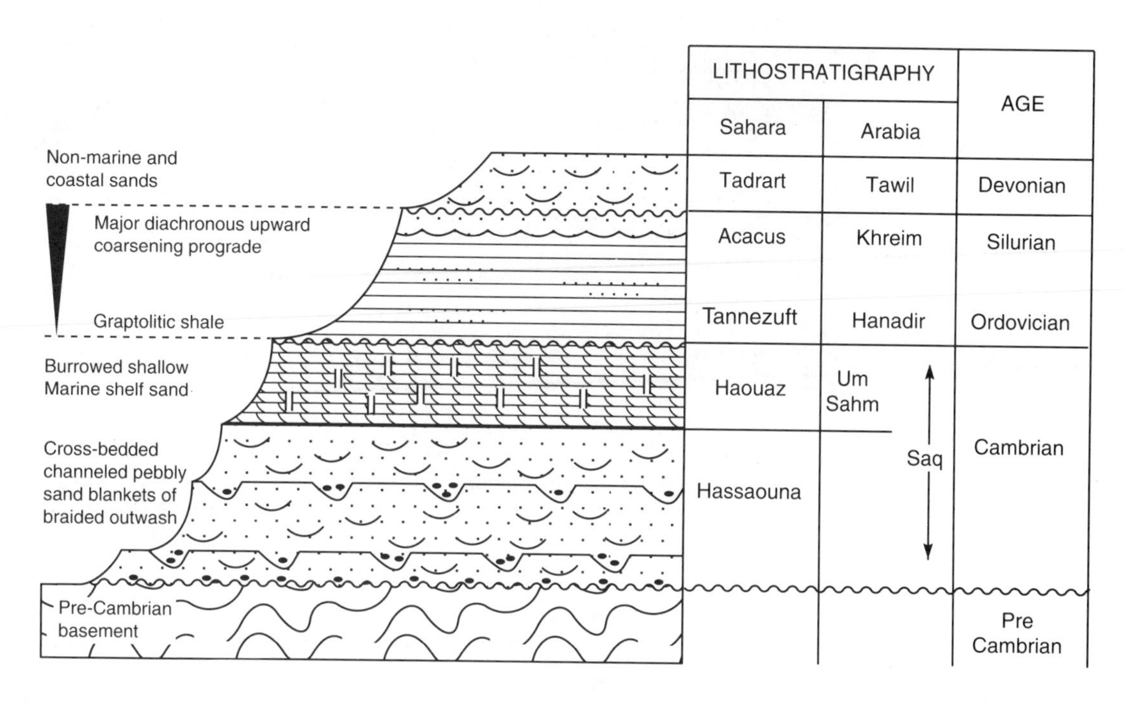

Figure 3.6 Stratigraphic section of the Lower Palaeozoic succession of North Africa and Arabia.

The earliest part of this sequence is barren of fossils, and may well include formations of PreCambrian age, but the graptolitic shales indicate Ordovician and younger ages. Though remarkably uniform in facies the sequence is strongly diachronous, and locally interrupted by erosive events and associated deposits attributed to glaciation (Powell *et al.*, 1994). The Cambro-Ordovician sandstones are of some economic importance both as aquifers (Lloyd and Pim, 1990), and as oil and gas reservoirs in Algeria, Libya, Jordan and Saudi Arabia (McGillivray and Husseini, 1992). The graptolitic shales commonly, but not invariably, act as the source for the petroleum. These Lower Palaeozoic sediments will be used in this book as case histories to illustrate ancient examples of fluvial and other environments.

Pebbly channel sand facies: description

The lowest facies is composed of coarse pebbly channel sands. This will now be described and diagnosed as of fluvial origin. Formations of this facies have essentially sheet geometries with little regional thickness variation. The base of the sequence is marked by a conspicuous unconformity with PreCambrian igneous rocks (Figure 3.7).

The unconformity is a mature pediment surface with occasional residual inselbergs. The overlying sediments range from pebbles to silt, but are largely

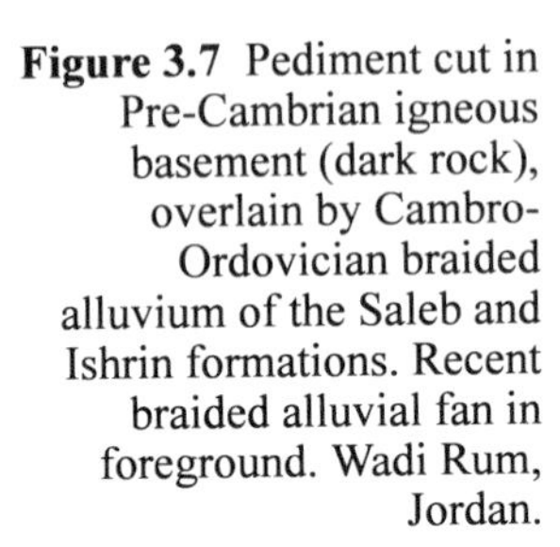

Figure 3.7 Pediment cut in Pre-Cambrian igneous basement (dark rock), overlain by Cambro-Ordovician braided alluvium of the Saleb and Ishrin formations. Recent braided alluvial fan in foreground. Wadi Rum, Jordan.

of medium to very coarse sand grade. Intraformational conglomerates of reworked siltstone occur throughout the sequence. Petrographically these sands include arkoses, especially at the base of the sequence. But they often pass up in to more mature quartz arenites. They are normally red coloured, though many formations are bleached during subsequent meteoric flushing, but the early red colour is retained by impermeable claystones and siltstones.

This facies is composed of a series of superimposed channel complexes. Three main types may be recognized. Two types are broad and shallow, either with heterogenously infilled sandstones, or shale infilled. The third type of channel has steep sided walls, and contains a sequence that fines up from a basal intraformational conglomerate, via sand back to shale.

Channels of the first type comprise about 90% of the facies. They are typically some 300 m wide and 5 m deep and infilled by sandstones with no apparent regular vertical arrangement of grainsize and structure. The channels are floored by a thin conglomeration of pebbles, and are infilled by various types of cross-bedding. Cross-bedding dips are markedly unimodal at any one locality. When plotted regionally they indicate palaeocurrents flowing off the basement shields, generally northerly in the Sahara, and easterly in Arabia. Flat bedding is also present, sometimes with scattered pebbles. Quicksand deformation structures are common, including both recumbent foresets as well as convolute bedding.

Channels of the second type are rare. Though similar in scale and profile to the sand-infilled channels, internally they are quite distinct. The channel floors are marked by an extraformational pebble lag, abruptly overlain by siltstone, whose laminae drape over the pebbles. The fill is almost entirely composed of laminated micaceous siltstone. These shale-infilled channels are overlain by the more common sand-infilled channels. Where the shale channels can be mapped they exhibit a braided morphology (Figure 3.8).

Body fossils are normally absent from this facies. Obscure unidentifiable trails and tracks occur. Bilobate trails attributable to *Cruziana* are sometimes found in the shale-filled channels.

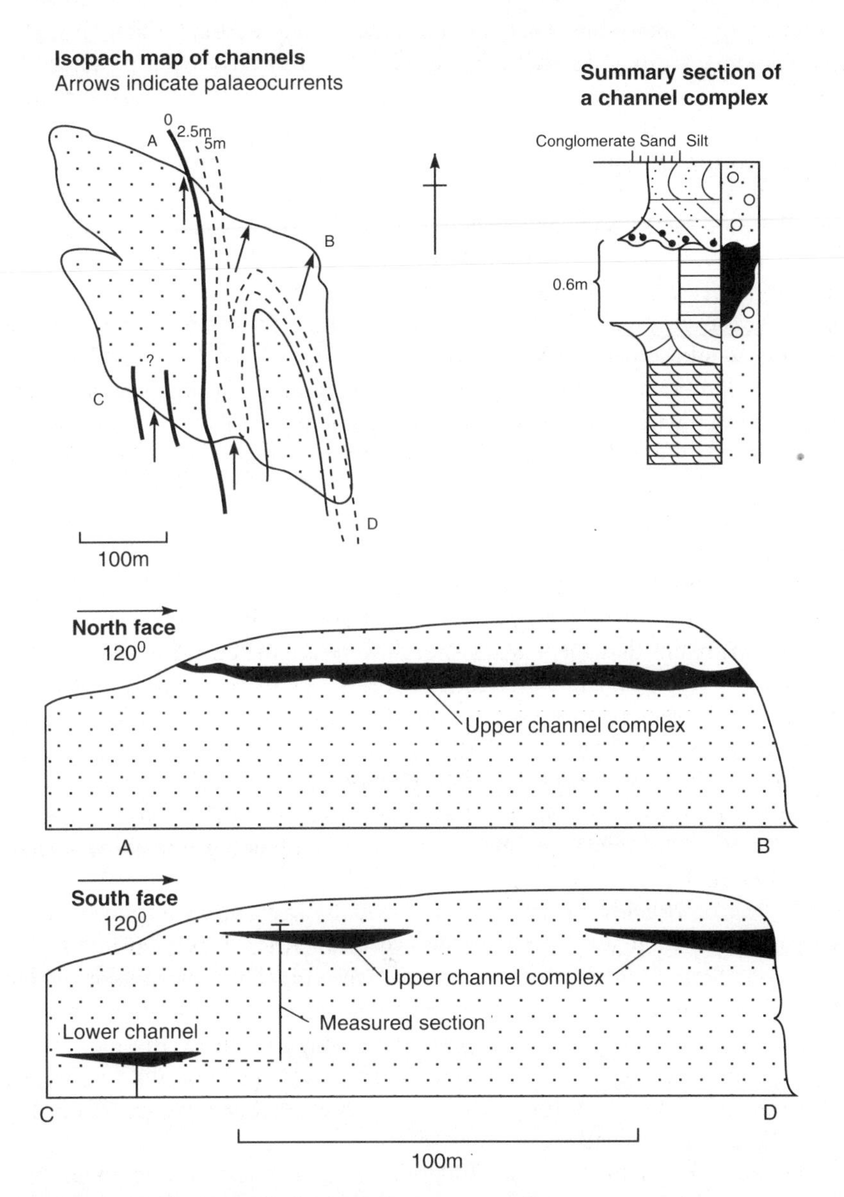

Figure 3.8 Map, field sketches and log of shale-infilled abandoned braided channels in the Disi Formation of Jordan (from Selley, 1972).

Pebbly channel sand facies: interpretation

From the preceding account it is clear that this facies was deposited in channels. The coarse texture and cross-bedding indicates sedimentation from the bed load of extremely powerful currents. The small variability of cross-bed orientations and channel trends shows that the currents were unidirectional and regionally persistent. The siltstones clearly originated from the infilling of abandoned channels. The predominant red colour suggests early diagenesis above the water table. There are no unequivocal marine fossils. *Cruziana* is commonly attributed to trilobite activity, but may have been produced by freshwater crustaceans. (Camel flies produce *Cruziana*-like trails on modern sand dunes.) The combination of sedimentologic parameters described above conform to the braided sedimentary model described earlier. Figure 3.3 may serve as an illustration of the environment in which the pebbly channel sand facies was deposited.

It is difficult to see, however, how the steep gradients and high current velocity necessary for braided rivers could have been maintained down current over the vast tectonic shelves on which these formations were deposited. One would expect sediment to build up quickly to base level so that current velocity would be diminished and fine grained meandering alluvium would be deposited. An answer to this problem was suggested by Stokes (1950) in a study of the Shinarump and similar formations on the Colorado Plateau of the USA. Here there are several sheets of coarse sandstones and conglomerates each generally less than 100 m thick and each overlying a planar unconformity.

Stokes pointed out that, since these rocks contain terrestrial fossils, it is unlikely that they were laid down by a sea transgressing over a peneplain. It is more probable that these sand sheets were deposited on piedmont fans derived from scarps which retreated as they cut back across a pediplain. Williams (1969) put forward a similar explanation for PreCambrian Torridonian alluvial fans in north-west Scotland. This concept can be applied to the Cambro-Ordovician pebbly channel sand facies of the Saharan and Arabian Shields. In the Southern Desert of Jordan the 700 m thick Cambro-Ordovician pebbly sand facies is divisible into three formations (Figure 3.9).

The PreCambrian basement is unconformably overlain by a (penecontemporaneously?) weathered surface south of Wadi Rum. Down current to the north inselbergs protrude 35 m into the overlying Saleb Formation which thickens northwards from 30 m south of Wadi Rum to about 60 m in a distance of 30 km. The base of the overlying Ishrin Formation is marked by huge steep-sided channel complexes whose floors are lined with imbricate siltstone slabs over 1 m long. These were presumably reworked from the Saleb Formation beneath.

The Ishrin Formation is about 300 m thick and shows little regional thinning over hundreds of square kilometres. The top of the Ishrin Formation is itself incised by steep-sided channels over 5 m deep which again are sometimes completely infilled by siltstone slabs. These channels mark the base of the Disi Formation which, like the Ishrin Formation, is about 300 m thick and shows little regional thickness variation. The Disi Formation passes up abruptly, but apparently without erosion, into overlying marine shelf sands of the Um Sahm Formation (p. 199).

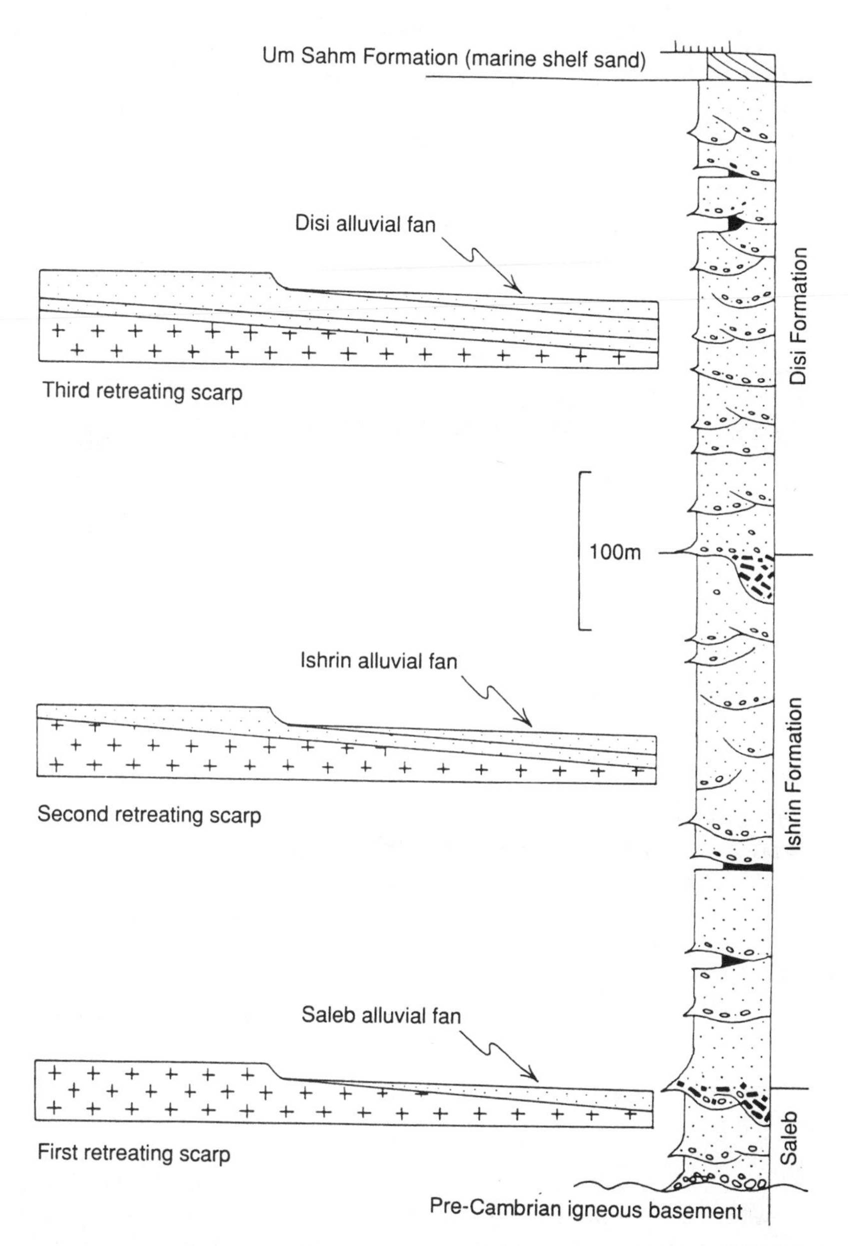

Figure 3.9 Section of Cambro-Ordovician braided alluvial sandstones in the Southern Desert, Jordan. Diagrams illustrate possible depositional mechanism of repeated rejuvenation, uplift, scarp retreat and pedimentation. Though the whole sequence is sedimentologically homogenous, sequence boundaries are delineated by regionally mappable planar surfaces dissected by steep sided channels.

Deposition from the braided fans of repeatedly retreating scarps is certainly a most attractive explanation for these extensive alluvial deposits. According to this concept the Arabian Shield was uplifted three times. Each phase of rejuvenation of the landscape caused scarps to retreat into the hinterland of the shield. The first scarp would have cut a pediment into the basement, on which the braided alluvium of the Saleb Formation was deposited. The next phase of

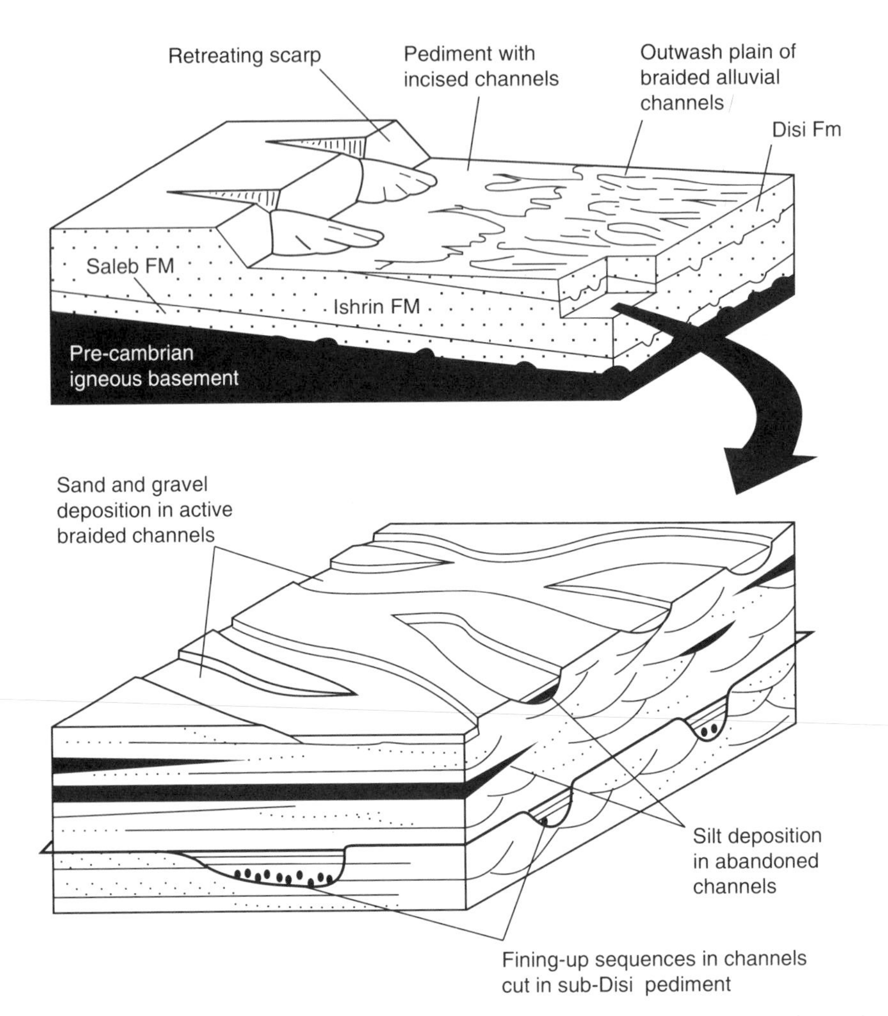

Figure 3.10 Geophantasmogram to illustrate the depositional model for Cambro-Ordovician braided alluvial sands of the Sahara and Arabia, and to illustrate the mechanism whereby laterally extensive blanket formations may be deposited by repeated rejuvenation, pedimentation and scarp retreat (from Selley, 1972).

uplift caused a second scarp to cut a new pediment into the Saleb deposits and bury it beneath Ishrin alluvium. A third repetition of this process deposited the Disi Formation (Figure 3.10).

The steep sides of the sub-Ishrin and sub-Disi channels indicate that they were cut into previously lithified sandstone, not in to soft penecontemporaneously deposited sands. An interesting alternative hypothesis is that the channels were cut in to permafrost; each formation being a periglacial outwash braid plain associated with discrete glacial episodes within the overall Ordovician-Silurian glaciation that is well documented both in Arabia and the Sahara (see Powell *et al.* (1994) and Biju-Duval (1974) respectively).

A process of repeated rejuvenation, pedimentation and scarp retreat can explain the continuity of the Cambro-Ordovician braided sand sheets of the Arabian and Saharan deserts, and elswhere in time and space. Some 40 years old, this concept of continental sequences being bounded by rejuvenation surfaces has been rejuvenated in the modern sequence stratigraphic style (Shanley and McCabe, 1994).

ECONOMIC ASPECTS

Alluvial deposits are of economic interest in the search for petroleum, uranium, metals and coal. Alluvial deposits can make good petroleum reservoirs where there are adjacent source beds. Since these may be absent, especially in arid continental basins, an unconformity generally separates petroliferous alluvial reservoirs from their source rocks. A major distinction must be made between the petroleum potential of the sand blankets of braided channel systems, and the shale-enclosed isolated channel sands of meander belts. The former give rise to giant structurally trapped fields, while the latter host small stratigraphically trapped fields.

Braided channel systems can deposit thick blankets of porous permeable sands with few impermeable shale beds. Subsequently the region may subside, often accompanied by faulting, and be drowned by a marine transgression. This may result in the deposition of organic rich muds. After burial petroleum may migrate from the shales and become trapped in truncated braided alluvial sands on the crests of anticlines or tilted fault blocks (Figure 3.11).

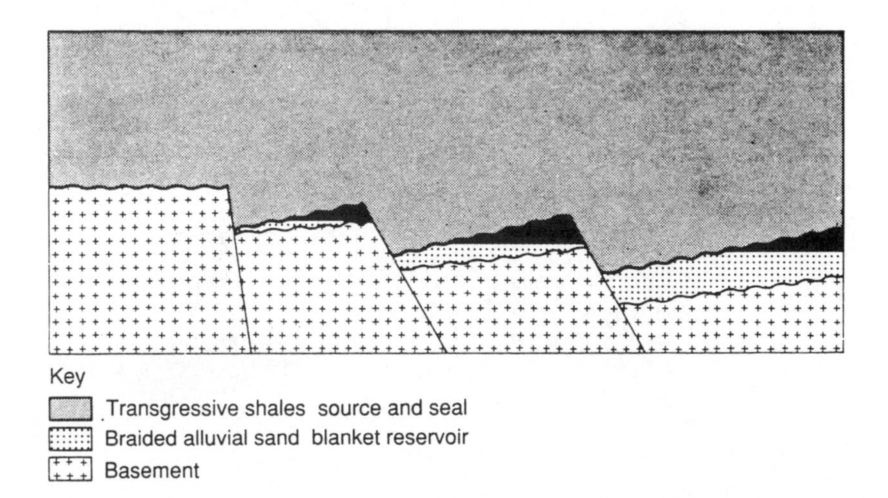

Figure 3.11 Cartoon to show how braided alluvial sand sheets may form major petroleum reservoirs in tilted fault blocks when transgressed by organic rich shales. Examples cited in text. Oil shown in black.

Meteoric flushing beneath the unconformity may have lead to the development of secondary solution porosity. This sequence of events has lead to the development of giant fields at Prudhoe Bay, Alaska (Jamison, Brockett and McIntosh, 1980; Atkinson *et al.*, 1989), Hassi Messaoud, Algeria (Balducci and Pommier, 1970) and the Sarir and Messla fields of Libya (Sanford, 1970, Clifford *et al.*, 1980) and Statfjord in the North Sea.

Because of the high ratio of shale to sand the alluvium of ancient meander channel flood plain systems can seldom host giant structural petroleum traps. Smaller stratigraphic accumulations are the norm. The limits of such fields may conform solely to the boundaries of a sand-filled channel. Usually however there is an element of closure due to regional tilt.

Figure 3.12 illustrates four common types of trap in alluvial flood plain deposits. Sometimes a sand-filled meander channel has been tilted in the direction of termination of point-bar sands where they are sealed laterally by overbank shales (Figure 3.12a). The Recluse field of Wyoming may be an example of this type of trap (Forgotson and Stark, 1972). It is important to

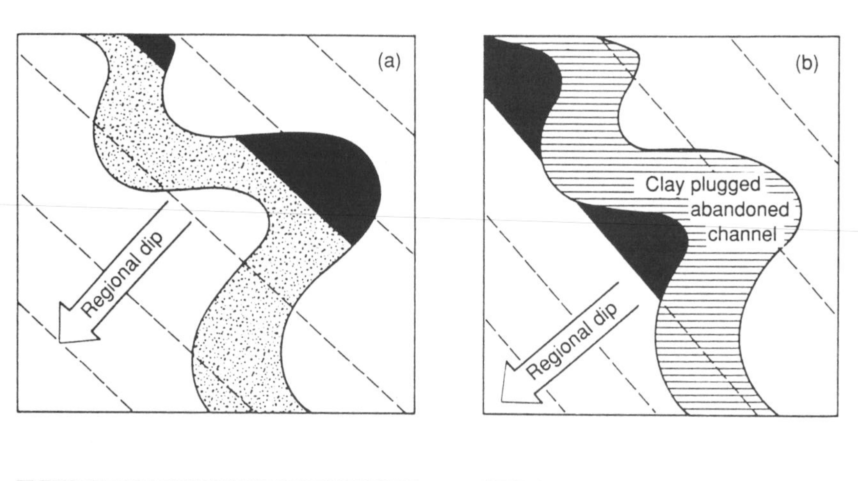

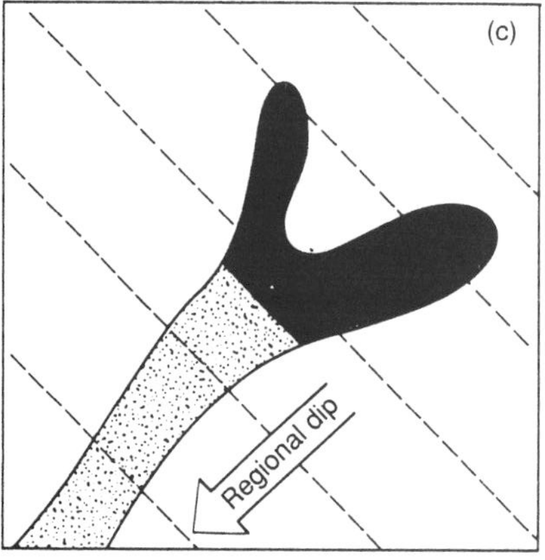

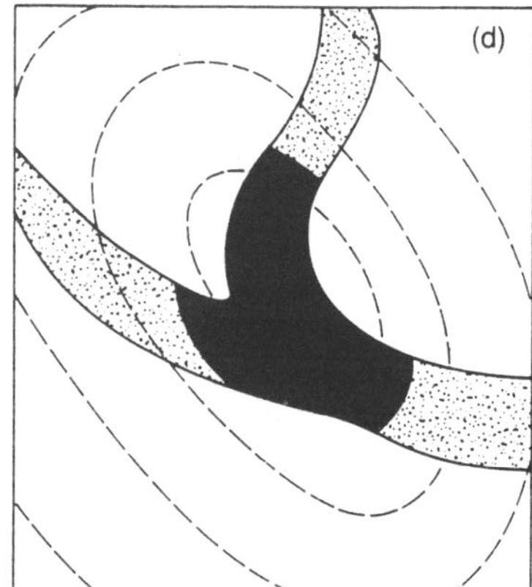

Figure 3.12 Cartoons to show how petroleum may be trapped in alluvial flood plain deposits. Note that, though in most cases the reservoir is provided by sand within the channel, in case B the channel is clay-plugged and acts as a seal. Examples of the various types of trap are cited in the text.

remember that a channel is an environment for transporting sand, but is not always an environment in which sand is deposited. Some channels become abandoned and clay plugged. Thus there are cases where abandoned channel shales have provided an updip seal to petroleum migration (Figure 3.12b).

Examples of this type of trap include the Coyote Creek and Miller Creek fields of Wyoming (Berg, 1968). If tilting is in the direction of the palaeodip then oil may migrate far up a sand-filled channel and become trapped where it finally pinches out between impermeable beds above and below (Figure 3.12c). The Clareton and Fiddler Creek fields of Wyoming are of this type. Lastly channels may cross-cut anticlines giving rise to very elusive accumulations (Figure 3.12d). The Pikes Peak field of Saskatchewan illustrates this type of trap (Putnam, 1983). Whereas the production of petroleum from braided sand blankets is fairly easy, production from discrete channel sands cut in to flood plain shales can be very challenging. It is very difficult to predict the continuity of fluvial sand bodies, and to establish which sands are in fluid communication with one another (North and Prosser, 1993).

Aside from oil and gas, alluvial deposits can also be metalliferous. The gold deposits of Witwatersrand in South Africa are a case in point. These have been extensively described and discussed in the literature (e.g. McCarthy, 1990). The Rand basin lies on the PreCambrian basement of South Africa. It measures about 250 km from north-east to south-west and 170 km from north-west to south-east. It is infilled by over 8 km of PreCambrian clastics which coarsen upwards and north-westwards towards their presumed source. There has been continuous debate about the depositional environment of the Witwatersrand deposits and of the genesis of the gold and uranium that they contain. Mapping of pebble size and palaeocurrent directions has shown that the sediments were deposited in a series of fans that radiated out into the basin from several points along the western margin (Figure 3.13).

Current opinion favours a braided alluvial outwash plain depositional environment. Regardless of whether the ore is a detrital placer, or of syngenetic origin, there is no doubt about the close correlation between its distribution and sedimentary facies. Regionally the ore is richest near fan heads, but on a local scale is concentrated in conglomerate channels some 600 m wide and 60 m deep.

Uranium mineralization occurs in Triassic-Jurassic alluvium in the Colorado Plateau and in Eocene alluvium in Wyoming, USA. Again there is debate as to the origin of the ores, but widespread agreement about the close correlation between mineralization and sedimentary facies. On a regional scale ore bodies occur in arcuate zones, halfway down ancient alluvial fans. There appears to have been a critical ratio of permeable sands to impermeable shales which favoured mineral precipitation (Crawley, 1982). This has been noted for example in the Uraven mineral belt within the Morrison alluvial fan of Colorado (Fischer, 1974), and in the Puddle Springs fan of

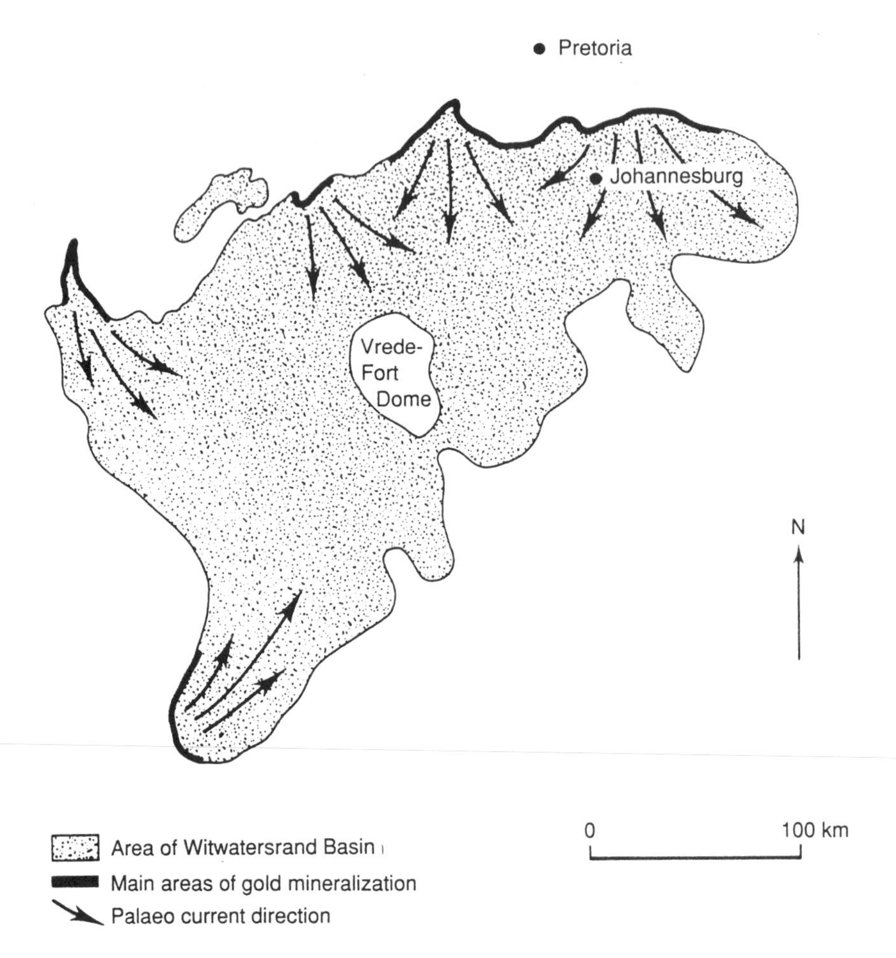

Figure 3.13 Map of the Witwatersrand basin (Pre-Cambrian), South Africa, showing the correlation between gold and braided alluvial fan heads. (For authentication see McCarthy, 1990.)

Wyoming (Galloway *et al.*, 1979). On a local scale the ore occurs in meniscus shaped bodies termed 'roll fronts' within fluvial channels. Precipitation of the ore probably occurred at the interface of mixing of connate and uranium-enriched meteoric waters (Figure 3.14).

Coal occurs in ancient alluvial deposits. For convenience sake the application environmental interpretation to coal will be dealt with in Chapter 6, which is concerned with deltas.

DIAGNOSIS OF FLUVIAL DEPOSITS

Since the deposits of braided and meandering channel systems are so different, it is appropriate to consider them separately, reviewing the five parameters of geometry, lithology, sedimentary structures, palaeocurrents (including borehole dip motifs) and palaeontology for each in turn.

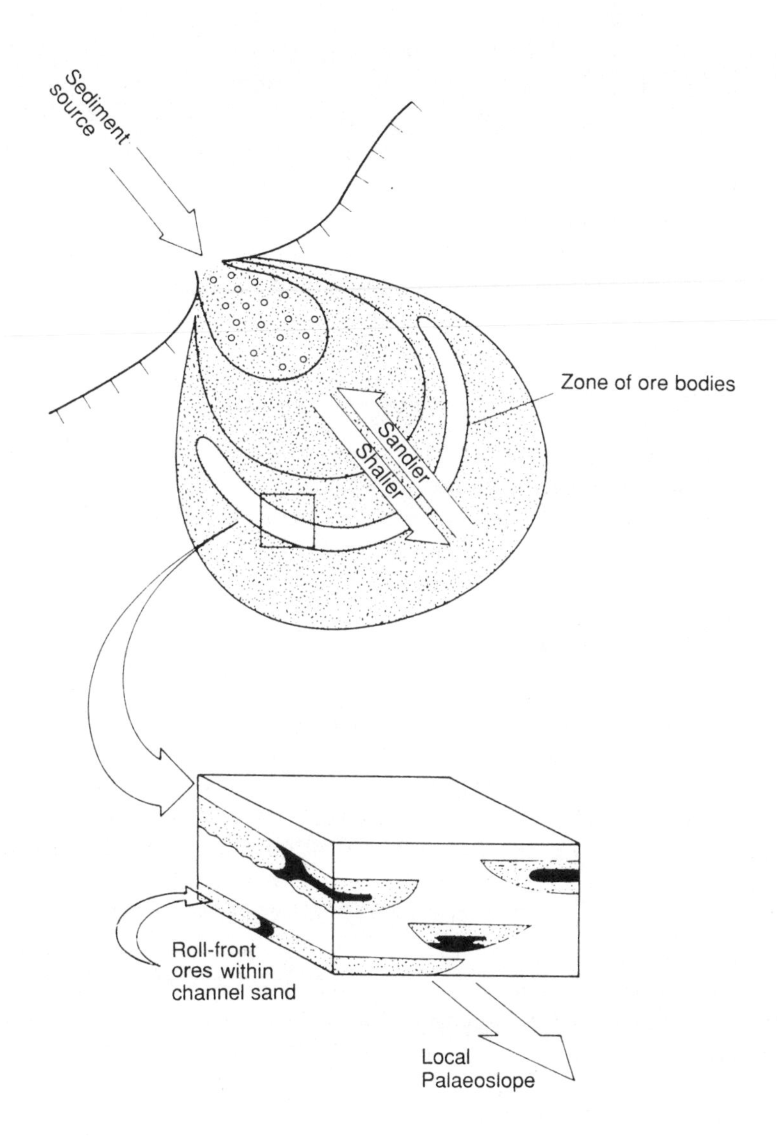

Figure 3.14 Cartoons to illustrate the sedimentological control on the distribution of uranium in alluvial fans. For examples see text.

Diagnosis of braided fluvial deposits

The geometry of braided fluvial deposits tends to be sheet-like, thick, laterally extensive, and often overlying an irregular or pedimented unconformity surface. In the subsurface the upper and lower boundaries of braided alluvial deposits may show up as seismic reflectors where they generate sufficient velocity contrast with the adjacent rocks. Blanket or fan geometries may be mappable. Because of the uniform sandy nature of braided alluvial deposits they contain few internal velocity variations and thus seldom show any internal reflectors.

The lithology of braided fluvial deposits consists almost entirely of conglomerates and coarse sands, with only minor amounts (some 10%) of fine sands and siltstones. Glauconite is absent, naturally. Carbonaceous organic matter, and even beds of coal, may occur in braided alluvium formed in periglacial or humid climates, but is absent in those that formed in arid climates. Braided alluvial sediments tend to be red in colour, due to the presence of red ferric oxide cement. This reflects early diagenesis above the water table, as discussed in Chapter 2 (refer back to Figure 2.7).

Thus braided alluvial fan sands tend to be red coloured, as do some other deposits of arid or semiarid continental environments. There are exceptions to these general statements (naturally). Some deep marine oozes are red (see pp. 283–284) and secondary reddening occurs beneath some unconformities. Red beds may become grey-green in colour if they are flushed by strongly reducing connate fluids. This phenomenon is especially common adjacent to petroleum accumulations. The characteristic sedimentary structure of braided alluvial deposits is channelling, though this cannot be seen in a single well log or core in the subsurface. The channels are infilled with cross-bedded and flat-bedded sands, with occasional recumbent foresets and disturbed bedding. One of the most characteristic features is shale-infilled abandoned channels (refer back to Figure 3.3).

In cores the presence of these may be inferred from double erosion surfaces above and below shale units. These abandoned channel sequences are diagnostic of braided deposits. Recognition of these channels in the subsurface is important. They indicate that the shales have shoestring geometries, and that they will not act as barriers to fluid migration in hydrocarbon reservoirs or aquifers.

Cross-bedding in braided channel systems tends to show a well-developed down current vector mode (Smith, 1972). Sedimentological logs may show large vertical changes in mean direction of dip, reflecting interbedding of channels where adjacent alluvial fans coalesce and interfinger. Regionally radiating fans may be mappable, and it may be possible to demonstrate that these originate from penecontemporaneously active fault scarps (Figure 3.15).

In the subsurface dipmeter motifs in braided deposits tend to be complex. Green patterns indicative of structural dip may be present in the abandoned channel shales. Bag o'nails motifs are common in the channel conglomerates and sands, but 20–25° foreset beds may be present which dip in the direction of current flow. Braided alluvial deposits tend to be palaeontologically barren because of the oxidizing nature of the environment. There may, however, be rare vertebrate tracks and trails, including orgasmoglyphs produced by rutting dinosaurs.

These then are the diagnostic criteria to look for in ancient braided alluvial deposits. Figure 3.16 shows the gamma profile, dip log pattern and sedimentary core log characteristics to be anticipated in a borehole penetrating such a facies.

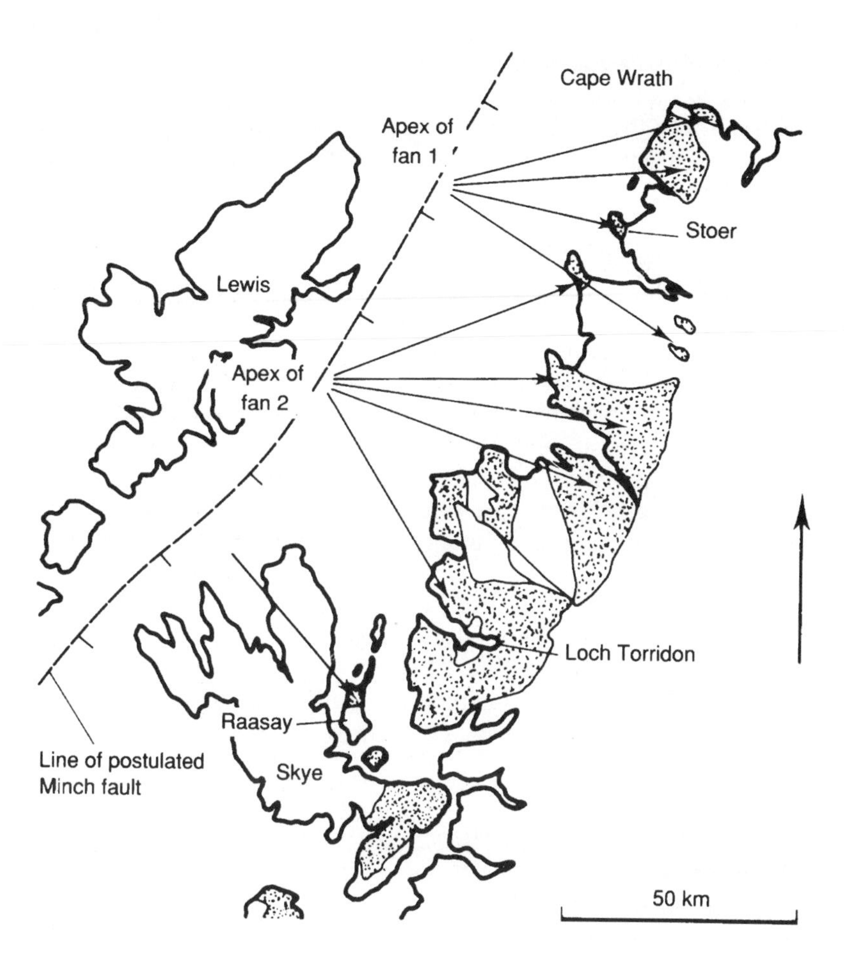

Figure 3.15 Distribution of Torridonian (PreCambrian) sediments, north-west Scotland. Arrows indicate red facies palaeocurrents. Note how these describe two radiating fans whose apices lie along the Minch fault. Simplified from Williams (1969), Fig. 12; by courtesy of the Journal of Geology.

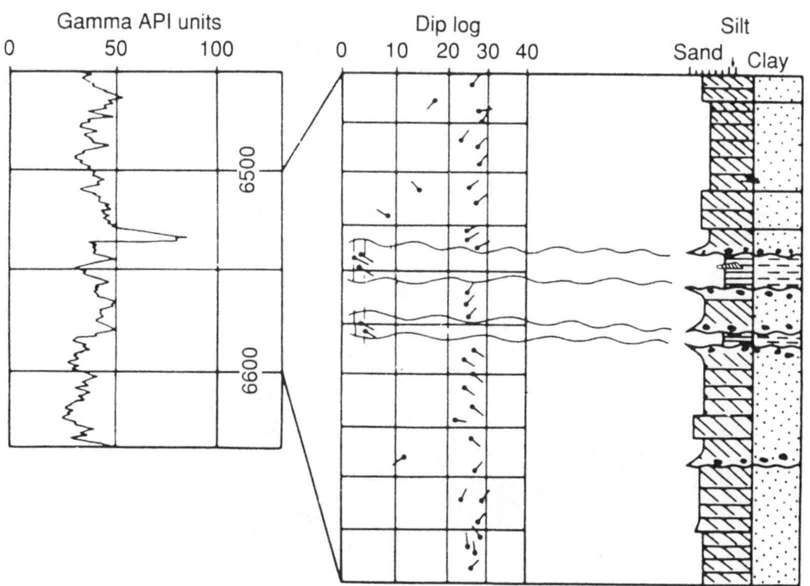

Figure 3.16 Log motifs for a borehole penetrating braided alluvial deposits. The gamma log shows that there is a more or less uniform sandy sequence, with a single shale unit. The dip log shows low angle structural dips in the shales, and higher angle dips, possibly cross-bedding, in the sands. Core log shows mainly cross-bedded channel sands, with abandoned channel shales with erosional surfaces above and below.

Diagnosis of meandering fluvial deposits

In the same way that braided channels pass into meandering ones, so their deposits show a similar gradation. True meander flood plain sediments differ from braided outwash plains principally in their sand:shale ratio. Whereas braided outwash plains are almost exclusively composed of coarse sand and gravel, the alluvium of meander channel flood plains consists of about equal parts channel sand and flood plain shale. The external geometry of the two types may be comparable, but the internal characters are very different.

Though channel cutting and filling occurs extensively on a meandering alluvial flood plain, the geometry of the resultant sands seldom actually has a discrete channel geometry. The to and fro migration of a channel across a flood plain may deposit sands that range from isolated channels to laterally extensive blankets, with all possible geometric permutations in between. This geometric variation will be controlled by the availability of sand, the discharge of water, and the rate of subsidence. In the subsurface the chanelled nature of flood plain sediments may generate a series of impersistent seismic reflectors. Since channels commonly have an erosional base and a gradational top detailed analysis suggests that these signals are generated by the channel floors (Figure 3.17).

Some fluvial channel stratigraphic traps are very subtle indeed (Figure 3.18).

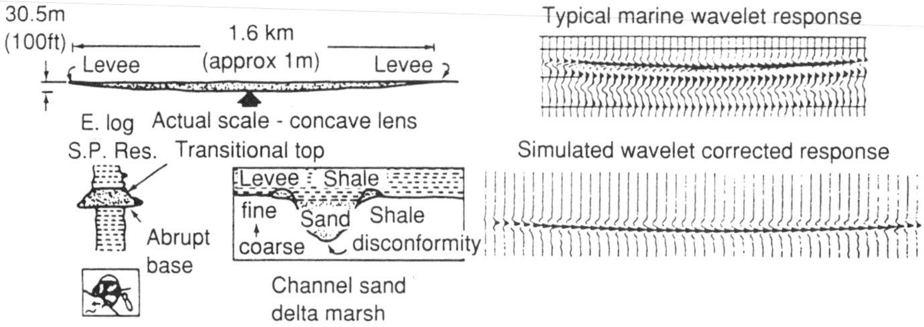

Figure 3.17 Geological model and seismic response for an upward-fining channel (from Schramm *et al.*, 1977, by courtesy of the American Association of Petroleum Geologists).

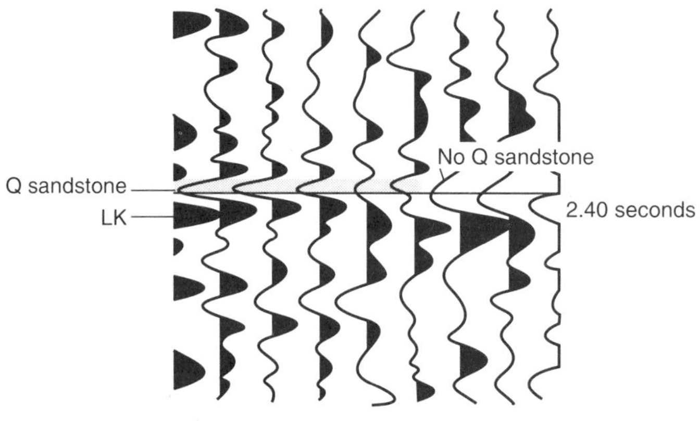

Figure 3.18 Part of a seismic line illustrating the Little Creek oil field in the mid-Tuscaloosa (Upper Cretaceous), Mississippi, USA. This occurs in a 20 m thick meander channel sand.This is what many geologists might term a subtle trap. (From Werren *et al.*, 1990.)

Three-dimensional seismic surveys may be able to delineate a whole meander belt in which channels and ox-bow lakes can be differentiated from the overbank shales (Figure 3.19).

Figure 3.19 Three-dimensional seismic survey from the Gulf of Thailand, showing alluvial flood plain meander belt. When seismic is able to directly image the geometry of potential petroleum reservoirs, then environmental diagnosis becomes of only academic interest. (From Brown *et al.*, 1982, by courtesy of the American Association of Petroleum Geologists.)

Lithologically meandering fluvial deposits have an overall finer grainsize, with sand: shale ratios of about 50:50. Conglomerates are rare, apart from those of intraformational origin. The sands tend to be relatively fine-grained. Red coloration may be present in semi-arid alluvial deposits, and may be associated with nodular carbonate caliches. Drab-coloured sand and shales, reflecting waterlogged alluvial plains, occur however, and these are often associated with coal beds, and disseminated carbonaceous detritus in sands and shales alike. Outcrop sections or cored intervals should show the suites of sedimentary structures associated with upward-fining grain-size sequences described previously. A channel floor erosion surface may be followed by cross-bedded point-bar sands which grade into cross-laminated finer sands and desiccation-cracked overbank shales (Figure 3.20).

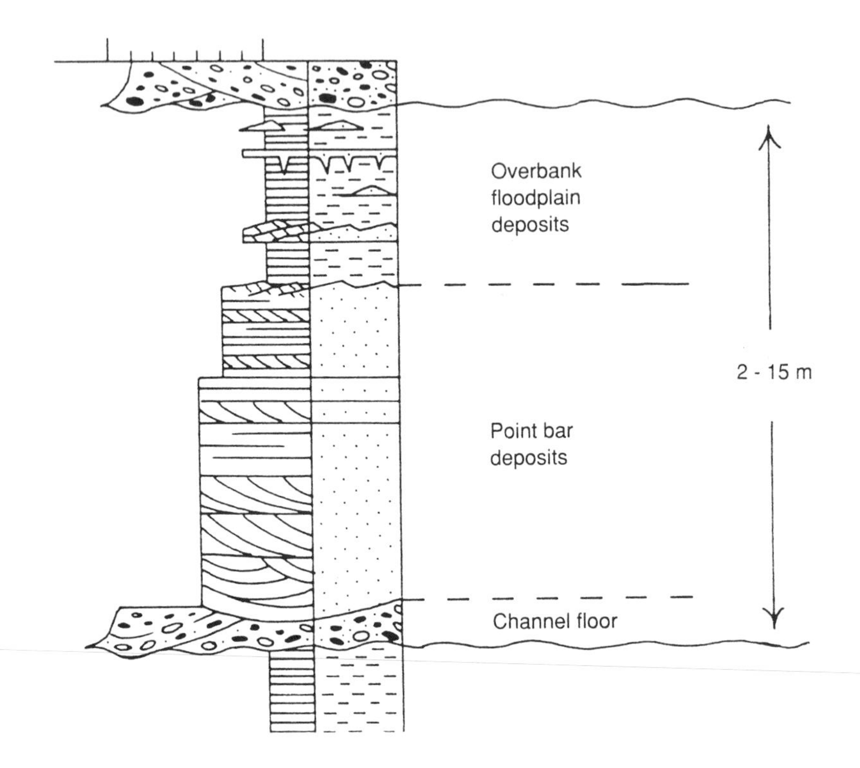

Figure 3.20 Typical genetic increment for meandering alluvial deposits of the Old Red Sandstone (Devonian) of South Wales. (For authentication see Allen, 1964 and 1970, and Allen and Williams 1982.)

Cross-bedding plots out with a much wider arc of dips than for braided channels, reflecting the greater sinuosity of meander channels. In the subsurface dipmeter patterns are of two types. Green motifs indicative of structural dip occur in the shales, while upward-decreasing red motifs characterize channels sands. Theoretically detailed analysis of the red patterns might show a bimodal pattern: one mode pointing towards the channel axis due to major point-bar bedding surfaces, the second mode, at 90° to the first, pointing down channel, reflecting cross-beds. In reality it is often hard to demonstrate convincingly such a distribution, particularly in homogenous sediments where there is little vertical variation in formation resistivity for the log to delineate. This dip motif is illustrated and discussed in more detail in the context of deltaic distributaries with which it is analogous (p. 158).

The deposits of meandering river systems are more fossiliferous than braided ones, because the waterlogged nature of the sediments favours the preservation of organic remains. Plant and animal matter may be fossilized yielding a fauna of terrestrial and subaqueous vertebrates, and invertebrates. These will be unidentifiable in the subsurface, where non-marine pollen and spores may be of more environmental and biostratigraphic importance.

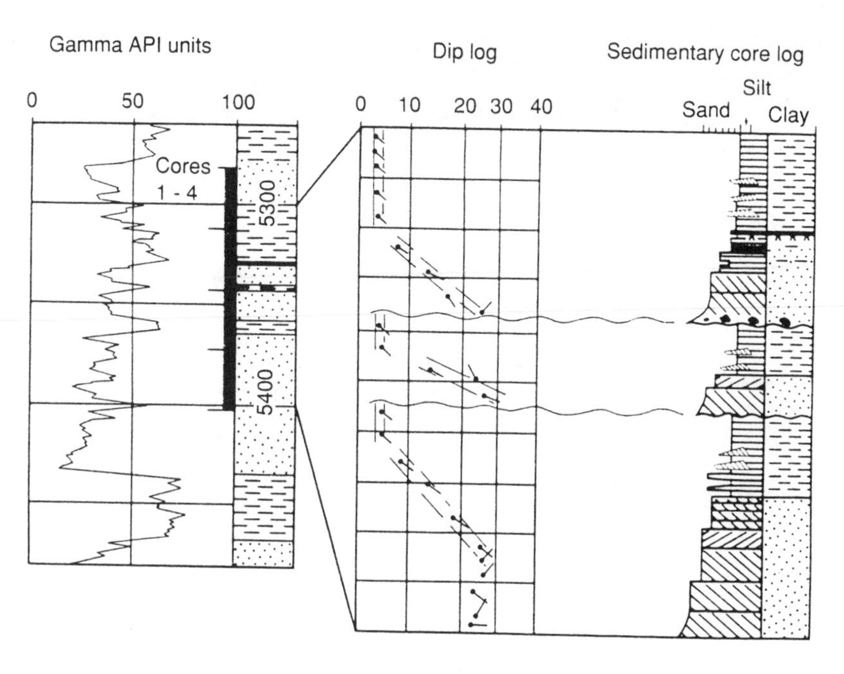

Figure 3.21 Log motifs for a borehole penetrating alluvium of meandering fluvial systems. Regular upward-fining motifs are not readily apparent on the gamma log, but the high ratio of shale to sand differentiates these deposits from continental beds of braided channel or eolian origin. Note the low angle structural dips in the shales and the upward-declining dip motifs possibly related to channel fill. Upward-fining grain-size profiles are readily seen in the core.

Figure 3.21 shows the log motifs which may be anticipated in boreholes which penetrate the deposits of meandering fluvial systems and Figure 3.22 illustrates a well log from the North Sea that penetrated several increments of channel sands and overbank coals and shales.

REFERENCES

The account of the Cambro-Ordovician sandstones of the Sahara and Arabia was based on the author's own fieldwork and:

Amireh, B. S., Schneider, W. and Abed, A. M. (1994) Evolving fluvial-transitional-marine deposition through the Cambrian sequence of Jordan, *Sed. Geol.*, **89**, 65–90.

Balducci, A. and Pommier, G. (1970) Cambrian oil field of Hassi Messaoud, Algeria. In M. T. Halbouty (ed.), *Geology of Giant Petroleum Fields*, Amer. Assoc. Petrol. Geol., Mem. No. **14**, 477–88.

Biju-Duval, B. (1974) Examples de depots fluvio-glacières dans l'Ordovician supérieur et le Precambrian supérieur du Sahara central, *Soc. Pet. Nat. d'Aquitaine (SNPA) Bull. Cent. Rech. Pau.*, **8**, 209–26.

Doughty, C. (1888) *Travels in Arabia Deserta*, Cambridge University Press.

Husseini, M. I. (1990) The Cambro-Ordovician Arabian and adjoining plates: a glacio-eustatic model, *Journ. Petrol. Geol.*, **13**, 267–88.

—— (1991) Tectonic and Depositional Model of the Arabian and Adjoining Plates During the Sihurian - Devonian, *Bull. Amer. Assoc. Petrol. Geol.*, **75**, 108–20.

Lloyd, J. W. and Pim, R. H. (1990) The hydrogeology and groundwater resources development of the Cambro-Ordovician sandstone aquifer in Saudi Arabia and Jordan, *J. Hydrogeol.*, **121**, 1–20.

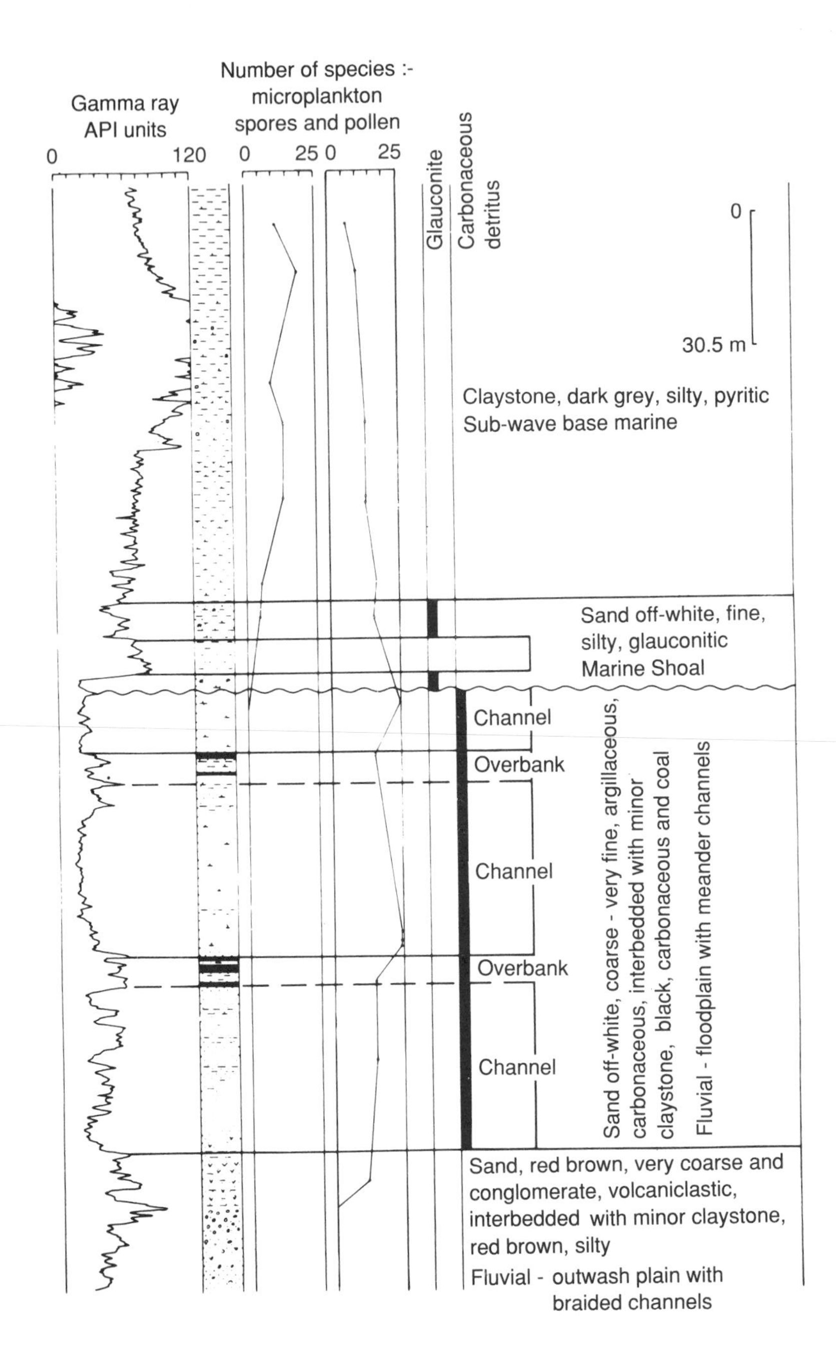

Figure 3.22 Borehole from the North Sea showing fluvial deposits unconformably overlain by marine shales. There are several cycles in which channel sands with abrupt erosional bases pass up with a steady or upward shaling gamma log motif into overbank shales and coals. (From Selley (1976), Figure 5, by courtesy of the American Association of Petroleum Geologists.)

McGillivray, J. G. and Husseini, M. I. (1992) The Palaeozoic petroleum geology of central Arabia, *Bull. Amer. Assoc. Petrol. Geol.*, **76**, 1475-90.
Selley, R. C. (1970) Ichnology of Palaeozoic sandstones in the Southern Desert of Jordan: a study of trace fossils in their sedimentologic context. In J. C. Harper and T. P. Crimes (eds), *Trace Fossils*, Lpool. Geol. Soc., 477–88.
—— (1972) Diagnosis of marine and non-marine environments from the Cambro-ordovician sandstones of Jordan, *Jl. Geol. Soc. Lond.*, **128**, 109–17.

Other references cited in this chapter were:

Allen, J. R. L. (1964) Studies in fluviatile sedimentation: six cyclothems from the Lower Old Red Sandstone, Anglo-Welsh basin, *Sedimentology*, **3**, 163–98.
—— (1970) Studies in fluvial sedimentation: a comparison of fining-upwards cyclothems, with special reference to coarse member composition and interpretation, *J. Sediment. Petrol.*, **40**, 298–323.
—— (1974) Studies in fluviatile sedimentation: lateral variation in some fining-upwards cyclothems from the Red Marls, *Geol. Joul.*, **9**, 1–16.
—— and William, B. P. J. (1982) The architecture of an alluvial suite: rocks between the Townsend Tuff and Pickard Bay Tuff Beds (Early Devonian), S. W. Wales, *Phil. Trans. Ser. B*, **297**, 51–89.
Atkinson, C. D., McGowan, J. H., Bloch, S., Lundell, L. L. and Trumbly, P. N. (1989) Braidplain and deltaic reservoir, Prudhoe Bay Field, Alaska. In J. W. Barwis, J. G. McPherson and J. R. J. Studlick (eds), *Sandstone Petroleum Reservoirs*, Springer-Verlag, Berlin, 7–30.
Berg, W. A. (1968) Point bar origin of Fall River sandstone reservoirs, northeastern Wyoming, *Amer. Assoc. Petrol. Geol. Bull.*, **52**, 2116–22.
Best, J. L. and Bristow, C. S. (eds) (1993) *Braided Rivers,* Geol. Soc. Sp. Pub. **75**.
Bevan, K. and Kirkby, M. J. (eds) (1993) *Channel Network Hydrology*, J. Wiley, Chichester.
Billi, P., Hey, R. D., Thorne, C. R. and Tacconi, P. (eds) (1992) *Dynamics of Gravel Bed Rivers*, J. Wiley, Chichester.
Brown, A. R., Graebner, R. J. and Dahm, C. G. (1982) Use of horizontal seismic sections to identify subtle traps. In M. T. Halbouty (ed.), *Amer. Assoc. Petrol. Geol.*, **32**, 47–56.
Callow, P. and Petts, G. E. (eds) (1992) *The Rivers Handbook*, 2 vols, Blackwell Scientific, Oxford.
Cant, D. J. (1982) Fluvial facies models. In P. A. Scholle and D. Spearing (eds), *Sandstone Depositional Environments*, Amer. Assoc. Petrol. Geol. Tulsa, Mem. No. **31**, 115–38.
Clifford, H. J., Grund, R., and Musrati, H. (1980) Geology of a stratigraphic giant: Messla oil field, Libya. In M. T. Halbouty (eds), *Giant Oil and Gas fields of the Decade 1968–78,* Amer. Assoc. Petrol. Geol., Mem. No. **30**, 507–22.
Costello, W. R. and Walker, R. G. (1972) Pleistocene sedimentology Credit river, Southern Ontario: a new component of the braided river model, *J. Sediment. Petrol.*, **42**, 389–400.
Crawley, R. A. (1982) Sandstone uranium deposits in the U.S.A., *Energy Exploration and Exploitation*, **1**, 203–43.
Dahm, C. G. and Graebner, R. J. (1982) Field development with three-dimensional seismic methods in the Gulf of Thailand, *Geophysics*, **47**, 161–72.
Fielding, C. R. (ed.) (1993) *Current Research in Fluvial Sedimentology*. Sed. Geol., **85**.
Fischer, P. P. (1974) Exploration guides to new uranium districts and belts, *Econ. Geol.*, **65**, 778–84.
Forgotson, J. M. and Stark, P. H. (1972) Well-data files and the computer, a case history from the northern Rocky Mountains. *Bull. Amer. Assoc. Petrol. Geol.*, **56**, 1114–27.

Friend, P. F. (1993) Control of river morphology by the grain size of sediment supplied. In C. R. Fielding (ed.), *Current Research in Fluvial Sedimentology*, Sed. Geol., **85**, 171–8.

Galloway, W. E., Kreitler, C. W. and McGowan, J. H. (1979) Depositional and ground water flow systems in the exploration for Uranium, *Bur. Econ. Ecol. Univ. Texas*, Austin Research Colloquium.

Jamison, H. C., Brockett, L. D. and McIntosh R. A. (1980) Prudhoe Bay: a ten year perspective. In M. T. Halbouty (ed.), *Geology of Giant Petroleum Fields*, Amer. Assoc. Petrol. Geol., Mem. No. **30**, 289–314.

McCarthy, T. S. (ed.) (1990) Geological studies related to the origin and evolution of the Witwatersrand basin and its mineralization, *S. African Jl. Geol.*, **93**.

Miall, A. D. (1977) A review of the braided river depositional environment, *Earth Sci. Rev.*, **13**.

—— (1981) Sedimentation and tectonics in alluvial basins, *Geol. Soc. Can.*, Sp. Pap. **23**.

Newsom, M. (1994) *Hydrology and the River Environment*, Oxford University Press, Oxford.

Nilsen, T. H. (1982) Alluvial fan deposits. In P. A. Scholle and D. Spearing (eds), *Sandstone Depositional Environments*, Amer. Assoc. Petrol. Geol., Tulsa, 49–86.

North, C. P. and Prosser, D. J. (1993) *Characterization of Fluvial and Aeolian Reservoirs*, Geol. Soc. Lond. Sp. Pub. **73**.

Powell, J. H., Moh'd, B. K. and Masri, A. (1994) Late-Ordovician-Early Silurian glaciofluvial deposits preserved in palaeovalleys in South Jordan, *Sed. Geol.*, **89**, 303–14.

Putnam, P. E. (1983) Fluvial deposits and hydrocarbon accumulations: examples from the Lloydminster area, Canada. In J. D. Collinson and J. Lewin (eds), *Modern and Ancient Fluvial Systems*, Internat. Assn. Sedol. Sp. Pub. **6**, 517–32.

Rachocki, A. H. and Church, M. (eds) (1990) *Alluvial Fans*, J. Wiley, Chichester.

Sanford, R. M. (1970) Sarir oil field, Libya□Desert surprise. In M. T. Halbouty (eds), *Geology of Giant Petroleum Fields*, Amer. Assoc. Petrol. Geol., Mem. No. **14**, 449–76.

Schramm, M. W., Dedman, E. V. and Lindsey, J. P. (1977) Practical stratigraphic modelling: an interpretation. In C. E. Payton (ed.), *Seismic Stratigraphy – Applications to Hydrocarbon Exploration*, Amer. Assoc. Petrol. Geol., **26**, 477–502.

Selley, R. C. (1965) Diagnostic characters of fluviatile sediments of the Torridonian Formation (PreCambrian) of North-West Scotland, *J. Sediment. Petrol.*, **35**, 366–80.

—— (1969) Torridonian alluvium and quicksands, *Scott. J. Geol.*, **5**(4), 328–46.

—— (1976) Sub-surface environmental analysis of North Sea sediments, *Bull. Amer. Assoc. Petrol. Geol.*, **60**, 184–95.

Shanley, K. W. and McCabe, P. J. (1994) Perspectives on the sequence stratigraphy of continental strata, *Amer. Assoc. Petrol. Geol.*, **78**, 544–68.

Smith, N. D. (1972) Some sedimentological aspects of planar cross stratification in a sandy braided river, *J. Sediment. Petrol.*, **42**, 624–34.

Stanistreet, I. G. and McCarthy, T. S. (1993) The Okavango fan and the classification of subaerial fan systems. In C. R. Fielding (ed.), *Current Research in Fluvial Sedimentology*, Sed. Geol., **85**, 115–34.

Stokes, W. I. (1950) Pediment concept applied to Shinarump and similar conglomerates, *Bull. Geol. Soc. Amer.*, **61**, 91–8.

Williams, G. E. (1969) Characteristics and origin of a PreCambrian pediment, *J. Geol.*, **77**, 183–207.

Woncik, J. (1972) Recluse field, Campbell County, Wyoming. In R. E. King (ed.), *Stratigraphic Oil and Gas Fields*, Amer. Assoc. Petrol. Geol., Mem. No. **16**, 376–82.

4 Wind-blown deposits

INTRODUCTION

Ancient wind-deposited sediments are often very hard to distinguish from those laid down by running water. As this chapter shows, however, a number of criteria have been proposed to distinguish eolian from aqueous sediments. Some of these are of debatable merit when used in isolation. The ultimate decision of whether a sediment is of eolian origin must be based on a critical evaluation of all the available data. Fortunately this is helped by studies of Recent wind blown deposits, notably by Bagnold (1941), McKee (1966; 1979), Glennie (1970; 1987), Wilson (1972; 1973), Ahlbrandt and Fryberger (1981; 1982) and Pye (1993).

When discussing sedimentary environments in an earlier section it was pointed out that most non-marine environments are areas of erosion. Most deserts are equilibrial surfaces, that is to say regions of the earth's crust where there is neither erosion, nor deposition taking place; though there may be a considerable amount of sediment being reworked and transported across desert surfaces. Unfortunately the Hollywood image of deserts as regions of sand dunes is a distortion of the facts, probably only about 25% of deserts are composed of sand dunes. The major surface areas are flat sand, gravel or bedrock covered plains.

RECENT EOLIAN SEDIMENTS

When the wind blows across a desert silt and clay are carried up into the atmosphere and may be transported far away ultimately to settle out in the seas. Any dust that does resettle in the desert may form beds of loess, as in periglacial realms, or be carried by floods into ephemeral desert lakes. Sand, by contrast, is transported close to the ground largely by saltation. Wind velocities are seldom sufficiently high to transport gravel, fortunately. The bed forms of sands transportation are asymmetric ripples and megaripples whose geometry resembles that of subaqueous ones (Figure 4.1).

The physics of eolian and aqueous transportation are basically similar, concerning a granular solid within a fluid. This is probably the main reason why the eolian environment is the hardest one to recognize in the subsurface. There are many different types of dune shape. Arbitrarily four main types can be defined though there are transitional varieties and some minor ones (Figure 4.2).

Figure 4.1 Photograph of a modern desert dune. The dune in the distance deposits large scale cross-bedding. Notice the ripples in the foreground that will deposit cross-lamination. This illustration shows how eolian and waterlain traction deposits have similar bed forms and broadly comparable internal sedimentary structures. Hence the problems of differentiating the two types of deposit.

Probably one of the best documented is the barchan or lunate dune. As with most dunes they have a gentle windward slope and a steep accretionary foreset directed downwind at either end of the slipface. Barchans are very photogenic and photos are common in many text books. Their internal structures have also been studied in some detail. This effort is probably largely unnecessary because it is unlikely that barchan dunes actually get preserved. Barchans are isolated dunes out away from the big sand seas. They generally occur migrating across flat equilibrial surfaces of rock, gravel and sand. It is easy to drive around a barchan, take a few pictures and dig some holes with a reasonable certainty of returning to camp in the evening and surviving to write a paper for one of the learned journals.

The second major dune type is the pyramidal, stellate, or Matterhorn variety. This is a very beautiful dune to look at, sometimes attaining a height of over a hundred metres. Stellate dunes are relatively rare, occurring in tangled patterns within sand seas, or where sand seas but against jebels or escarpments. The third type of dune is the longitudinal or seif variety. These are generally gregarious, many seifs being aligned parallel to one another and to the prevailing wind direction. Detailed study of seifs shows that they are complex bed forms in which the basic longitudinal geometry is punctuated by a series of barchanoid or pyramidal culminations. Seif dunes are like barchans, however, in that they overly flat desert surfaces. They are bedforms in which sand is reworked and transported but from which little net deposition actually takes place.

The transverse dune is the fourth variety to consider. This is seldom

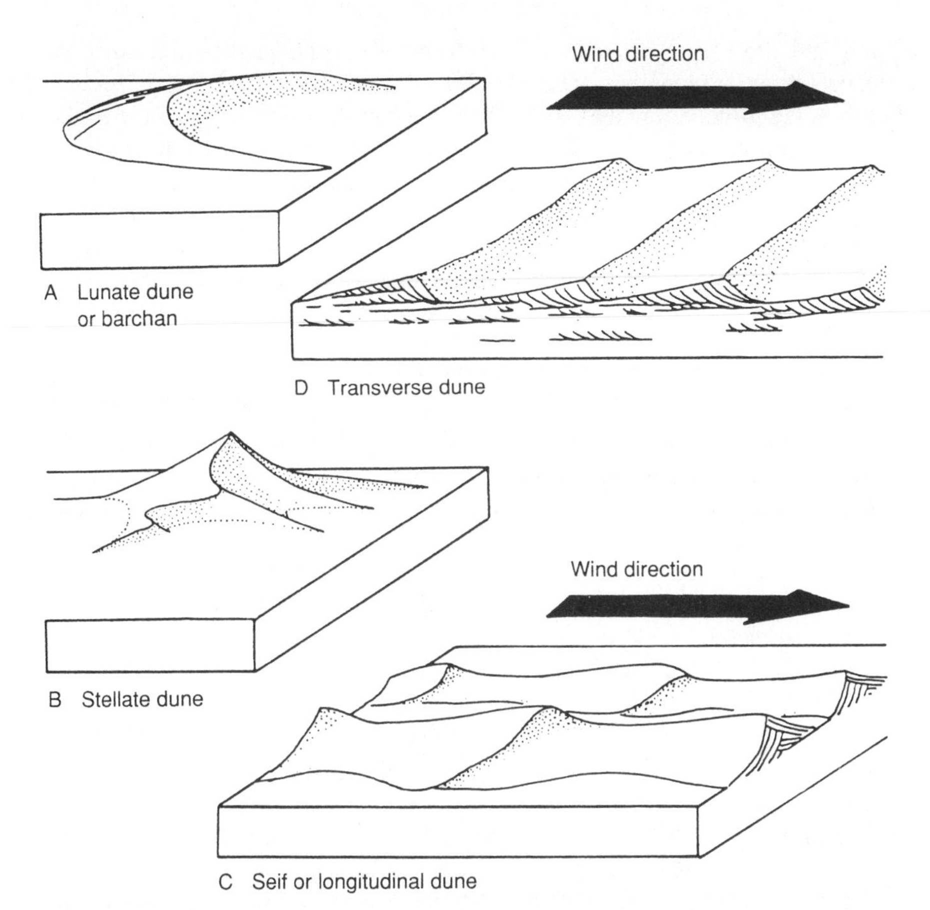

Figure 4.2 Cartoons of the four common types of dune seen in recent deserts. Probably only the transverse dune is responsible for significant deposition of sand. The other three types are commonly bed forms of sand reworking and transportation across equilibrial surfaces (from Selley, 1988).

described in the literature, but is probably the most important variety. Transverse dunes are not lonely dunes migrating across flat surfaces, but gregarious dunes that migrate over the backside of the next dune downwind. It is transverse dunes which actually deposit sands which may be buried and preserved in the stratigraphic record. They are seldom described in the literature because it is difficult to penetrate sand seas and to return to camp for a cool beer in the evening. Essentially transverse dunes are just like the straight-crested megaripples which deposit cross-bedded sands under water.

The interdune area has been defined as 'a geomorphic surface commonly enclosed or at least partially bounded by dunes or other eolian deposits such as sand sheets' (Ahlbrandt and Fryberger, 1981). As already discussed, many interdune areas are equilibrial environments where bed rock is exposed or lag gravels just lie there. In some instances, however, deposition takes place in interdune areas.

Ahlbrandt and Fryberger (1981; 1982) recognize three such types: dry, wet, and evaporitic. Dry interdune deposits consist of massive or flat bedded sand and granules which are generally coarser grained than the adjacent dunes and are bimodally sorted. Wet interdune deposits are rather different. They contain more clay and silt since these particles may be attached to the damp substrate. Similarly, sand may form adhesion ripples as it is blown across the interdune flat. Animal tracks and trails may be preserved, together with rhizoconcretions (Glennie and Evamy, 1968). Evaporitic interdune deposits develop where sand seas occur adjacent to sabkhas. Nodules, layers and intergranular cements of evaporites are associated with fine-grained flat-bedded terrigenous sediment. Adhesion ripples and shrinkage polygons occur.

Interdune deposits are of significance in two ways. First they may form permeability barriers which may slow down, and sometimes totally inhibit, fluid movement in eolian sands. Secondly, considerable amounts of organic matter have been discovered in recent evaporitic interdune sediments, and it has been suggested that this may act as a petroleum source. Ancient rocks which have been attributed to eolian deposition are rare in the stratigraphic record, but they are widely distributed and range in age from PreCambrian to Recent. Some of the best documented complexes of supposed ancient eolian sediments occur in the western USA. These will now be briefly described and the reasons for attributing them to eolian action stated.

EOLIAN DEPOSITS OF WESTERN USA: DESCRIPTION

Large areas of the western United States contain sandstones that are believed to be of eolian origin. This region was an important site of intermittent continental sedimentation through some 150 m.y. from the Pennsylvanian (Late Carboniferous) to the Late Jurassic. The regional geology is complicated but can often be resolved due to a combination of good exposure and dissected topography (Figure 4.3).

In general the Cordilleran fore-arc basin lay to the west and was an area of marine deposition through much of this time. The present Colorado Plateau was then an unstable shelf on which sedimentation took place in a number of basins which were intermittently connected to one another and to the sea. These include the Paradox, San Juan, Kaiparowits and Black Mesa basins. The stratigraphy of this region is complex therefore, with rapid lateral and vertical facies changes due to both tectonic and sedimentary causes. Three main facies can be recognized in the Pennsylvanian to Jurassic rocks of the Colorado Plateau:

1. easterly thinning sheets of limestones, shales, dolomites, and evaporites (thought to be shallow marine, sabkha and ephemeral lake);
2. wedges and fans of conglomerates, pebbly channel sands with small-scale cross-bedding, and red shales, generally best developed in the

Figure 4.3 Isolated jebel showing the excellent exposure of the Colorado Plateau, Monument Valley, Arizona. Cliff of eolian Wingate sandstone (Lower Jurassic) forms a steep cliff above a talus slope with ledges of fluvial Chinle sandstones and shales (Triassic). Photo by courtesy of W. F. Tanner.

east, e.g. the Moenkopi and Morrison Formations (thought to be fluviatile);

3. red and white sandstones with irregular sheet geometries, and large scale cross-bedding, e.g. the Entrada, Navajo, Wingate, Weber, De Chelly, and Coconino Formations (thought to be eolian).

The distribution of these facies through time enables successive palaeogeographies to be mapped (Figure 4.4), and the distribution of the ancient sand dunes to be delineated. The supposed eolian sandstones are interbedded with the other two facies with interfingering but generally abrupt margins. They are irregular in plan and seldom greater than 60 m thick, though some examples like the Navajo Sandstone approach 300 m. Lithologically these deposits are exclusively sand-grade sediment. Petrographically they are protoquartzites with minor amounts of feldspar, mica, chert, and red ferruginous clay. Cementation is by quartz and calcite.

In one or two exceptional cases, as in the upper part of the Todilto Formation, wind-blown gypsum sands occur. The grade of the sandstones range from very fine to coarse, but it is mainly fine. Sorting is moderate to good, occasionally bimodal, with coarse grains in a fine sand. Rounding of grains is moderate to good. Cross-bedding is the characteristic sedimentary structure of these rocks (Figure 4.5).

Both tabular planar and trough sets are present. Set heights vary from 1 m to 60 m and troughs range in width from 1 m to 60 m. Individual foresets

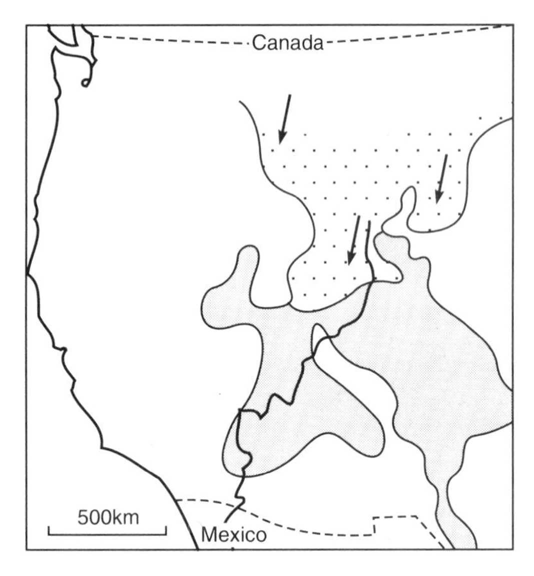

1. Pennsylvanian – Early Wolfcampian (Tensleep)

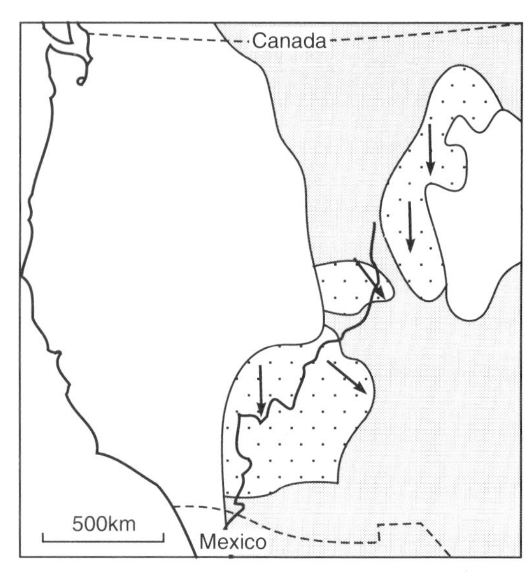

2. Wolfcampian (Weber)

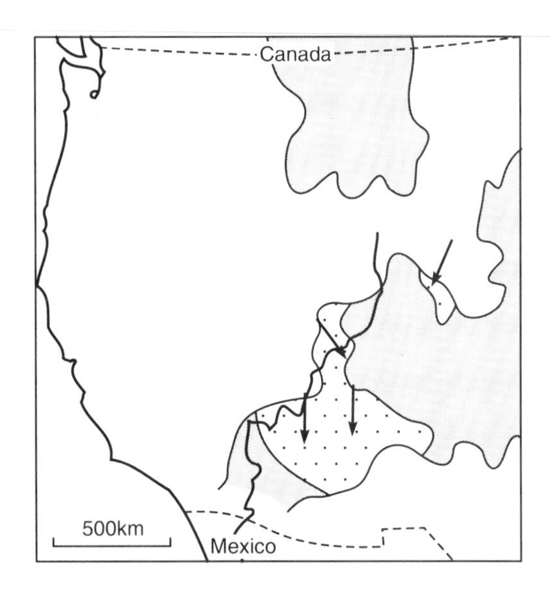

3. Leonardian (Late Permian)
(Coconino and de chelly)

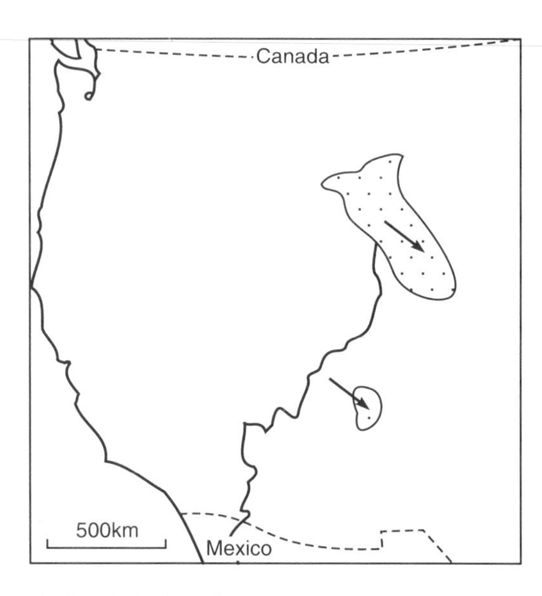

4. Mid – Late Triassic

Figure 4.4 Palaeogeographic maps to show the distribution of eolian sands in the western interior of the USA from the Pennsylvanian (Late Carboniferous) to the Late Jurassic. Simplified from Parrish and Peterson (1988).

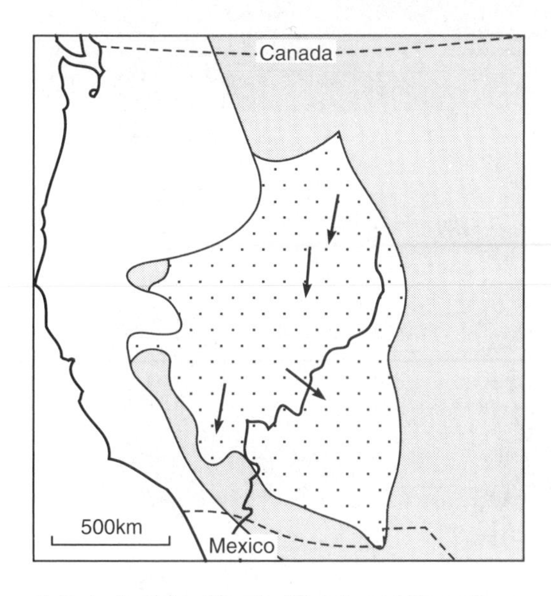

5. Early Jurassic (Navajo, Wingate and Nugget)

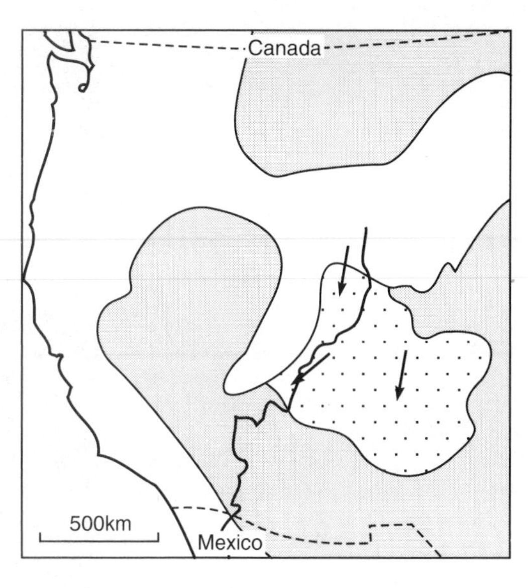

6. Middle Jurassic (Entrada)

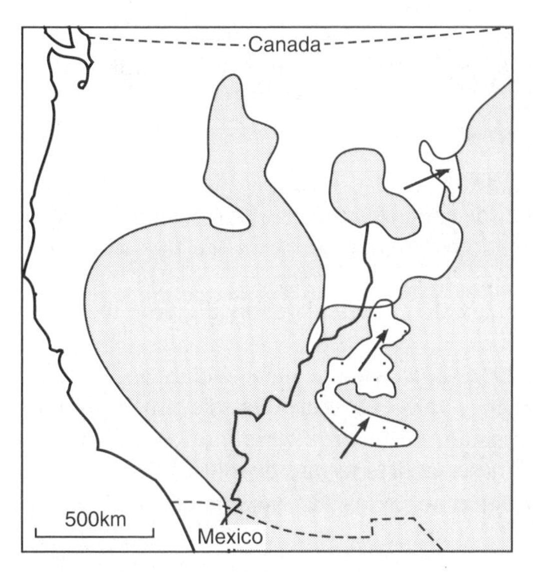

7. Late Jurassic

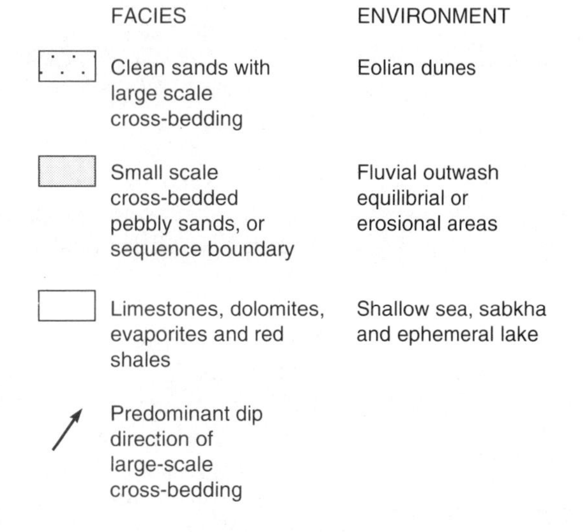

Figure 4.5 Eolian cross-bedding in Entrada Formation (Jurassic), Church rock, New Mexico. Photo by courtesy of W. F. Tanner.

dip at between 20° and 30° and are generally curved at the base. Low angle backsets also occur sub-parallel to erosion surfaces. Penecontemporaneous deformation of bedding is sometimes present. Some of the fine sandstones show long wavelength low amplitude asymmetrical ripples with RIs between 20 and 50 (the ripple index (RI) is calculated by dividing the wavelength by the amplitude).

Dip directions of cross-bedding show wide scatters within one outcrop. Throughout the majority of Pennsylvanian to Late Jurassic time they indicate transport generally to the south, with some variation to southeasterly and southwesterly directions. These sandstones are devoid of fossils except for occasional footprints attributed to terrestrial quadrupeds and bipeds, largely dinosaurs.

EOLIAN SANDSTONES OF WESTERN USA: DISCUSSION

There are two main lines of argument for the eolian origin of these sands.

Some arguments are negative, suggesting that these are not water-laid sands, others are positive, suggesting that they are wind-blown. It is unlikely that these are water-laid sands because of the absence of pebbles, which are generally too heavy to be wind-blown, and of clay, which is generally too light to come to rest on a windy earth. There is no sign of an aqueous biota either marine or non-marine. Conglomerate-lined channels suggestive of running water are absent.

Positively in favour of an eolian origin for these sandstones is their close comparison with Recent desert dunes. Points in common are the dominance of sand-grade sediment which is generally well sorted, matrix free, and well rounded. Cross-bedding of such vast height is unknown in Recent aqueous sediments, but is known from Recent terrestrial dunes. Ripple indices > 15 are recorded from Recent eolian ripples whereas aqueous ripples have RIs < 15 (Tanner, 1967). Recent dune country is generally lifeless, such plants and animals as live and die there are desiccated and destroyed by the shifting sands. The persistent (generally) southerly direction of sand transport is perpendicular to the (general) westerly palaeoslope of the Colorado Plateau and locally can be seen trending oblique to slope-controlled palaeocurrents of the associated fluviatile red beds.

Some doubt has been expressed as to the eolian origin of these sands. It has been pointed out that the large scale cross-bedding, the well-rounded, well-sorted texture and frosted grain surfaces of these sediments could also have formed in a marine shelf sea. Log-probability sorting curves of the sands are comparable to those of modern tidal-current environments. Green pellets have also been recorded, and interpreted as glauconite of marine origin (Stanley *et al.*, 1971; Visher, 1971; Freeman and Visher, 1975). It is difficult to pontificate in such a controversy, especially without field experience of the rocks in question. Two points can be made, however.

First, this dilemma points up the absence of any clear criteria for recognizing eolian deposits. Second, on a broad shelf where a tidal shallow sea transgresses and regresses a desert, eolian sands may inherit aqueous textural characteristics and vice versa. Due to the wealth of stratigraphic and sedimentologic data available it is possible to reconstruct the palaeogeography of the ancient Colorado Plateau area in some detail. In Pennsylvanian and early Permian time dunes extended intermittently along a coastal plain separating the Rocky Mountain fore-arc seas from alluvial piedmont fans to the east. There are no known eolian sands of Late Permian to Middle Triassic age. In Late Triassic and Early Jurassic time dune fields developed in inland basins away from the sea. Middle Jurassic eolian sandstones are unknown. In Late Jurassic time both coastal and interior basin dune fields developed (Figure 4.6).

Tanner (1965) has produced a particularly elegant detailed study of the Upper Jurassic palaeogeography of the 'Four Corners' region of the Colorado Plateau. Marine deposits with southwesterly directed longshore

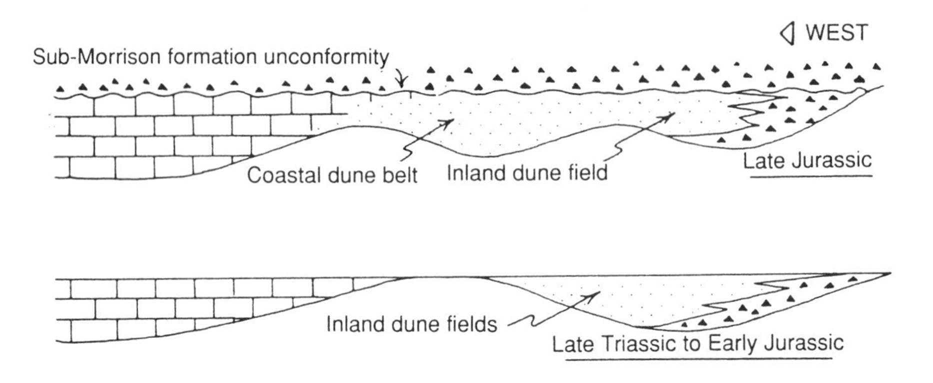

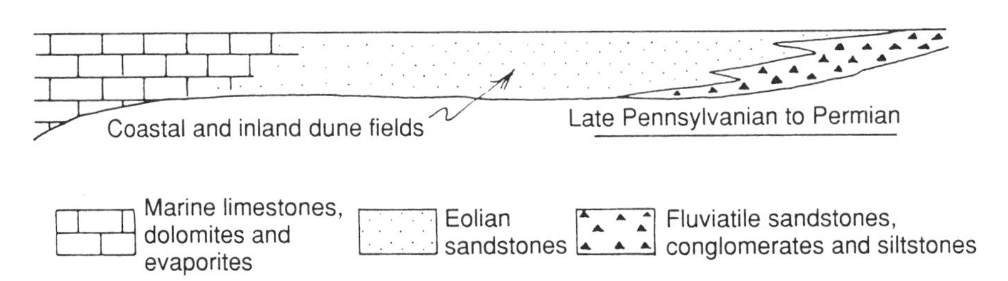

Figure 4.6 Diagrammatic cross-sections illustrating successive palaeogeographies of the western interior of the USA, showing shifting shorelines and dune fields.

palaeocurrents interfinger eastwards with the Entrada sandstone. This was deposited by a complex of southeasterly migrating coastal dunes which, for a time, separated the open sea from the hypersaline deposits of the Todilto Formation lake. Tongues of fluvial sediment within the eolian Entrada sandstone show northwesterly directed cross bedding indicating the direction of the palaeoslope as the river flowed down to the sea (Figure 4.7).

ECONOMIC SIGNIFICANCE OF EOLIAN DEPOSITS

Eolian sandstones are potentially of high porosity and permeability because they are typically well-rounded, well-sorted, and generally only lightly cemented. Regional permeability is likely to be good due to an absence of shale interbeds. Because of these features eolian sandstones can be important aquifers and hydrocarbon reservoirs. In general, however, eolian sandstones are rated as poor hydrocarbon reservoir prospects. This is because they frequently lie within continental basins far from marine shales which could be potential source rocks. Exceptions to this general rule are found where eolian sandstones are brought into contact with organic-rich source beds by unconformities or by tectonic movement.

This first situation is demonstrated by many gas fields in the southern North Sea (Parsley, 1990). Most of this gas is reservoired in Rotliegende

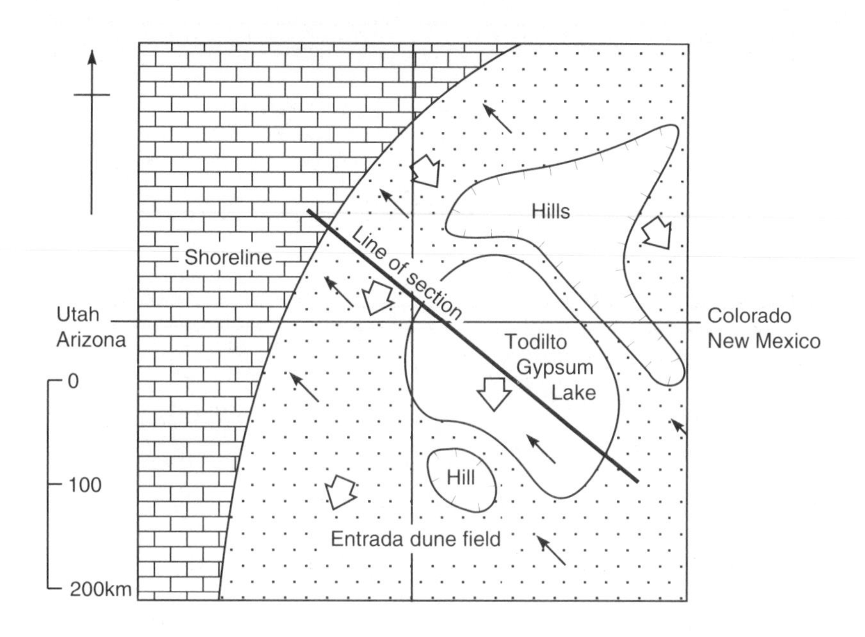

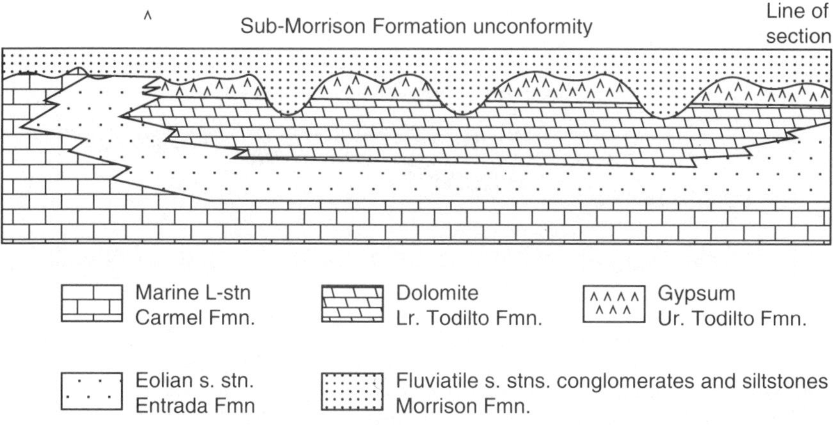

Figure 4.7 Reconstruction of Entrada sandstone (Middle Jurassic) palaeogeography. Cross-section about 100 m thick. Thin arrows: slope controlled seaward flowing fluvial palaeocurrents. Thick arrows: slope independent eolian palaeocurrents. Modified from Tanner (1965), Figure 7.

(Permian) sands that unconformably overlie Upper Carboniferous (Pennsylvanian) Coal Measures. Deep burial of coal beneath the younger sediments may have caused gas to be expelled during devolatization. The Permian sediments consist of a complex group of interbedded eolian, fluvial, playa lake and sabkha deposits, of which only the eolian sands produce high-quality reservoirs (Glennie, 1990). The traps occur in uplifted fault blocks that are sealed by the Zechstein (Late Permian) evaporites (Figure 4.8A).

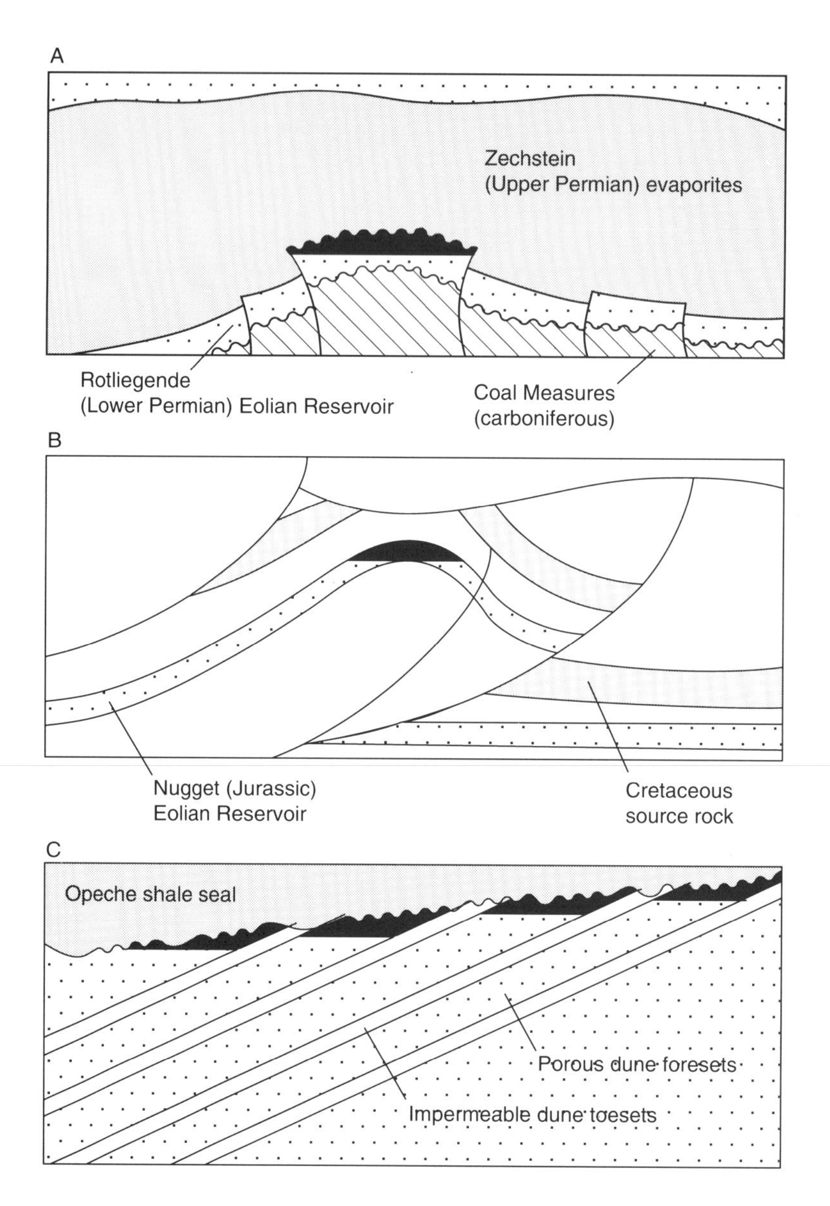

Figure 4.8 Cartoons to illustrate styles of petroleum occurrence in eolian reservoirs. A: Rotliegende (L. Permian) gas fields of the southern North Sea, sourced by unconformably underlying Coal Measures (Carboniferous). B: Nugget (Jurassic) sands thrust over Cretaceous source rocks in the Rocky Mountain thrust belt, Utah and Wyoming. C: Stratigraphically trapped petroleum in truncated Minnelusa (Permian) eolian sands of Wyoming. For authentication see Parsley (1990), Lindquist (1988) and Fryberger (1993) respectively.

The second way in which eolian sands can become in fluid communication with source rocks is by faulting. In Utah and Wyoming several oil fields produce from some of the Triassic and Jurassic eolian sands described earlier in this chapter. Here, within the foothills of the Rocky Mountains, several of the eolian formations have been involved in thrusting during the Laramide orogeny (Figure 4.8B). This has resulted in the porous permeable eolian sands being thrust over organic-rich Cretaceous shale petroleum source rocks (Lamb, 1980; Lindquist, 1983; 1988).

A third type of eolian petroleum play occurs where dune sands contain stratigraphically trapped hydrocarbons. Sometimes the trap takes the form of a simple truncation, as in the case of the Caprock field of New Mexico. This produces from a coastal dune field on the northwestern shelf of the Delaware basin (Malicse and Mazullo, 1989). A more complex situation is exemplified by the Minnelusa (Permian) eolian sands of the Powder River basin, Wyoming (Fryberger, 1993). Here multiple truncation traps occur where the toesets of individual dunes serve as regional permeability barriers (Figure 4.8C). Much work has been done on the permeability variations of eolian foresets and toesets on both exploration and production scales (e.g. Weber, 1987, Prosser and Maskall, 1993).

The recognition of eolian sediments, and the ability to differentiate them from water-lain ones, can also be important where the latter are mineralized and the former are barren. This situation is found in the distribution of uranium mineralization of the Colorado Plateau (p. 83) and in the Zambian Copper belt.

DIAGNOSIS OF ANCIENT EOLIAN SEDIMENTS

Because the processes of eolian and aqueous sedimentation are so similar these deposits are extremely difficult to differentiate. Even the classic Mesozoic eolian sands of the USA have been subjected to attempts to reinterprete them as just discussed. There is no single criterion which unequivocally indicates an eolian origin for a sandstone. Only an assemblage of criteria may suffice, backed by the absence of criteria which positively demonstrate a subaqueous environment. Subsurface diagnosis will now be reviewed under the five headings: geometry, lithology, palaeontology, sedimentary structures and palaeocurrent pattern.

Geometry

There is no distinctive shape to eolian formations. The original dune morphology of a sand sea is generally planed out before sediments of another facies are deposited. Basically a blanket is the final geometry. The environments of adjacent facies may be significant. Because deserts are equilibrial environments sand dunes often migrate across unconformities, cemented by ferrocrete, calcrete or silcrete and overlain by a veneer of pebbles which may include dreikanter, fierkanter and occasionally even the legendary funfkanter. Eolian sands may be expected to pass laterally into braided alluvial outwash sands, playa lake shales and evaporites or fanglomerates. There is no significant seismic response. Because of their uniform sandstone composition there are no significant internal reflectors and, as already discussed, dune palaeotopography is unlikely to be preserved.

Lithology

According to geological folklore eolian sands are well-rounded, well-sorted orthoquartzites with grain surfaces which are frosted when examined by a microscope, and show diagnostic markings under a scanning electron microscope. Statistical analysis of grain-size curves can differentiate recent eolian sands from subaqueous ones. Alas, it is not quite so simple. As already discussed granulometric analysis and surface texture are dangerous tools to use in environmental diagnosis because of the problems of inheritance and reworking, together with the difficulty of disaggregating lithified sandstone, and of surface textures being modified by solution and cementation. It is true that most Saharan and Arabian desert dunes are well-rounded, well-sorted orthoquartzites. But so they would be regardless of their present environment. In both deserts their recent dune sands are poly-poly-polycyclic (Figure 4.9).

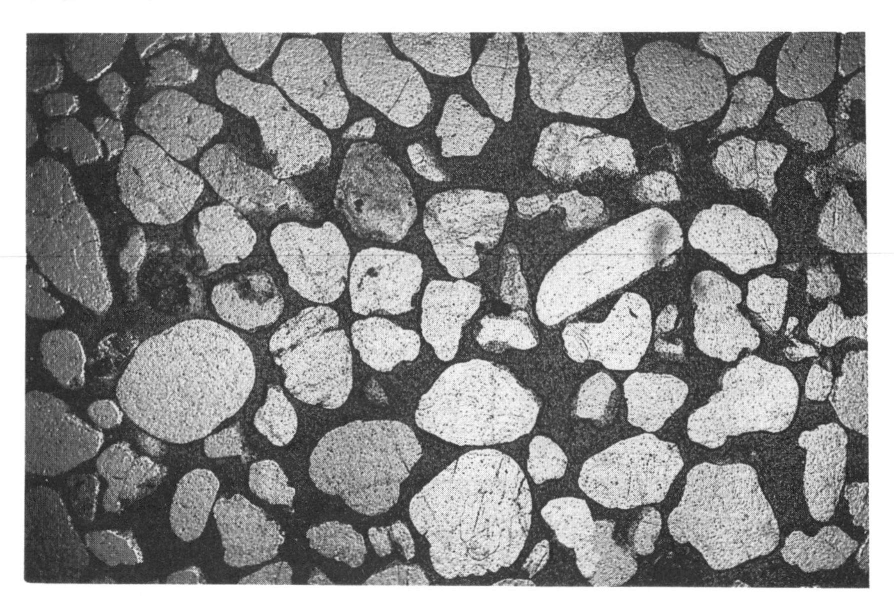

Figure 4.9
Photomicrograph of a thin section of a modern Saharan dune sand. This is a hypermature orthoquartzite with a polycyclic history of weathering, erosion, transportation and redeposition. Whereas all eolian sands are texturally mature, many are mineralogically immature, with well-rounded grains of unstable minerals.

They have been through many cycles of weathering, erosion, deposition, transportation and lithification. One will expect these sands to be texturally and mineralogically mature, regardless of their present environment. In point of fact though most eolian sands are texturally mature, they are often mineralogically very immature. Well-rounded feldspar grains occur in the recent Sahara and in the North Sea Permian dune sands. There are black dunes of volcaniclastic sand in Iceland and around Tertiary lava flows of the Sahara and elsewhere.

Many geologists have proposed textural criteria to distinguish wind-blown from water-laid sand. It is widely accepted that eolian sands are better sorted than aqueous ones. This is generally true, but there is no arbitrary dividing limit. The classic eolian Triassic Bunter Sandstones of England have a Trask sorting coefficient of 1.41 (Shotton, 1937). This is not so well-sorted as a beach sand cited by Krumbein and Pettijohn (1938, p. 232) which has a value of 1.22. It is widely accepted that eolian sands are positively skewed, with a tail of fines (Folk, 1966, p. 88; Folk, 1971). This factor can generally be used to distinguish them from beach sands, but not from fluvial ones (Friedman, 1961, p. 514). Though, as discussed on p. 40, textural criteria have been developed to distinguish eolian sands from aqueous ones in Recent sediments it is hard to use these in ancient deposits for technical reasons (see also Bigarella, 1973).

Eolian sands are widely believed to be very well-rounded and certainly experiments show that wind is much more efficient at rounding quartz sand than is running water (Kuenen, 1960). It is important to note, however, the polycyclic history of the eolian sandstones of the Colorado Plateau, the Permo Trias of Europe, and the present Sahara. These sands may be expected to be well-rounded and well-sorted, whether they were eolian or not. Eolian sand grains show a frosted, pitted surface under the optical microscope and, under the higher powers of the electron microscope, show a variety of characters which can be used to distinguish them from sands subjected to aqueous and glacial action (Krinsley and Funnel, 1965).

All these textural features of eolian sands must be carefully interpreted, remembering that dunes may be re-worked by running water, and alluvial fans may dry out and be re-worked by wind. Water-laid deposits may therefore inherit eolian textural parameters and vice versa. For this same reason wind-faceted pebbles (ventifacts, dreikanters) should be interpreted with care. Many recent beaches and barrier islands are capped by eolian dunes. These sands often contain marine shell debris and some are indeed composed of nothing but carbonate grains. Coasts around the world are rimmed by Pleistocene limestones with spectacularly large cross-bedding. These are variously termed Bahamite (in the Bahamas), Eolianite or Miliolite (in the large Gulf between Iran and Arabia).

Geologists have had many arguments about the eolian or marine origin of these rocks. Limestones may of course become dolomitized, so we must expect eolian dolomites to exist in the subsurface. The problem is worse than that. Some recent eolian dunes are composed of gypsum sands. These occur where sabkhas are deflated and gypsum and crystals eroded and abraded. Recent examples are known from both coastal dunes, like those of Abu Dhabi, and inland dunes, like those in White Sands National Park, New Mexico. The analogous Jurassic Todilto dunes have just been mentioned. When buried and heated of course, gypsum dehydrates to anhydrite. It is not surprising therefore that cross-bedded anhydrite and halite have been recorded in the subsurface, as in the Permian Zechstein of Germany (Zimmermann, 1908).

Not all eolian deposits are of sand grade. The Pleistocene loess silt deposits of North America, Europe, and China are widely interpreted as due to wind action in arid zones around the ice caps (Smalley, 1975). It is apparent therefore that it may be difficult to use lithology as a diagnostic criterion of eolian deposition. In a simple world we might look for well rounded, well-sorted medium to fine-grained sand of diverse mineralogical composition, ranging from quartzite and arkose to limestone, dolomite or evaporite. One would not expect to find pebbles or shale clasts in these sands. A red ferric staining should be anticipated, indicating early diagenesis above the water table as discussed on p. 42.

Palaeontology

From the preceeding account of eolian sand lithology it is apparent that these deposits may contain, and indeed be totally composed of, a diverse marine fauna. One would however expect the fossils to be extremely fragmented, abraded and rounded. The presence of abundant bioturbation would be one criterion which would serve to differentiate marine shoal sands from eolian ones; the problem being that these two facies are otherwise remarkably similar in lithology, texture and sedimentary structures. Marine vertical burrows should however be differentiated from rhizoconcretions. These are pipes of carbonate or gypsum cemented sand which form around plant roots (Figure 4.10).

Figure 4.10 Rhizoconcretions attributable to mineral precipitation around plant roots in arid continental Sherwood Sandstone (Triassic), Sidmouth, England.

As the plants draw moisture from the sand salts are precipitated around the rhizomes (Glennie and Evamy, 1968). Rhizoconcretions are not of course unique to eolian sands, but also occur in fluvial sands in arid environments. Intraformational conglomerates of fragmented concretionary tubes also occur.

Sedimentary structures

Sedimentary structures have often been used to distinguish wind-blown from water-laid sediment. As already noted (p. 102) long, low ripples have been recorded from wind-blown sediments and Tanner (1967) states that an RI of >15 can be used to distinguish wind ripples from those due to water, except for swash zone ripples on beaches. Tanner also empirically derived a number of other statistical criteria for distinguishing eolian and aqueously formed ripples. One might have supposed that contorted sand bedding was restricted to aqueous deposits. Unfortunately this is not so. It has been recorded in Recent dunes (McKee, 1966, p. 69). It occurs in the Colorado Plateau eolian sandstones, and is particularly widespread in the Navajo Formation (Stokes, 1961, p. 158). Very thick cross-beds are often held to be restricted to eolian sands (Figure 4.11).

However, insufficient is known of Recent marine sand bars to give the maximum set height of water-formed cross-bedding. Therefore no arbitrary limit between eolian and aqueous set heights may be given. Eolian foresets

Figure 4.11 Photograph of large curvaceous cross-bedding in red Permian sandstone, Dawlish, England. Many geologists accept this style of cross-bedding as characteristic of eolian deposition.

are often no higher than water-laid sets, and only a fraction of the height of the dune in which they occur (see McKee, 1966, Figure 8). Eolian dune foresets are often said to be asymptotically curved towards their bases, and to have steeper inclinations than water-laid foresets. An angle of about 30° is often quoted as critical for distinguishing water-laid from wind blown foresets. Angles in excess of this figure have been widely recorded from Recent dunes and supposed ancient ones (e.g. McKee, 1966; Mackenzie, 1964; McBride and Hayes, 1962; Laming, 1966; Poole, 1964). On the other hand Shotton (1937) recorded a mean value of only 24° for the Triassic Bunter dune sands of England, and Tanner (1965) recorded dips between 21° and 26° as common in the Entrada sandstone of the Colorado Plateau.

Care should also be taken in studying foreset inclination data from ancient rocks since it is hard to determine the horizontal datum at the time of their formation. Major bedding surfaces generally represent old erosional slopes of the dune surface, and could seldom have been horizontal (see also Potter and Pettijohn, 1963, pp. 78, 80 and 86).

Palaeocurrent patterns

Recent dunes show extremely varied morphologies. Transverse dunes whose foresets consistently dip downwind are generally quantitatively subordinate to barchan (lunate), seif (longitudinal), and stellate dunes. These types have much more complex geometries and correspondingly more varied foreset orientations. Barchan foresets curve through an arc of about 180°, while seif dunes show bipolar forest dip orientations perpendicular to mean wind direction (McKee and Tibbits, 1964) (Figure 4.12).

Despite these complexities ancient eolian deposits have produced consistent palaeocurrent trends, though with predictably high scatters of readings. Thompson (1967), in an extremely elegant account of the three-dimensional geometry of Triassic sand dunes in the English Midlands, measured many unimodal dip patterns. Similar results have been obtained from dipmeter logs of Permian Rotliegende dunes of the North Sea (Glennie, 1972). These often exhibit a clean low API gamma log with occasional upward asymmetric sawtooth patterns. The corresponding dip motif is a blue pattern in which the upward increase in dip occurs opposite to the short shaley interval of the gamma curve. Calibration of this log motif with cores shows that the subhorizontal dips in the silty sand correspond with the toeset of the dune on which clay and mica settle out. This passes up into the clean gamma interval with uniform high angle foreset dips (Figure 4.13).

Genetic increments with this motif occur associated with more regular gamma intervals with heterogenous dip motifs. These are interpreted as waterlaid fluvial intervals, in which eolian sands were deposited by episodic flash floods (Figure 4.14).

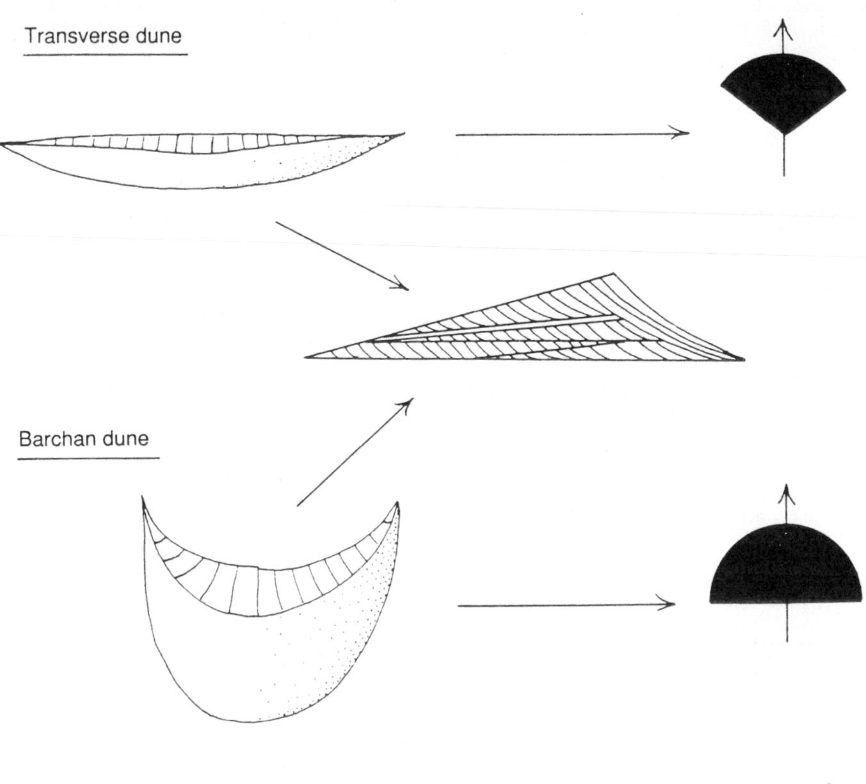

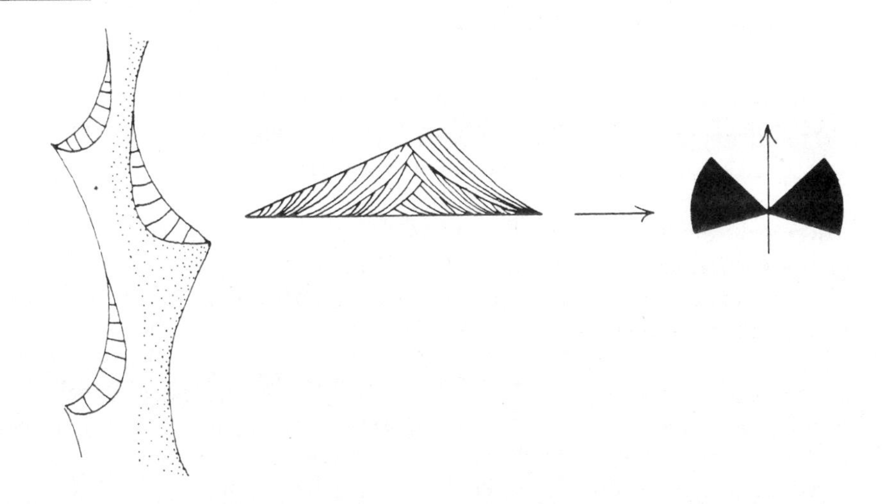

Figure 4.12 Diagrams illustrative of the relationship between dune morphology and cross-bedding orientation, shown plotted on adjacent azimuths. Wind blows up the page. Palaeocurrents will be unrelated to depositional slope. Barchan and seif dunes are generally ephemeral bed forms which are seldom preserved. Net eolian sedimentation comes from gregarious transverse dunes with straight crests that deposit tabular planar cross-bedding, and lunate crests that deposit trough cross-bedding.

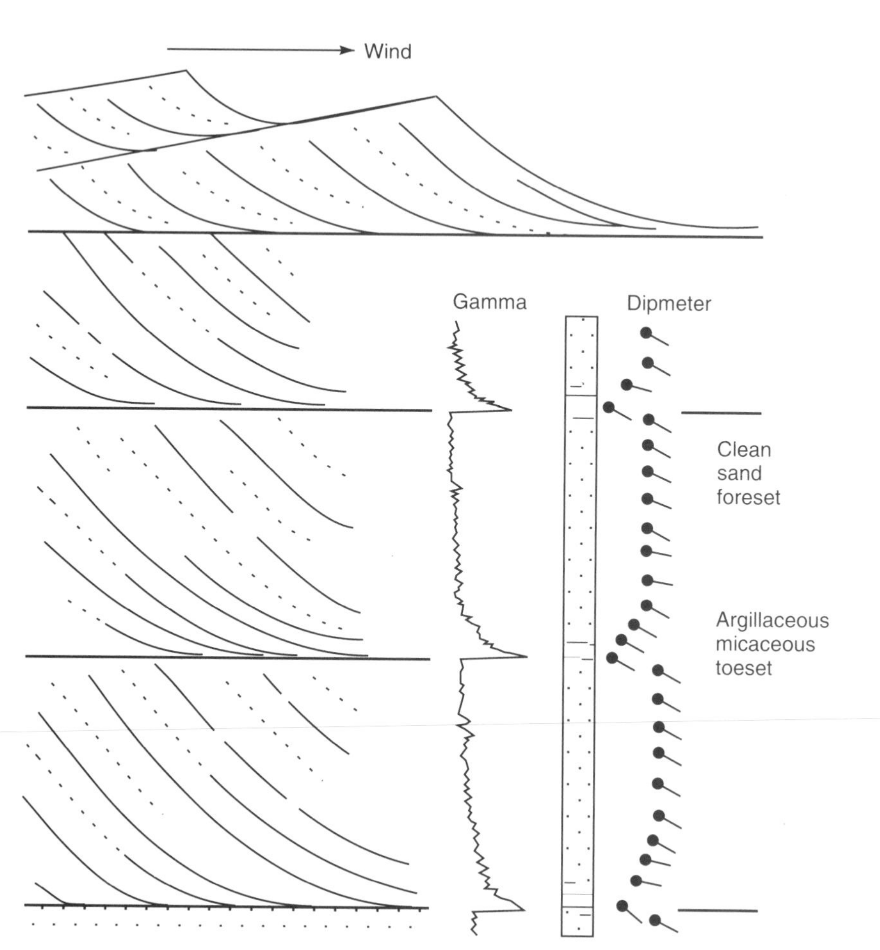

Figure 4.13 Diagram to illustrate the formation of the characteristic gamma and dip meter motifs seen in ancient eolian sands.

The distinctive motif outlined above has also been recognized in presumed eolian sands in the Botucatu Formation of Brazil, and in the Colorado Plateau eolian sands described earlier.

Regional studies have been used in attempts to reconstruct palaeoclimates, particularly the past distribution of high pressure air cells (Shotton, 1956; Opdyke, 1961, Parrish and Peterson, 1988). A further problem of eolian palaeocurrents is that they are not slope controlled like many (but not all) aqueous currents, as illustrated in the Entrada sandstone of the Colorado Plateau case history. In an elegant study of Permotrias desert deposits of south-west England, Laming has shown (1966) how sediment was alternately flushed down alluvial fans by floods and blown upslope again by wind.

In conclusion this discussion shows that it is not easy to recognize ancient wind-blown sediments. This is basically because wind and water both transport and deposit sand in ripples and dunes. Many textural and structural criteria have been proposed to distinguish the results of these processes; few, if

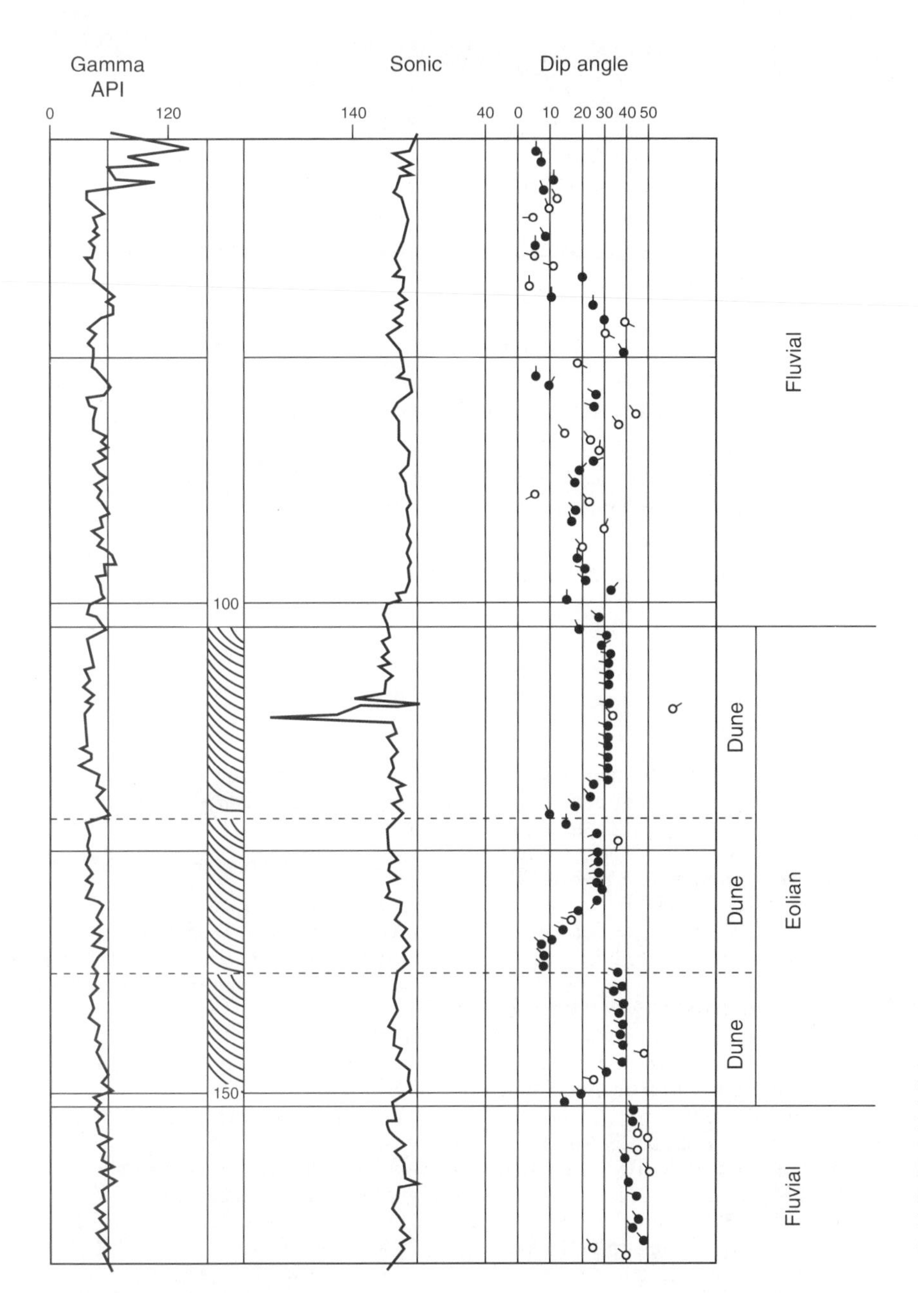

Figure 4.14 Gamma and dipmeter log of a Rotliegende (Lower Permian) well in the southern North Sea, showing the characteristic motif attributed to eolian dunes.

any, are foolproof. Recognition of an ancient eolian sediment must be based on a critical evaluation of all the available data. It may help to distinguish criteria which weigh against an aquatic origin separately from those positively in favour of eolian deposition. Finally it often helps to ask the question: 'If this facies is not eolian, what else could it be?'

EOLIAN DEPOSITS: SUMMARY

From the preceding account it will be apparent that it is very hard to prove conclusively the eolian origin of a facies at outcrop, and may be impossible in the subsurface. The underlying difficulty is because both in air and under water sediment is transported by traction currents whose bed forms of ripples and megaripples are broadly comparable, and whose resultant structures of cross-lamination, flat bedding and cross-bedding are likewise similar.

Though the geometries of modern desert dunes are distinctive, they actually deposit nondescript blanket formations with no diagnostic geometry, often gradationally interbedded with waterlain sands of fluvial origin. Eolian formations are composed of a wide range of lithologies. Though the classic eolian sandstone may be a pink well-sorted, well-rounded, medium fine grained quartzite, other eolian sediments include arkoses, volcaniclastic sands, evaporites, bioclastic limestones and red loessic siltstones. Palaeontologically, eolian dunes may contain a rich, if fragmented, marine biota.

REFERENCES

The account of the Colorado Plateau eolian sandstones was based on:

Baars, D. L. (1961) Permian blanket sandstones of the Colorado plateau. In J. A. Peterson and L. C. Osmond (eds), *Geometry of Sandstone Bodies*, Amer. Assoc. Petrol. Geol. Symposium, 179–207.

Freeman, W. E. and Visher, G. S. (1975) Stratigraphic analysis of the Navajo Sandstone, *J. Sediment. Petrol.*, **45**, 651–8.

Fryberger, S. G. (1993) A review of aeolian bounding surfaces, with examples from the Permian Minnelusa Formation, USA. In C. P. North and D. J. Prosser (eds), *Characterization of Fluvial and Aeolian Reservoirs*, Sp. Pub. Geol. Soc. London, **73**, 167–98.

Haun, J. D. and Kent, H. C. (1965) Geological history of Rocky Mountain Region, *Bull. Amer. Assoc. Petrol. Geol.*, **49**, 1781–1800

Kokurek, G. (ed.) (1988) Late Palaeozoic and Mesozoic Eolian deposits of the Western Interior of the United States, *Sed. Geol.*, **56**.

Lessentine, R. H. (1965) Kaiparowits and Black Mesa Basins: stratigraphic synthesis, *Bull. Amer. Assoc. Petrol. Geol.*, **49**, 1997–2019.

Opdyke, N. D. (1961) The paleoclimatological significance of desert sandstone. In A. E. M. Nairn (ed.), *Descriptive Paleoclimatology*, Interscience, NY, 45–59.

Peterson, F. (1988) Pennsylvanian to Jurassic eolian transportation systems in the westen United States. In G. Kokurek (ed.), Late Palaeozoic and Mesozoic Eolian deposits of the Western Interior of the United States, *Sed. Geol.*, **56**, 207–60.

Poole, F. G. (1964) Paleowinds in the Western United States. In A. E. M. Nairn (ed.), *Problems of Paleoclimatology*, Interscience, NY, 390–405.

Stanley, K. O., Jordan, W. M. and Dott, R. H. (1971) New hypothesis of Early Jurassic Paleogeography and sediment dispersal for western United States, *Bull. Amer. Assoc. Petrol. Geol.*, **55**, 10–19.

Stokes, W. L. (1961) Fluvial and eolian sandstone bodies in Colorado Plateau: In J. A. Peterson and J. C. Osmond (eds), *Geometry of Sandstone Bodies*, Amer.

Assoc. Petrol. Geol. Symposium, 151–78.
Tanner, W. F. (1965) Upper Jurassic paleogeography of the Four Corners Region, *J. Sediment. Petrol.*, **35**, 564–74.
Visher, G. S. (1971) Depositional processes and the Navajo sandstones, *Bull. Geol. Soc. Amer.*, **82**, 1421–4.

Other references cited in this chapter are:

Ahlbrandt, T. S. and Fryberger, S. C. (1981) Sedimentary features and significance of interdune deposits. In F. G. Ethridge and R. M. Flores (eds), *Recent and Ancient Non-marine Depositional Environments: Models for Exploration*, Soc. Econ. Pal. & Min. Sp. Pub., 31. 293-314.
—— and —— (1982) Eolian deposits. In P. A. Scholle and D. Spearing (eds), *Sandstone Depositional Environments*, Amer. Assoc. Petrol. Geol., Mem. No. **31**, 49–86.
Bagnold, R. A. (1941) *Physics of Blown Sand and Desert Dunes*, London, Methuen.
Bigarella, J. J. (1972) Eolian environments, their characteristics, recognition and importance: In J. K. Rigby and W. K. Hamblin (eds), *Recognition of Ancient sedimentary environments,* Soc. Econ. Pal. Min. Sp. Pub. **16**, 12–62.
—— (1973) Textural characteristics of the Botucatu sandstone, *Boletin Paranaense de Geociencies*, **31**, 85–94.
Brookfield, M. E. and Ahlbrandt, T. S. (1983) *Eolian Sediments and Processes*, Elsevier, Amsterdam.
Folk, R. L. (1966) A review of grainsize parameters, *Sedimentology*, **6**, 73–93.
—— (1971) Longitudinal dunes of the northwestern edge of the Simpson Desert, Northern Territory, Australia, 1, Geomorphology and grain size relationships, *Sedimentology*, **16**, 5–54.
Friedman, G. M. (1961) Distinction between dune, beach and river sands from their textural characteristics, *J. Sediment. Petrol.*, **31**, 514–29.
Glennie, K. W. (1970) *Desert Sedimentary Environments*, Elsevier, Amsterdam.
—— (1972) Permian Rotliegendes of northwest Europe interpreted in the light of modern sediment studies, *Bull. Amer. Assoc. Pet. Geol.*, **56**, 1047–8.
—— (1987) Desert sedimentary environments, present and past – a summary, *Sed. Geol.*, **50**, 135–65.
—— (1990) Lower Permian – Rotliegende. In K. Glennie (ed.), *Introduction to the Petroleum Geology of the North Sea*, 3rd edn, Blackwell, Oxford, 120–52.
—— and Evamy, B. D. (1968) Dikaka: Plant and plant root structures associated with eolian sand, *Palaeogeog. Palaeoclimatol. and Palaeoecol.*, **4**, 77–87.
Křinsley, D. H. and Funnel, B. M. (1965) Environmental history of sand grains from the Lower and Middle Pleistocene of Norfolk, England, *Quart. J. Geol. Soc. London*, **121**, 435–61.
Krumbein, W. C. and Pettijohn, F. J. (1938) *Manual of Sedimentary Petrography*, Appleton-Century-Crofts, NY.
Kuenen, Ph. (1960) Experimental abrasion: 4, Eolian action, *J. Geol.*, **68**, 427–49.
Lamb, C. F. (1980) Painter reservoir field: giant in the Wyoming Thrust Belt. In M. T. Halbouty (ed.), *Giant Oil and Gas Fields of the Decade 1968–78*, Amer. Assoc. Petrol. Geol., **30**, 281–8.
Laming, D. J. C. (1966) Imbrication, paleocurrents and other sedimentary features in the Lower New Red Sandstone, Devonshire, England, *J. Sediment. Petrol.*, **36**, 940–57.
Lindquist, S. J. (1983) Nugget formation reservoir characteristics affecting production in the Overthrust Belt of Southwestern Wyoming, *Jl. Pet. Tech.*, **35**, 1355–65.
—— (1988) Practical characterisation of eolian reservoirs for development: Nugget Sandstone, Utah-Wyoming thrust belt. In G. Kokurek, G. (ed.), *Late Palaeozoic*

and Mesozoic Eolian Deposits of the Western Interior of the United States, Sed. geol., **56**, 315–40.

McBride, E. F. and Hayes, M. O. (1962) Dune cross-bedding on Mustang Island, Texas, *Bull. Amer. Assoc. Petrol. Geol.*, **46**, 546–52.

McKee, E. D. (1966) Structures of dunes at White Sands National Monument, New Mexico (and a comparison with structures of dunes from other selected areas), *Sedimentology*, **7**(1) (Sp. Issue).

—— (1979) *A Study of Global Sand Seas*, US Geol. Surv. Prof. Pap. 1052.

—— and Tibbitts, G. C. (1964) Primary structures of a seif dune and associated deposits in Libya, *J. Sediment. Petrol.*, **34**, 5–17.

Mackenzie, F. T. (1964) Bermuda Pleistocene eolianites and paleowinds, *Sedimentology*, **3**, 52–64.

Malicse, A. and Mazullo, J. (1989) Reservoir properties of the Desert Shattuck Member, Caprock Field, New Mexico. In J. W. Barwis, J. G. McPherson and J. R. J. Studlick (eds), *Sandstone Petroleum Reservoirs*, Springer-Verlag, Berlin, 133–52.

North, C. P. and Prosser, D. J. (eds) (1993) *Characterization of Fluvial and Aeolian Reservoirs*, Geol. Soc. Sp. Pub., **73**.

Parrish, J. T. and Peterson, F. (1968) Wind directions predicted from global circulation models and wind direction determined from eolian sandstones of the western United States – a comparison, *Sed. Geol.*, **56**, 261–82.

Parsley, A. J. (1990) North Sea hydrocarbon plays. In K. Glennie (ed.), *Introduction to the Petroleum Geology of the North Sea*, 3rd edn, Blackwell, Oxford, 362–88.

Potter, P. E. and Pettijohn, F. J. (1963) *Paleocurrents and Basin Analysis*, Springer Verlag, Berlin.

Prosser, D. J. and Maskall (1993) Permeability variation within aeolian sandstone: a case study using core cut sub-parallel to slipface bedding, the Auk Field, Central North Sea, UK. In C. P. North and D. J. Prosser (eds), *Characterization of Fluvial and Aeolian Reservoirs*. Geol. Soc. Sp. Pub., **73**, 377–97.

Pye, K. (ed.) (1993) *The Dynamics and Environmental Context of Aeolian Sedimentary Systems*, Geol. Soc. Lond. Sp. Pub., **72**.

Sanders, J. E. and Friedman, G. M. (1967) Origin and Occurrence of Limestones. In G. V. Chilingar, H. J. Bissell and R. W. Fairbridge (eds), *Carbonate Rocks, Part A*, Elsevier, Amsterdam, 169–265.

Shotton, F. W. (1937) The Lower Bunter Sandstones of North Worcestershire and East Shropshire, *Geol. Mag.*, **74**, 534–53.

—— (1956) Some aspects of the New Red Desert in Britain, *Lpool. Manchr. Geol. J.*, **1**, 459–65.

Smalley, I. J. (1975) *Loess: Lithology and Genesis*, J. Wiley & Sons, Chichester.

Tanner, W. F. (1967) Ripple mark indices and their uses, *Sedimentology*, **9**, 89–104.

Thompson, D. B. (1967) Dome-shaped Aeolian dunes in the Frodsham Member of the so-called 'Keuper' sandstone formation (Scythian-Anisian: Triassic) at Frodsham, Cheshire, (England), *Sedimentary Geology*, **3**, 263–89.

Weber, K. J. (1987) Computation of original well productivities in aeolian sandstone – the basis of a geological model, Leman Gas Field. UK North Sea. In W. Tilman and K. Weber (eds), *Reservoir Sedimentology*, Soc. Econ. Pal. Min. Sp. Pub., **40**, 333–54.

Wilson, I. G. (1972) Aeolian bed forms□their development and origins, *Sedimentology*, **19**, 173–210.

—— (1973) Ergs. *Sed. Geol.*, **10**, 76–106.

Zimmermann, E. (1908) Steinsalz mit wellenfurchen, Shlitz in Hessen, *Zeit. d. Deut. Geol. Gessell.*, **60**, 70–5.

5 Lake deposits

INTRODUCTION

Lakes are landlocked bodies of non-marine water. Recent examples may be hundreds of kilometres long, as are the Canadian Great Lakes, and of widely varying depths. They may be permanent bodies of water or only of short duration. For example, the playa lakes of central Australia last for only weeks or months every few years after torrential rainstorms. Lakes occur in mountainous regions, as in the Alps, and on low-lying continental platforms, like Lake Chad. Their waters range from fresh in temperate and periglacial climate, to hypersaline in arid regions. Lakes may be caused by tectonic subsidence and faulting, by glacial erosion, and by damming due to ice, lava and, recently, man.

Because of their widely variable environmental settings Recent lacustrine sediments are of many kinds. We must expect ancient lacustrine deposits to show a similar diversity. It is beyond the scope of this book to examine all the different kinds and, after briefly reviewing recent lakes, one case history must suffice. This is followed however by a review of the diagnostic features of some other ancient lake deposits. Both recent and ancient lakes have been intensively studied for economic as well as aesthetic reasons. Detailed accounts of lakes ancient and modern have been given by Fouch and Dean (1982), Anadon *et al.* (1991) and Katz (1991).

RECENT LAKES

The chemistry, physics and biology of modern lakes have been extensively studied. But rather less is known about their sedimentology. Most of this work has concentrated on the permanent lakes of temperate climates. These studies have shown that the two main controlling parameters of lacustrine sedimentation are the degree of aridity and the amount of sediment input. The latter is controlled partly by climate (temperature, and the type, frequency and amount of precipitation) and partly by the relief and rock type of the hinterland. These variables have been used as a basis for classifying lakes.

Figure 5.1 presents a simple classification of lakes that can be recognized both on the surface of the earth today and in the stratigraphic record. Around the lake shores rivers may build out deltas from which wind-generated currents can transport sand in beaches and bars. Where there is no influx of terrigenous detritus swamps may deposit peat. On some lake

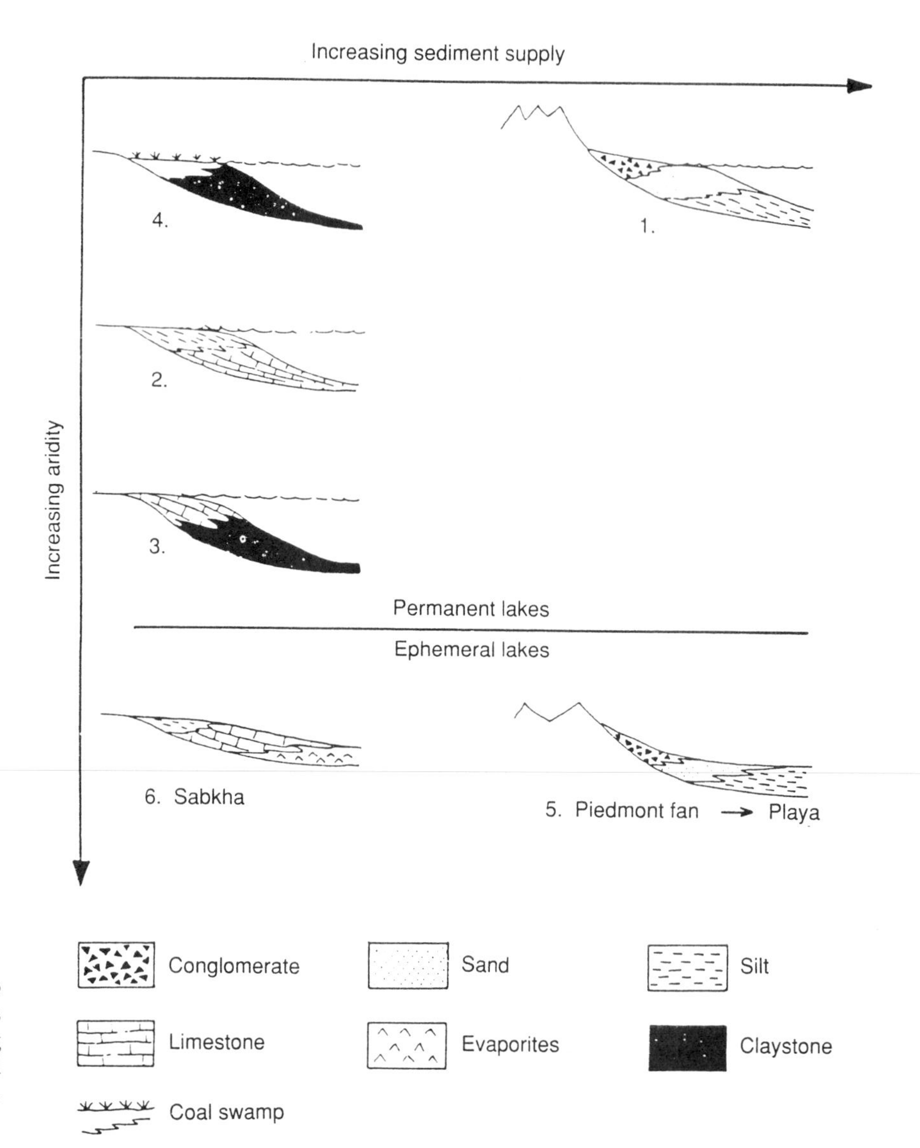

Figure 5.1 Cartoons to illustrate the various types of lake classified according to sediment supply and aridity.

shores lime is deposited. This is largely of biogenic origin, ranging from algal reefs to detrital sands of mollusc shells and algal debris, but lacustrine oolite sands are known.

Turbidity currents may transport detritus from the shoreline far out into the basin centre where the finest sediment comes to rest. This may include not only terrigenous detritus but also transported lime and humic organic matter. Extensive algal growth within the shallow photic zone in the centres of lakes may give rise to the deposition of sapropelic muds. In many lakes a stratification of water develops, with warm shallow water overlying cooler denser water.

The activity of phytoplankton in the upper layer generates oxygen which supports many other forms of life. If a water of the lake has a density stratification then the oxygen in the lower layer becomes depleted, and cannot be renewed because it is too dark for photosynthesis. In this situation the lower layer becomes anoxic and organic matter settling from above may be preserved. These two layers are referred to as the Epilimnion and the Hypolimnion respectively (Figure 5.2).

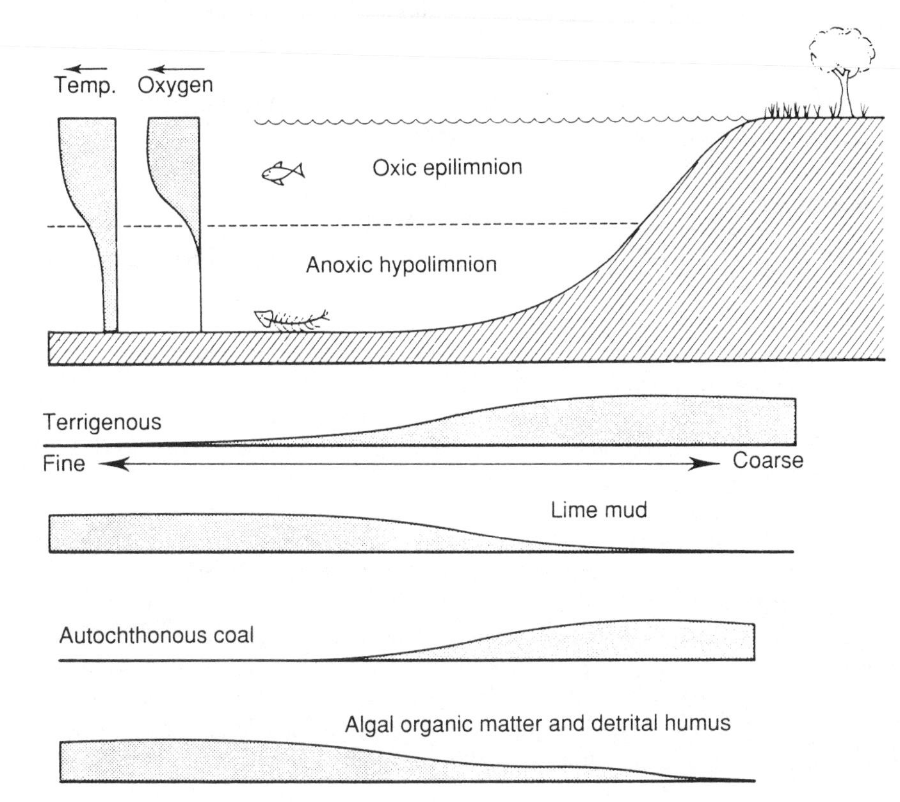

Figure 5.2 Cartoon to illustrate the relationship between sediment type and some critical physical and chemical parameters in a lake with thermally induced stratification of the water column.

Intermittent storms may mix the two water masses. This often leads to catastrophic mortality of organisms in the upper layer whose bodies may then be preserved on the lake floor as anoxic conditions re-establish themselves in the Hypolimnion. This is why many ancient lake deposits contain oil shales and petroleum source beds in their centres. Regular repetition of this process imparts a cyclic lamination to the lake sediments. Cyclicity is also commonly seen in glacial lakes where laminae of terrigenous detritus from melt water streams are interlayered with lime mud and organic matter. These regular layers called 'varves' are of course annual.

Lake basins in arid mountainous terrain are often referred to as playas. Red clays and silts in the centre of the lake basin generally pass via alluvial outwash plains into fanglomerates adjacent to mountains. Many playa lakes are only flooded for a few months every twenty years or so as a result of torrential storms and flash floods (it has been remarked that more people are drowned in the Sahara than die of thirst). For the rest of the time the lake basins are dry flat surfaces with intermittent sand dunes around their edges and occasional huge desiccation cracks in their centres. Some lakes in arid regions are areas of evaporite sedimentation. Recent examples include the Great Salt Lake of Utah, the Dead Sea and Lake Eyre, Australia.

TORRIDONIAN (PRECAMBRIAN) LACUSTRINE SEDIMENTS, NW SCOTLAND

Description

The Torridonian rocks crop out along the northwest coast of Scotland and on adjacent islands of the Hebrides (Figure 5.3). These sediments overlie the Lewisian gneiss with an irregular unconformity, and are themselves separated by a planar unconformity from overlying Lower Cambrian sandstones and limestones. The Torridonian is divisible into two groups by an unconformity which separates westerly dipping red sandstones and shales of the Stoer Group from the overlying sub-horizontal Torridon Group. This consists of about 3 km of interbedded red conglomerates, sandstones, and shales.

In the north these deposits are separated from the Lewisian gneiss by a planar pre-Torridonian weathered surface. Traced southward, they overlie the Stoer Group. Still further south around the type area of Loch Torridon these deposits overlie and infill a dissected topography cut into the gneiss basement with a relief of at least 100 m. In low-lying parts of this surface breccias banked against the flanks of pre-Torridonian mountain pass laterally into grey shales and sandstones. Such a situation can be seen on the island of Raasay whose Torridon Group sediments will now be described. These are divisible into three major facies defined as follows:

1. basal facies: red and grey breccias and granulestones present adjacent to the buried gneiss topography (interpreted as fanglomerate);
2. grey facies: grey sandstones and shales, overlying and laterally equivalent to the basal facies (interpreted as lacustrine);
3. red facies: coarse red pebbly sandstones and siltstones overlie the previous two facies (interpreted as braided alluvial outwash).

The relationship of the three facies are summarized in Figure 5.4. The sub-Torridonian unconformity (after allowing for later tectonic tilting) is a dissected plateau with a steep westerly facing scarp some 60 m high. This topography has largely been stripped of its Torridonian cover. Basal facies deposits can still be found infilling gullies cut into the scarp, and as fan-shaped deposits banked against its lower slope. These sediments are boulder

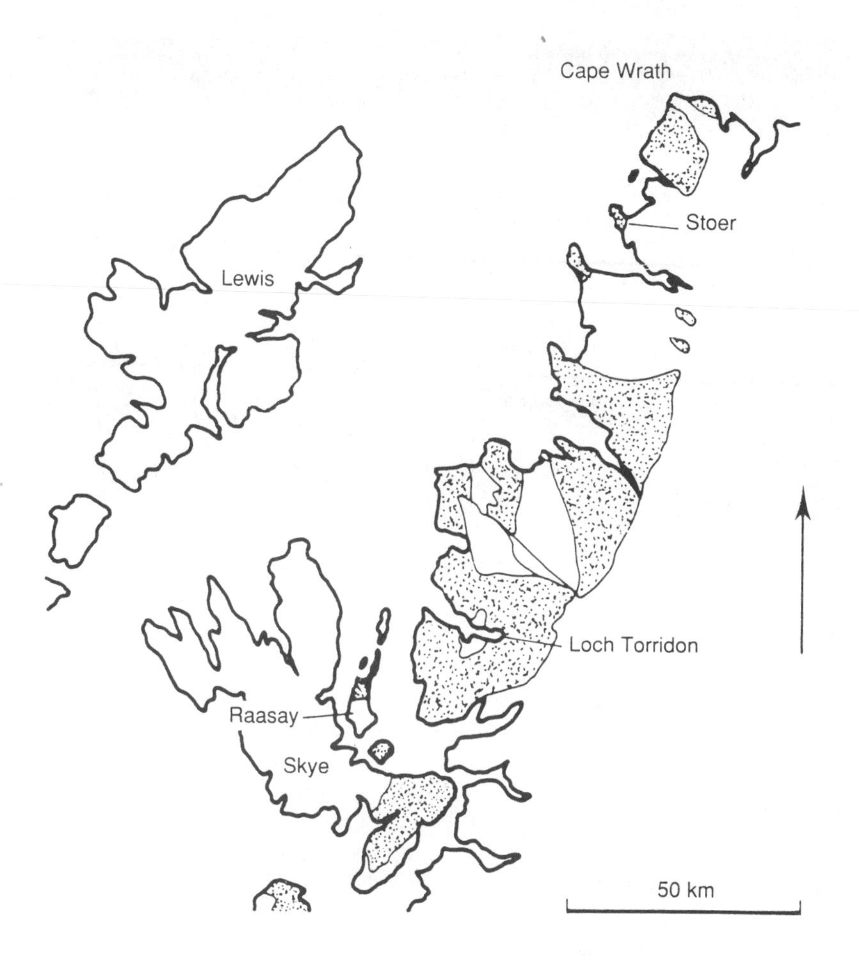

Figure 5.3 Distribution of Torridonian (PreCambrian) sediments of north-west Scotland. Most of the outcrop is of the red facies (braided alluvial outwash). The grey facies (lacustrine) crop out intermittently along the eastern limit of the Torridonian, notably around Loch Torridon, and on the isles of Raasay and Skye.

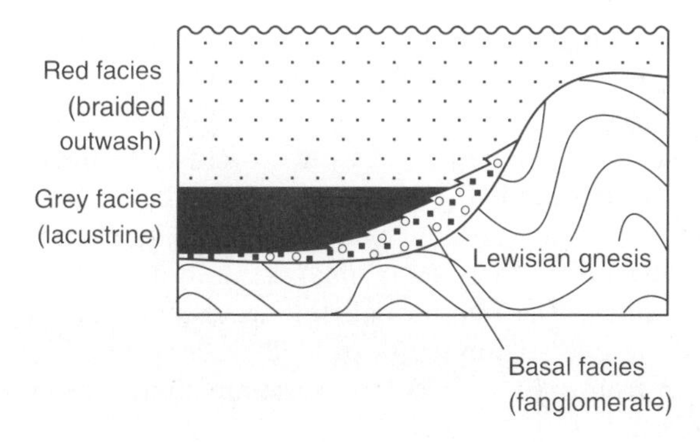

Figure 5.4 Sketch cross-section to show the relationships between the various facies of the Torridon Group.

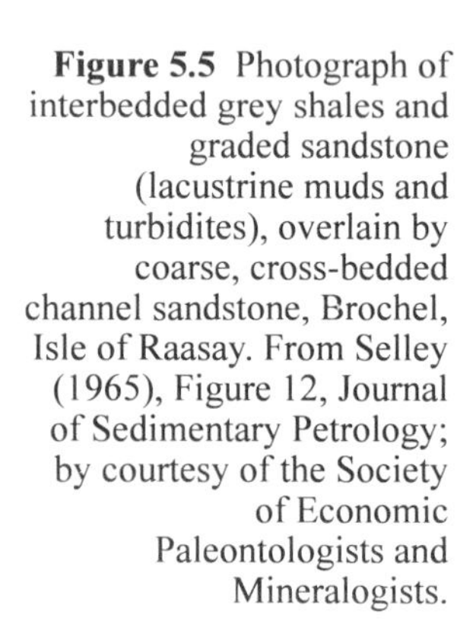

Figure 5.5 Photograph of interbedded grey shales and graded sandstone (lacustrine muds and turbidites), overlain by coarse, cross-bedded channel sandstone, Brochel, Isle of Raasay. From Selley (1965), Figure 12, Journal of Sedimentary Petrology; by courtesy of the Society of Economic Paleontologists and Mineralogists.

beds, breccias, and granulestones. The coarse grainsize, angularity, poor sorting and petrography of the basal facies deposits clearly show that they were derived locally from the Lewisian gneiss. Their geometry and the radiating depositional dips leave little doubt that they are ancient piedmont fans, as described in Chapter 3 (refer back to Figure 3.2).

The basal facies is overlain by, and also pass laterally into, the grey facies. The grey facies consists of over 100 m of three interbedded rock types: thick beds of coarse sandstone, thin beds of medium and fine sandstone, and shales (Figure 5.5).

Shales are the most abundant rock type, composing up to 80% of the grey facies in individual sections (Figure 5.6).

They are poorly sorted fissile grey laminated argillaceous micaceous siltstones. Thin clay and sand laminae are frequent. Within these shales are lenses and isolated ripples of very fine micro-cross-laminated sand. Desiccation cracks infilled with red medium well-rounded (wind-blown?) sands are common. Interbedded with the shales are 10–15 cm thick beds of medium and fine-grained sandstones.

These are of two types. Some are clean micro-cross-laminated arkoses. Others are poorly sorted graded greywackes, whose bases are often erosional and overlain by a thin layer of granules and shale clasts. Internally these sands are generally massive, laminated, or cross-laminated, with occasional convolutions. The interbedded fine sandstones and shales pass up gradationally in to coarse grey cross bedded arkose channel sandstones (Figure 5.5). Apart from their drab colour these are similar to those of the overlying red facies.

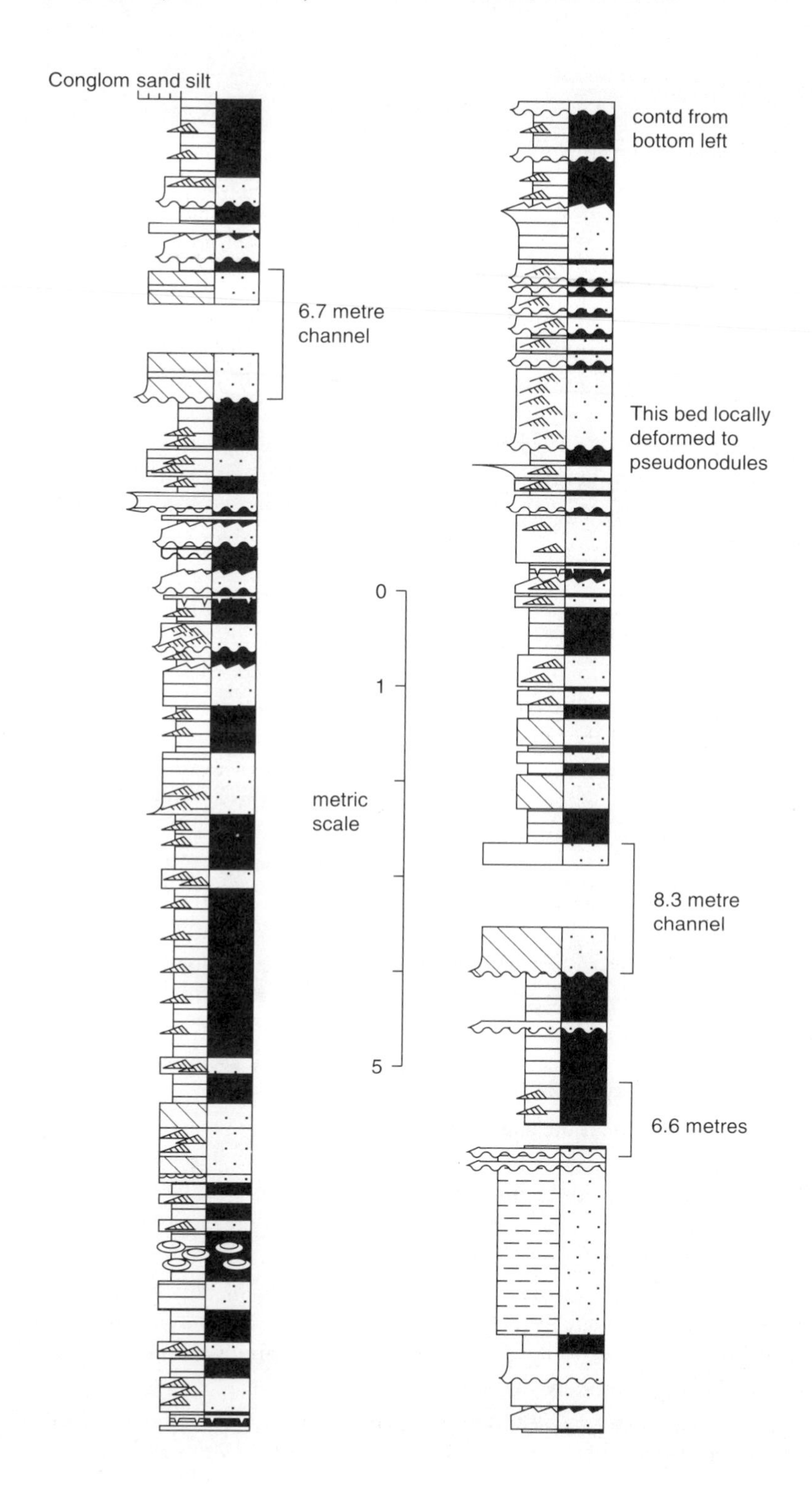

Figure 5.6
Sedimentological log of grey facies section, Brochel, Isle of Raasay. Note desiccation cracks, graded sands (interpreted as turbidites), cross-laminated sands, and cross-bedded channel sands.

Interpretation

The abundance of laminated grey shale suggests that this facies originated in a low energy environment where fine-grained sediment settled out of suspension. The presence of desiccation cracks shows that the water occasionally receded or evaporated and was therefore shallow. Intermittent higher energy conditions are indicated by the beds of medium and fine sandstone. The well-sorted micro-cross-laminated sands were probably deposited by low velocity traction currents. The poorly-sorted graded greywackes, however, show many of the typical features of turbidites (refer to Chapter 12).

The coarse cross-bedded channel sands, that gradually increase in abundance up the section, herald the advance of the overlying red facies. This has all the sedimentological features that characterize a braided alluvial environment, as outlined in Chapter 3 (p. 85). Regionally radiating cross-bedding in the red facies suggests that the braided channels flowed down a series of alluvial fans initiated by uplift of the Minch Fault (Figure 3.15). These fans gradually prograded out over the waters of the grey facies, their advance being heralded by the turbidites, and succeeded by their feeder channels.

Considered overall therefore the grey facies was deposited mostly under quiet shallow water which from time to time receded or evaporated. Sands, deposited both by traction currents and turbidity flows, indicate higher energy conditions which ultimately dominated the area as it was buried beneath the braided alluvial outwash of the red facies.

One of the problems of the grey facies, however, is whether it was deposited within enclosed lake basins, or in the bays of a dissected marine coastline. For Phanerozoic rocks this problem would be quickly solved by palaeontology. Though the grey facies shales do contain microfossils their environmental significance is uncertain. Likewise the sedimentological evidence is equivocal. Desiccation cracks can form on both tidal flats and dried out lakes. Turbidity flows have been recorded both from Recent lakes and fjords, as well as in the sea.

Palaeocurrents in the grey facies are unipolar, but even bipolar palaeocurrents would not be significant since they have been recorded from ancient lakes and are not therefore exclusive to marine tidal deposits. Exposure is nowhere good enough to show whether the grey facies is restricted to isolated hollows in the gneiss basement, or whether the facies extends in the subsurface around the headlands of a dissected marine shoreline. There is, however, no evidence for mud infilled channels, such as those that occur on modern tidal flats.

Geochemical data support a non-marine origin, the boron content of illites in the shales being in the order of 100 ppm, whereas illites from known marine shales commonly show more than twice that amount (Stewart and Parker, 1979). It is thus reasonable to conclude that the grey facies of the Torridonian was deposited in a lacustrine environment.

ECONOMIC ASPECTS OF LAKE DEPOSITS

Ancient lakes are of considerable economic importance for the production of petroleum, coal, and evaporite minerals. There are many ancient lake basins that contain organic-rich shales in their centres which have generated petroleum now trapped in sands around the old lake basin shoreline (Fleet *et al.*, 1988). The Green River Shale Formation has generated oil in several of the Tertiary basins of Utah and Wyoming (Chatfield, 1972; Randy Ray, 1982). Other petroliferous lake basins occur in China, Brazil and Australia (Chin Chen, 1980; Guangming and Quanheng, 1982; Grunau and Gruner, 1978 and Evans, 1981).

Many oil shales are of lacustrine origin (Yen and Chilingarian, 1976). The Green River Formation (Eocene) of ancient Lake Uinta in Utah is one of the best-known lacustrine oil shale formations (Dean and Fouch, 1983). Other examples occur in the Permo-Carboniferous of New South Wales and in the Carboniferous deposits of 'Lake Cadell' in the Midland Valley of Scotland (Cater, 1987).

Important coal fields also occur in ancient fluvio-lacustrine swamps (Falini, 1965). This type of environmental setting is widespread in Permian coal fields of the continents which are interpreted as once having formed Gondwanaland. These include those of the Karoo System in Zambia, South Africa, and Zimbabwe; the Damuda System of India, and equivalent strata in Australia and Antarctica. These coal deposits are all of about the same (Permian) age, they often overlie sediments of possible glacial origin and occur in thick cyclic non-marine clastic sequences. The majority of the coals seem to have originated in swamps on alluvial plains peripheral to lakes. The Oligocene Bovey Tracey basin of south Devon is an example of an ancient lake which contains both lignite and kaolin clays (Edwards, 1976). Other economic minerals formed in lacustrine deposits include iron ore (generally limonite), kiesulguhr (diatomaceous earth) and evaporites.

DIAGNOSIS OF ANCIENT LAKE DEPOSITS

Basically lacustrine sediments can easily be identified by the combination of low energy aqueous deposits and the absence of a marine fauna. However, as seen in the case of the grey facies of the Torridonian, in unfossiliferous PreCambrian sediments the distinction between lacustrine and marine sediments may be difficult. Though the majority of lakes are realms of fine grained sedimentation these may be either argillaceous, calcareous, or evaporitic.

A second feature typical of ancient lakes is cyclicity. This is often on a scale of millimetres, with alternating laminae of silt and clay, organic matter and carbonate. These laminae are commonly termed 'varves' and are believed to reflect annual depositional cycles. They were first recognized in the Pleistocene glacial lakes of northern Europe where they have been used to build up a detailed stratigraphy calibrated to Milankovitch cycles and climatic changes. Varves are not restricted to glacial lakes, though, and are well documented in the Green River Formation of the Eocene Lake Uinta in Utah mentioned earlier (Dean and Fouch, 1983).

Figure 5.7 Rhythmic lamination in Pleistocene lacustrine sediments, Libyan Sahara.

Figure 5.7 illustrates varved Pleistocene lacustrine sediments from the Sahara. A third feature of some lakes is the presence of evaporites. These also occur in Recent lakes such as the Dead Sea, and in their ancient deposits (Figure 5.8).

The source of lacustrine evaporites is often debatable. They may be derived from evaporitic rocks that crop out in the lakes' catchment area. The lake may have formed from a cut-off arm of the sea. Alternatively the salts may have come in the atmosphere from the oceans, being precipitated with rain and concentrated by rivers into ephemeral salt lakes. When attempting to find the source for lacustrine evaporites (where even their lacustrine origin may be in doubt) each case must be judged on its own merits.

Figure 5.8 Contorted interlaminated aragonite:gypsum couplets in the Lisan Marls (Pleistocene), on the shores of the Dead Sea.

Thus in the case of the Jurassic Todilto evaporites of the Colorado Plateau it may be significant that they are only a short distance from time equivalent marine strata (p. 105), whereas the Triassic Cheshire salt deposits of northern England, whose origin is in doubt, are far from time equivalent marine limestones and dolomites. It is more likely that they were reworked from Zechstein (L Permian) evaporites that cropped out contemporaneously. Such cases as these need careful observation and interpretation. Even the presence of marine micro-fossils is no proof of a marine origin. Holland (1912) records foraminifera being blown 200 miles inland from the Runn of Kutch to be deposited with evaporites in the Great Sambhar Lake of the Rajputana Desert.

This is a useful cautionary note on which to bring to a close this discussion on the diagnosis of ancient lacustrine environments sediments. The preceding account of lacustrine deposits shows that they are virtually indistinguishable from mariner ones in terms of lithology and sedimentology. The main diagnostic criterion is an absence of marine macrofossils. (Marine microfossils may be blown in to lakes, as just discussed.) There are also some differences between the geochemistry of lacustrine and marine evaporites.

REFERENCES

The description of the Torridonian grey facies sediments was based on field work by the author and:

Selley, R. C. (1965) Diagnostic characters of fluviatile sediments of the Torridonian Formation (PreCambrian) of North-West Scotland, *J. Sediment. Petrol.*, **35**, 366–80.

Stewart, A. D. (1969) Torridonian rocks of Scotland reviewed. In M. Kay (ed.), *North Atlantic Geology and Continental Drift*, Amer. Assoc. Petrol. Geol., Tulsa, 595–608.
—— (1975) 'Torridonian' rocks of western Scotland, *Spec. Rep. Geol. Soc. London*, **6**, 43–51.
—— (1982) Late Proterozoic rifting in North-west Scotland: the genesis of the Torridonian, *J. Geol. Soc. Lond.*, **139**, 413–20.
—— and Parker, A. (1979) Palaeosalinity and environmental interpretation of red beds from the Late Precambrian ('Torridonian') of Scotland, *Sedimentary Geology*, **22**, 29–41.
—— (1991) Torridonian. In G. Y. Craig (ed.), *The Geology of Scotland*, 3rd edn, Geol. Soc. Lond., 66–86.
Williams, G. E. (1969) Characteristics and origin of a PreCambrian pediment, *J. Geol.*, **77**, 183–207.

Other references cited in this chapter were:

Anadon, P., Cabrera, L. L. and Kelts, K. (eds) (1991) *Lacustrine Facies Analysis*, Internat. Assn. Sedol. Sp. Pub., **13**.
Cater, J. M. L. (1987) Sedimentology of the Lower Oil Shale Group (Dinantian) sequence at Granton, Edinburgh, *Trans. Roy. Soc. Edinburgh, Earth Sci.*, **78**, 29–40.
Chatfield, J. (1972) Case history of Red Wash Field, Uinta County, Utah. In R. E. King (ed.), *Stratigraphic Oil and Gas Fields*, Amer. Assoc. Petrol. Geol., **16**, 342–53.
Chin Chen (1980) Non-marine setting of petroleum in the Sungliao Basin of Northeastern China, *Jl. Pet. Geol.*, **2**, 233–64.
Dean, W. E. and Fouch, T. D. (1983) Lacustrine. In P. A. Scholle, D. G. Bebout and C. H. Moore (eds), *Carbonate Depositional Environments*, Amer. Assoc. Petrol. Geol., **33**, 97–130.
Edwards, R. A. (1976) Tertiary sediment and structure of the Bovey Basin, south Devon, *Proc. Geol. Assn. London*, **87**, 1–26.
Evans, P. R. (1981) The petroleum potential of Australia, *Jl. Pet. Geol.*, **4**, 123–46.
Falini, F. (1965) On the formation of coal deposits of Lacustrine origin, *Bull. Geol. Soc. Amer.*, **76**, 1317–46.
Fleet, A. J., Kelts, K. and Talbot, M. R. (1988) *Lacustrine Petroleum Source Rocks*, Geol. Soc. Sp. Pub., **40**.
Fouch, T. D. and Dean, W. E. (1982) Lacustrine environments. In P. A. Scholle and D. Spearing (eds), *Sandstone Depositional Environments*, Amer. Assoc. Petrol. Geol., **31**, 87–114.
Grunau, H. R. and Gruner, U. (1978) Source rock and origin of natural gas in the Far East, *Jl. Pet. Geol.*, **1**, 3–56.
Guangming, Z. and Quanheng, Z. (1982) Buried hill oil and gas pools in the North China Basin. In M. T. Halbouty (ed.), *The Deliberate Search for the Subtle Trap*, Amer. Assoc. Petrol. Geol., **32**, 317–36.
Holland, Sir T. H. (1912) The origin of Desert Salt deposits, *Proc. Lpool. Geol. Soc.*, **11**, 227–50.
Katz, B. J. (1991) Lacustrine basin Exploration, *Case Studies and Modern Analogs*, Amer. Assoc. Petrol. Geol., **50**.
Keighin, C. W. and Fouch, T. D. (1981) Depositional environments and diagenesis of some non-marine Upper Cretaceous Reservoir Rocks, Uinta basin, Utah. In F. G. Ethridge and R. M. Flores (eds), *Recent and Ancient Non-marine*

Depositional Environments: Models for Exploration, Soc. Econ. Pal. and Min. Sp. Pub., **31**, 233–59.

Randy Ray, R. (1982) Seismic stratigraphic interpretation of the Fault Union Formation, western Wind River Basin: example of subtle trap exploration in a nonmarine sequence. In M. T. Halbouty (ed.), *The Deliberate Search for the Subtle Trap*, Amer. Assoc. Petrol. Geol., **32**. 169–180.

Yen, T. F. and Chilingarian, G. V. (eds) (1976) *Oil Shale*, Elsevier, Amsterdam.

6 Deltas

INTRODUCTION

The development of shoreline sedimentary environments and their concomitant facies is a function of many variables, including rate of influx of land-derived sediment, tidal regime, current system, climate, and the relative movements of land and sea. Recent shorelines are sometimes areas of net erosion or net deposition. The former do not concern us since they leave no sediment in the geological record. The products of depositional shorelines, on the other hand, compose a large percentage of the world's sedimentary cover. The nature of shoreline deposits basically reflects the relationship between the amount of input of land-derived sediment, and the ability of marine processes to re-distribute it.

The interplay of these two processes produces a continuous spectrum of shoreline types ranging from the complete dominance of the shoreline by terrestrial sediment, to the other extreme where marine processes dominate over a negligible influx of detritus. With declining terrestrial sediment supply, carbonate sedimentary environments migrate closer and closer inshore. These concepts can be used as a basis on which arbitrarily to classify this continuous spectrum of shorelines into a number of types, as shown in Table 6.1.

Table 6.1 A classification of depositional shoreline deposits. This chapter deals with deltas, Chapters 7, 8 and 9 deal with linear coasts.

1. LOBATE SHORELINES – DELTAS

A. FLUVIALLY DOMINATED
with radiating 'birdfoot' distributary channels

B. MARINE DOMINATED

(i) WAVES DOMINANT
with radiating channels truncated by an arc of barrier islands

(ii) TIDES DOMINANT
with sub-parallel braided channels

2. LINEAR SHORELINES – BEACH/BARRIER-LAGOON COMPLEXES
include terrigenous, carbonate and mixed terrigenous/carbonate varieties

This chapter describes lobate (deltaic) shorelines. The next three chapters are concerned with terrigenous, mixed and carbonate shorelines respectively.

RECENT DELTAS

Herodotus applied the Greek letter delta Δ to the triangular area where the distributaries of the Nile deposit their load into the Mediterranean (*ibid.*, 454 BC). This term has gained widespread usage by geographers. More recently Lyell (1854) re-defined a delta as 'alluvial land formed by a river at its mouth'. This second definition is perhaps safer for geologists to use since it lays no stress on the deltaic geometry of such an area. Indeed, many Recent river mouths which are accepted as deltaic are far from triangular in shape. Nevertheless, it is of extreme economic importance to be able to distinguish lobate deltaic shorelines from linear ones. Fortunately this can often be achieved from the study of a single seismic section without drilling a single borehole.

Deltas are one of the most studied and written about of all Recent environments (Elliott, 1986). These studies show that deltas form where a river brings more sediment into the sea than can be re-deployed by marine currents. In such a case the river deposits progressively finer sediment in progressively deeper water as its current velocity diminishes seawards. The channel advances seaward between two raised banks of its own deposits, termed levees.

Ultimately the channel lays down so much sediment at its mouth that this is higher than the proximal, landward end of the levees. As the channel chokes in its own detritus, it breaks through the levees on its margin. From the crevasse thus formed a new channel flows seaward between two new levees. Thus an alluvial floodplain builds out to the sea, lobate in plan and wedge-shaped in vertical cross-section. The top of the delta consists of a radiating network of distributary channel sands flanked by levee silts. Finer silts and clays are laid down in the flood basins and lagoons of the interdistributary areas. Swamps often form and are the site of peat deposition.

Seaward of the distributary channel mouth, sediment is deposited on the sub-aqueous delta platform. Rippled interlaminated sand and mud are formed, often with bioturbation. If sedimentation is relatively slow, the platform facies may be re-worked by marine currents. The mud may be winnowed out to leave cross-bedded or flat-bedded sands. The sub-aqueous delta platform is separated from the prodelta by the delta slope. On this surface most of the mud finally comes to rest. From time to time this slope often becomes so steep that the mud slumps and slides down slope to be re-deposited at its foot. Slumping often appears to generate and move down gullies in the delta slope. Some deltas, ancient and modern, have submarine fans of turbidite sands at their base.

Thus a vertical section of a delta reveals a fine-grained marine deposit, which passes up transitionally into coarser freshwater sediments; these facies boundaries will be diachronous seaward as the delta builds out. Vertical repetition of this sequence may be expected, due to repeated crevassing and delta switching.

In searching for ancient deltas, therefore, we must look for sequences showing repeated cycles of upward-coarsening grainsize. Each cycle should begin, at the base, with a marine shale which passes up through silts into coarser freshwater channel sands at the top. In plan the channels should show a radiating shoestring pattern and be cut into freshwater shales and coals. Coals may also be present on the top of the channels (Figure 6.1).

Examination of modern deltas shows that they are in fact rather more complex than this simple picture. Two major types of delta may be differentiated: fluvially dominated deltas, and those dominated by marine processes (refer back to Table 6.1). The delta of the Mississippi River is the classic example of the first type. It is in fact a very unusual delta with a very low ratio of sand to mud (1:9), and it is also unusual in that it debouches into a sheltered embayment with a low tidal range. Thus distributary channels can radiate far out to sea. Deltas of this fluvially dominated 'birdfoot' pattern are uncommon except in lakes. Most deltas are subjected to extensive marine reworking. An important distinction must be made between deltas dominated by wave or tidal action.

The Nile delta is an excellent example of the first type. In this delta sand is no sooner deposited at the mouth of a distributary than it is reworked by wave action and redeposited on an arc of barrier islands around the delta periphery. Deltas in areas of high tidal range however have a quite different geomorphology. These have wide expanses of tidal flat fine sands and muds, often colonized by mangrove swamps. These are crosscut by braided distributary channels scoured by tidal currents. Instead of radiating from a point these channels tend to be subparallel. Tidally dominated deltas of this type are widespread in South-East Asia, examples include the Ganges: Brahmaputra, the Klang and the Mekong.

The differentiation between these three types of delta is important in subsurface facies analysis because of the different distribution and orientation of potential hydrocarbon reservoirs (Figure 6.2).

Ancient deltaic deposits are worldwide in their distribution and are known throughout geologic time. A case history of an ancient deltaic sequence will be described to illustrate their characteristic features.

CASE HISTORY: THE ORDOVICIAN-SILURIAN SANDSTONES OF THE SAHARA AND ARABIA

As noted in Chapter 3 geologists have long remarked on the remarkable uniformity of Lower Palaeozoic stratigraphy from the Atlantic coast of North Africa east to the shores of the large gulf between Arabia and Iran. Laterally

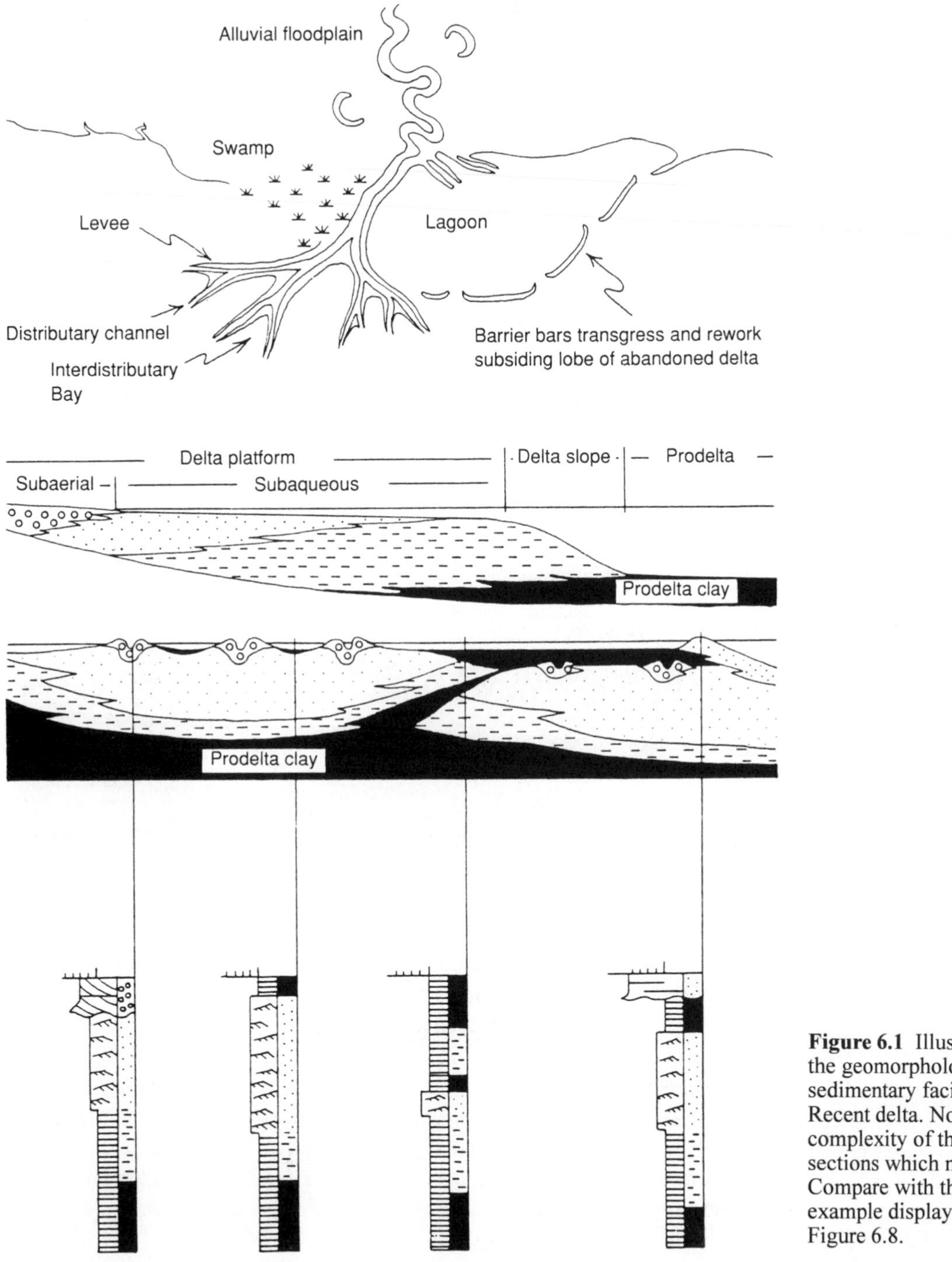

Figure 6.1 Illustration of the geomorphology and sedimentary facies of a Recent delta. Note the complexity of the vertical sections which may result. Compare with the real example displayed in Figure 6.8.

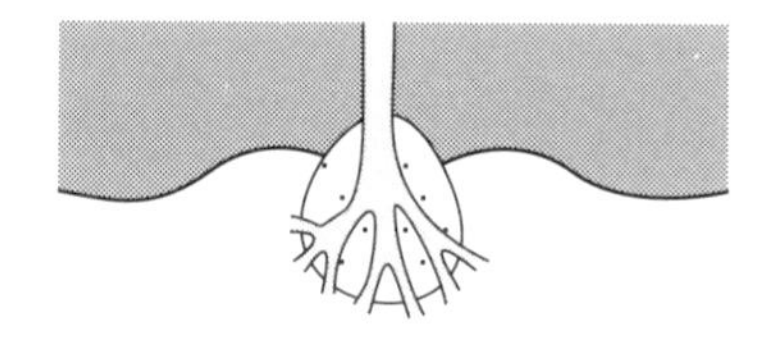

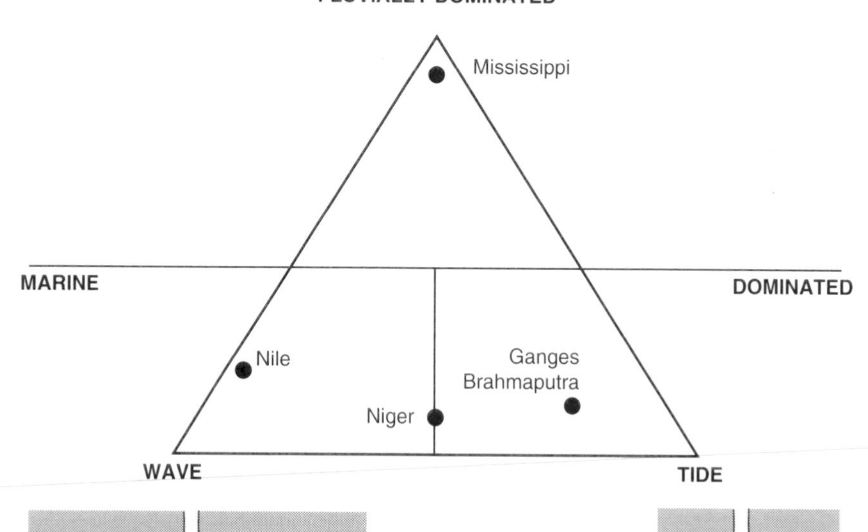

Figure 6.2 Sketch maps to illustrate the basic types of delta. For examples see text. Note how process controls the distribution and trend of sand bodies which may be potential hydrocarbon reservoirs.

extensive blankets of sandstone and shale crop out around the PreCambrian shields of the Sahara and Arabia, and can be traced intermittently northward and eastward towards the Mediterranean and Gulf coasts respectively (Figure 6.3).

These formations show a remarkably uniform vertical sequence of facies. The sequence normally commences with coarse pebbly channeled sands (attributable to fluvial environments). These pass up into better sorted finer sands, commonly bioturbated (attributable to shallow marine environments). This facies passes up in to graptolitic shales, which in turn pass up, often via turbidite sands, in to prograding deltaic, and finally fluvial sands (Figure 6.4).

The first facies was described as a braided outwash plain in Chapter 3, the second facies will be diagnosed as a shallow marine sand in Chapter 9. This chapter uses the Ordovician to Silurian sediments as a case history of deltaic deposits.

The sequence to be described overlies shallow marine shoal sands. There

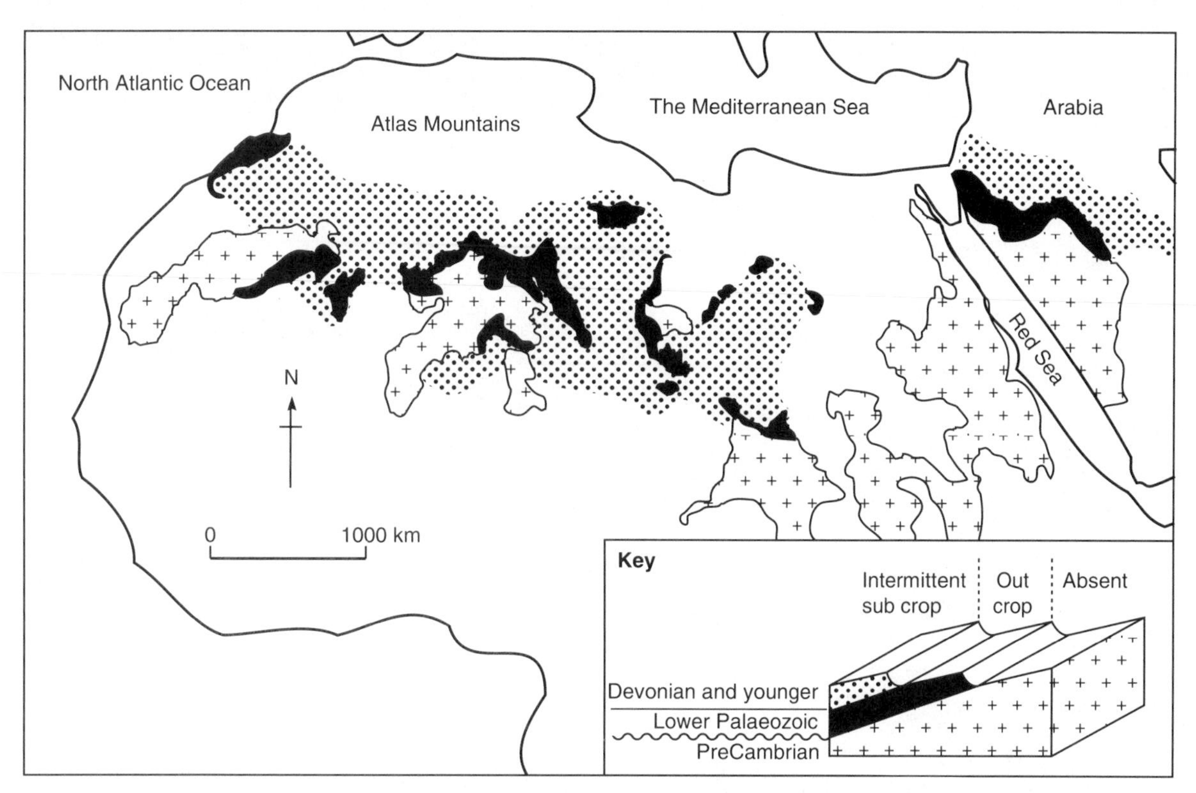

Figure 6.3 Map to show the distribution of the Lower Palaeozoic rocks of the Sahara and Arabia. The deltaic case history is concerned with the Tannezuft shale and Acacus sandstones of the Sahara, and the Khreim Group, and Hanadir Shale of Arabia.

is often a regional depositional break. Above this the section is essentially regressive and progradational. It begins with organic-rich graptolitic black shales, the Tannezuft shale of the Sahara and the Hanadir shale of Arabia. These are both petroleum source rocks. The Tannezuft shale is the source for petroleum in the Hassi Messaoud and other fields of Algeria (Tissot *et al.*, 1975), and the Hanadir Shale, together with the younger Quasaiba, are sources for petroleum in Arabia (Al-Husseini, 1991a).

The black shales pass up into a sequence of upward-coarsening shale to sand increments in which the sand percentage gradually increases upward. The upper sand is termed the Acacus sandstone in the Sahara, the Khreim Group in Jordan, and the Quasiba in Saudi Arabia (Mahmoud *et al.*, 1992). These formations are proven or potential petroleum reservoirs in all of these areas. Regional biostratigraphic studies using graptolites show that the shale–sand transition is strongly diachronous down the depositional slope both in the Sahara and Arabia (Figure 6.5).

In both areas the transition from the lower shales in to the upper sand formations is often marked by interbedded sequences of shales and thin sands with erosional bases, graded bedding, and fragmentary Bouma sequences (Figure 6.6).

These are the characteristic features of turbidites (sands deposited from

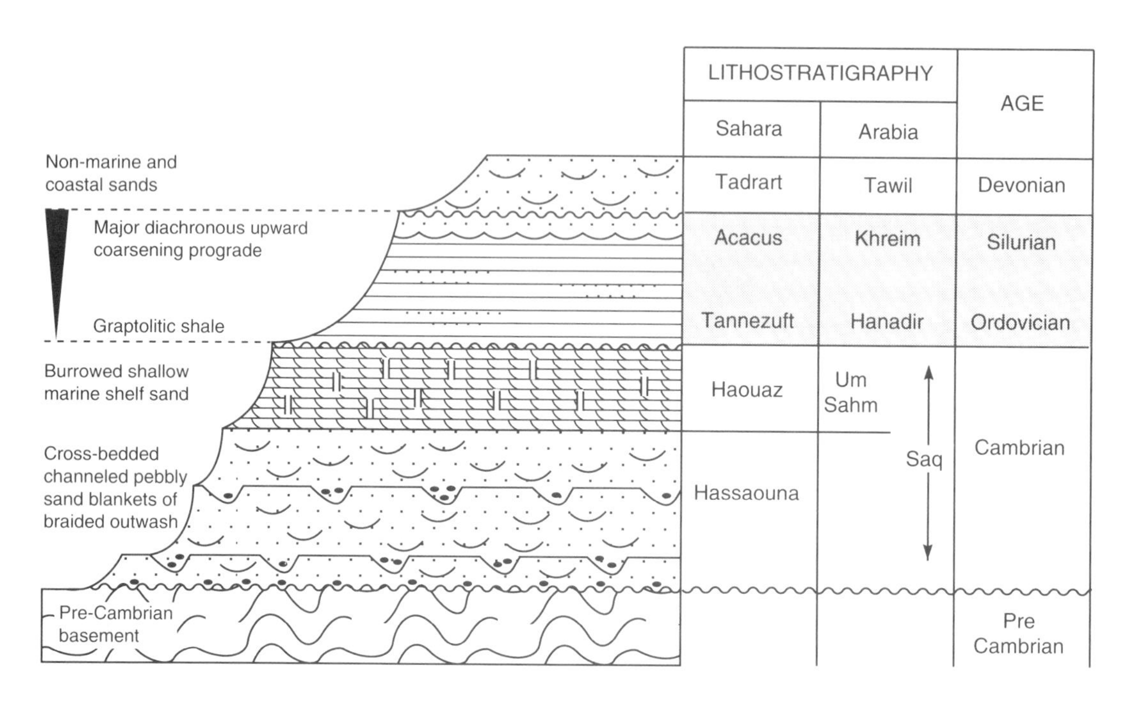

Figure 6.4 Section to show the Lower Palaeozoic stratigraphy of the Sahara and Arabia, including the Ordovician–Silurian deltaic sediments described in this chapter.

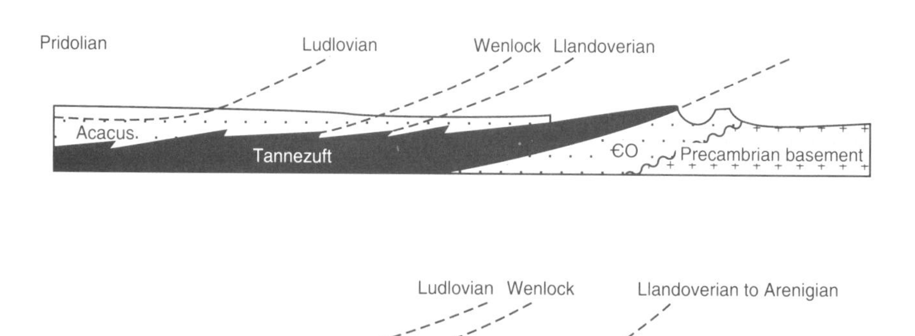

Figure 6.5 Cross-sections through the Ordovician-Silurian sediments of the Sahara (Upper) and Arabia (Lower) to show the diachronous nature of the prograding deltaic sequence. For authentication see Bellini and Massa (1980) and Mahmoud *et al.* (1992) respectively.

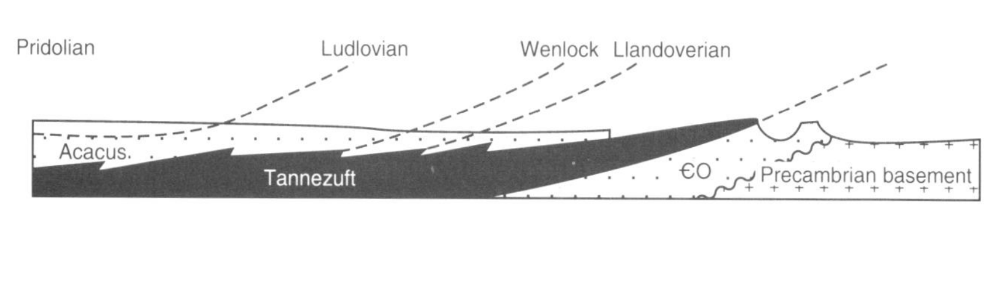

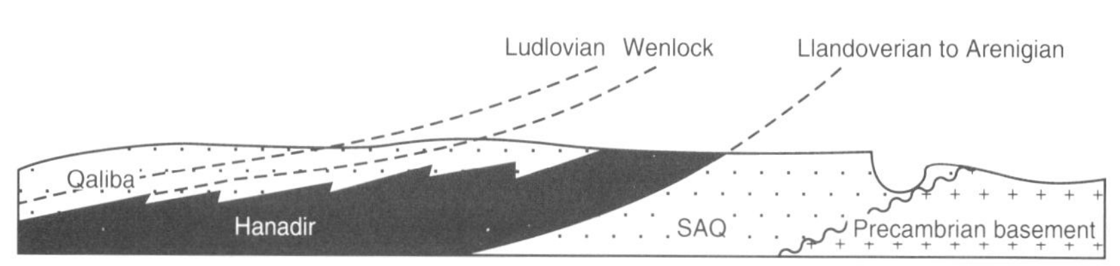

Figure 6.6 Graded sandstone, interpretable as a turbidite, in the Tannezuft shale–Acacus Sand transition zone (Silurian), Murzuk Basin, Libya.

turbidity currents, described and discussed more fully in Chapter 12, p. 254). The Khreim Group of Jordan will now be used as a case history of ancient deltaic sedimentation. It will first be described and then the reasons for attributing it to a deltaic depositional environment explained.

Khreim Group (Ordovician-Silurian), Southern Desert of Jordan: description

Figure 6.7 shows measured sections through the Khreim Group in the southern Desert of Jordan. This displays a genetic sequence composed of over five genetic increments in which three facies are repeated repeatedly. Sequence boundaries are marked by extensively bioturbated horizontal surfaces. These exhibit a range of burrows, together with trilobite trails (*Cruziana*) and parking bays (*Rusophycus*). The sequence boundaries are abruptly overlain by khaki-coloured laminated shales that grade up in to thinly interlaminated shales and fine sands. The sands are flaser bedded, with clay drapes and bimodal bipolar cross-lamination. This sub-facies is also commonly bioturbated by the vertical burrow known in Arabia as *Sabellarifex*, in the Sahara as *Tigillites*, and elsewhere variously as *Scolithos* or *Monocraterion*. The Bedu of the Howetat tribe call these burrows 'angel's tears'.

This sub-facies is cut in to and overlain by channel sands. The channels scour into the underlying facies, and their floors are lined with intraformation shale pebble conglomerates. The channels are infilled by cross-bedded

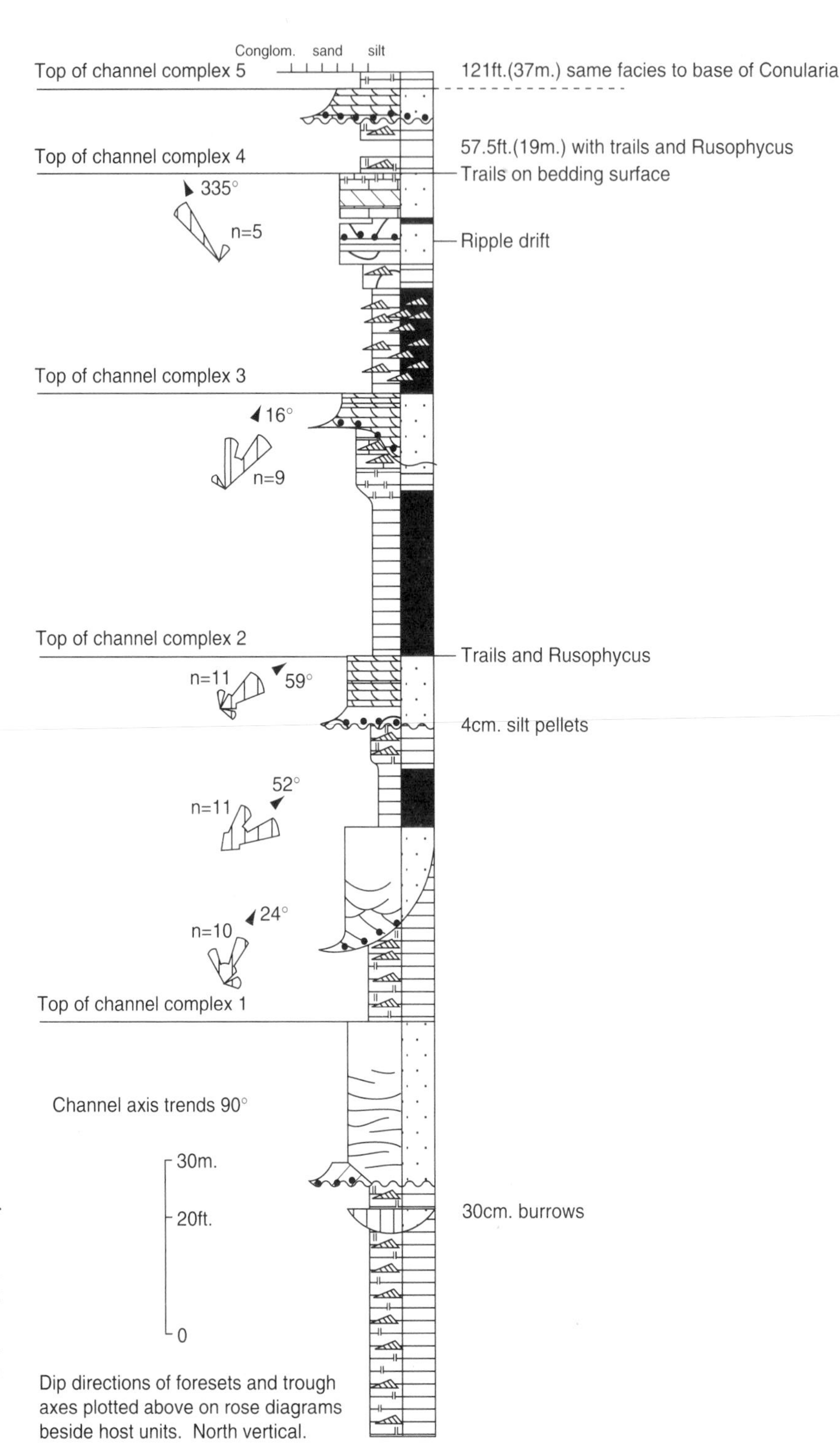

Figure 6.7 Sedimentological log through part of the Khreim Group (Ordovician–Silurian) southern desert, Jordan. Note the repetition of upward-coarsening genetic increments. This section does not include the lowest increment, in which the Hanadir shale passes up via a sequence of turbidites, into a channel complex.

sands, that pass up with decreasing set height into cross-laminated sand. Within any one channel the cross-bedding is markedly unimodal, though there is a wide arc of dips, and of presumed flow directions, if all the dips are plotted together.The channels are commonly sufficiently abundant to coalesce to form laterally extensive sheets, the tops of which form the sequence boundaries previously described.

Though it has been stated that the sequence consists of a series of upward-coarsening genetic increments that are continuously repeated, careful examination of Figure 6.7 shows that this is not always correct. Sometimes the shale unit is missing. Figure 6.8 illustrates why.

This shows a laterally surveyed cliff-section that displays the complex

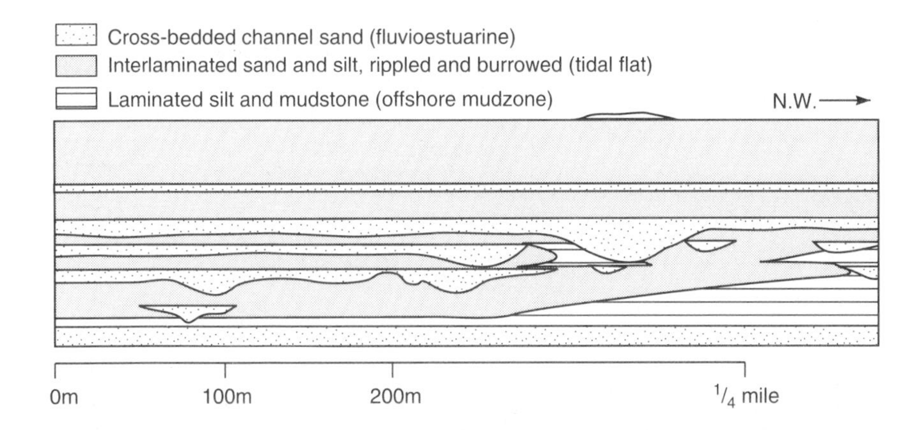

Figure 6.8 Photo and surveyed plan of a cliff in the southern desert, Jordan, to show the geometry of the upward-coarsening deltaic genetic increments of the Khreim Group (Ordovician–Silurian). Note rapid lateral facies variation.

nature of the facies boundaries, in particular the erratic vertical and lateral distribution of the channels. Because of the dissected nature of the topography it is, in fact possible to map out the geometry of some of the channel sands (Figure 6.9).

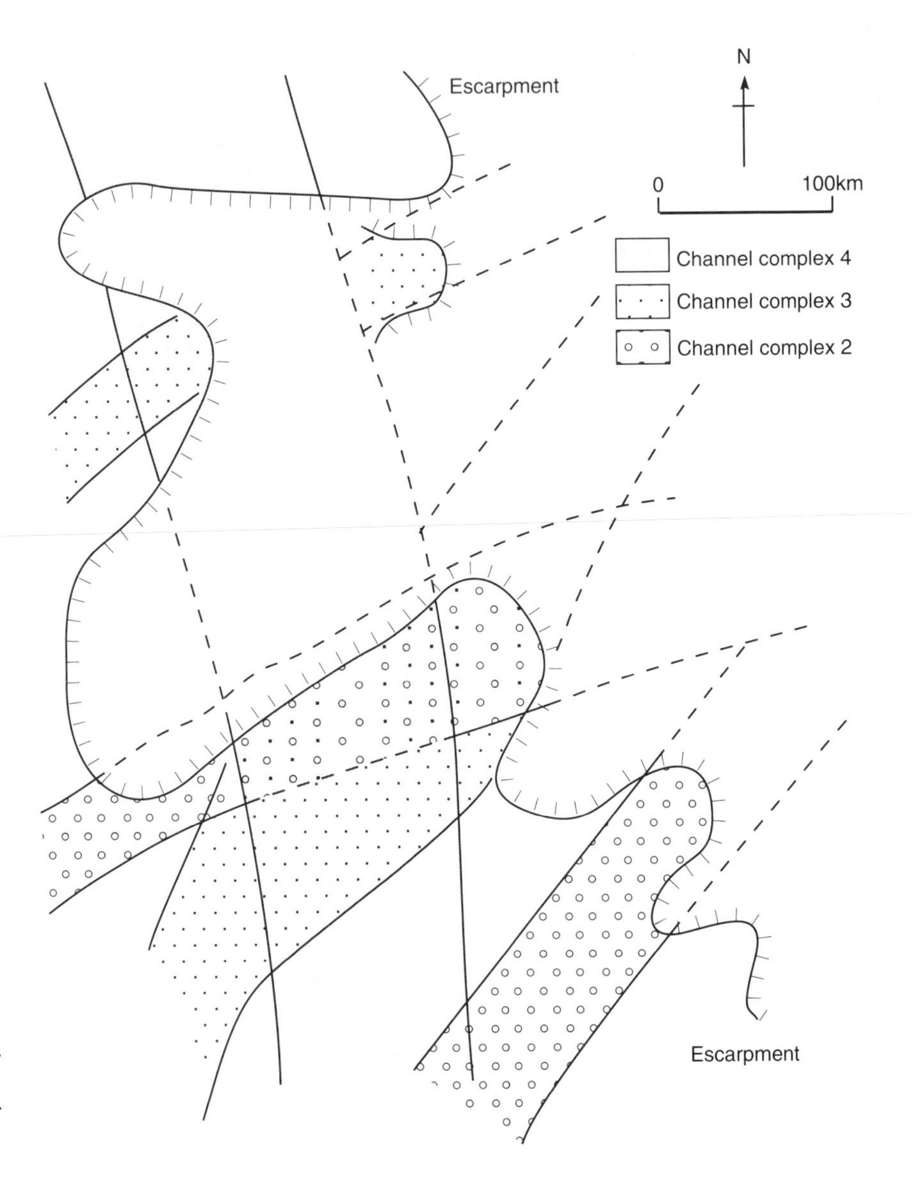

Figure 6.9 Map of dissected scarp and jebels to show the geometry of delta distributary channel sands in the Khreim Group (Ordovician–Silurian) in the southern desert, Jordan.

Plate 1
Though the lithology of a facies may best be studied from the rocks themselves, their chemical composition can be remotely sensed by mineralogical borehole logs. These are computed by integrating the data from as many as a dozen separate logging tools, measuring electric, radiometric, acoustic and other physical and chemical properties of the rock formations. Courtesy of Schlumberger.

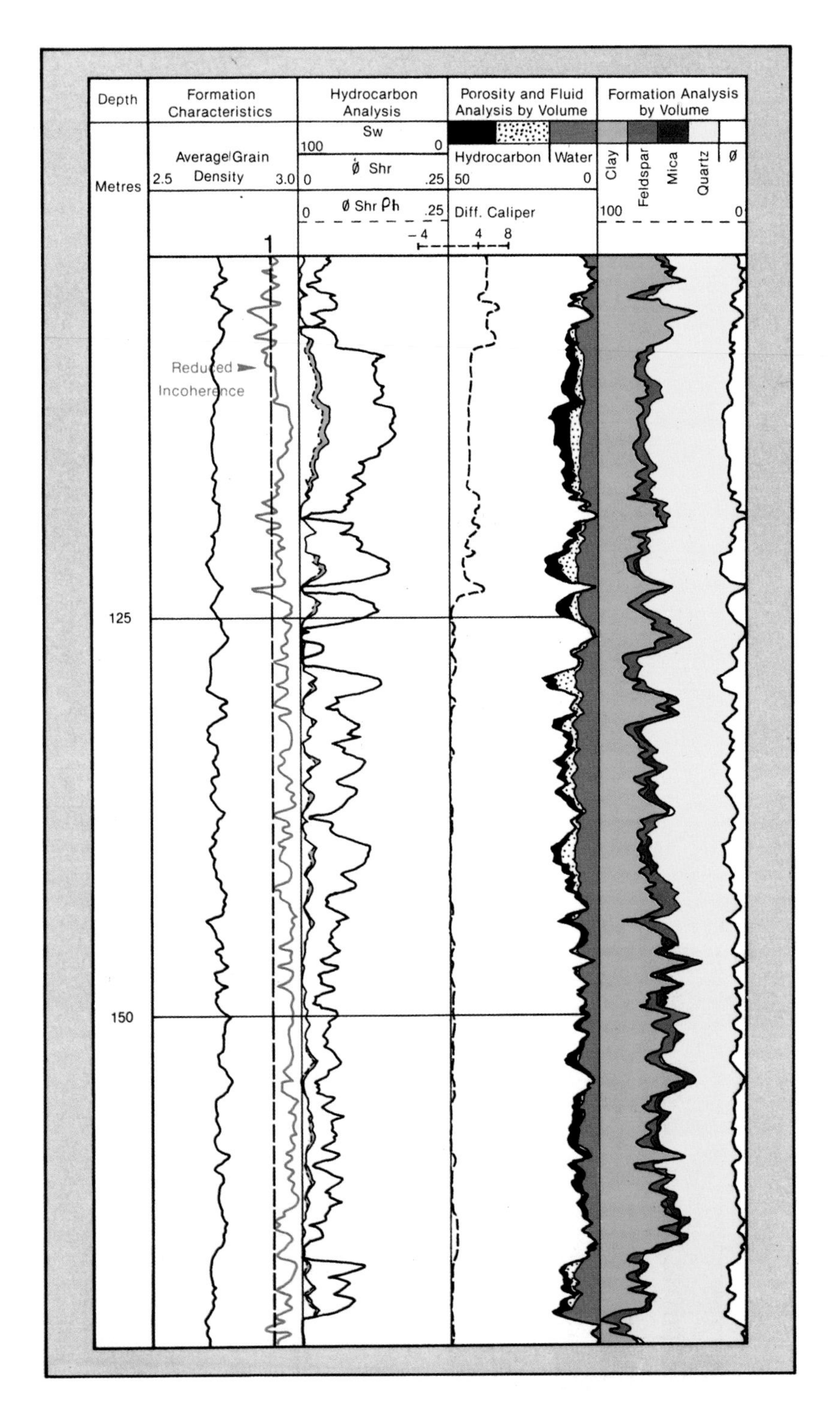

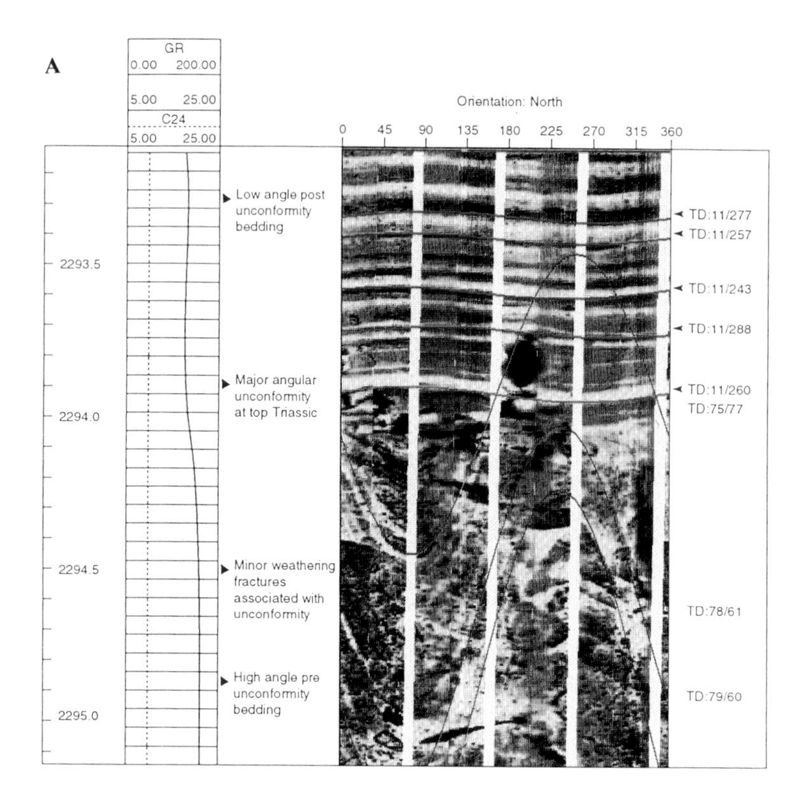

Plate 2

Images produced by Schlumberger's Formation MicroImager (FMI). A: An unconformity, courtesy J. Roestenburg . B: cross- bedding. C: A coal with plant roots beneath.

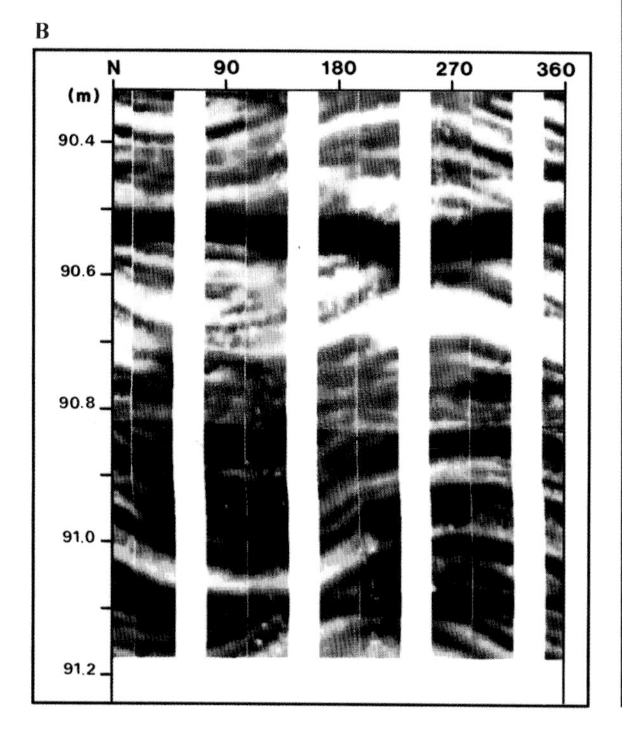

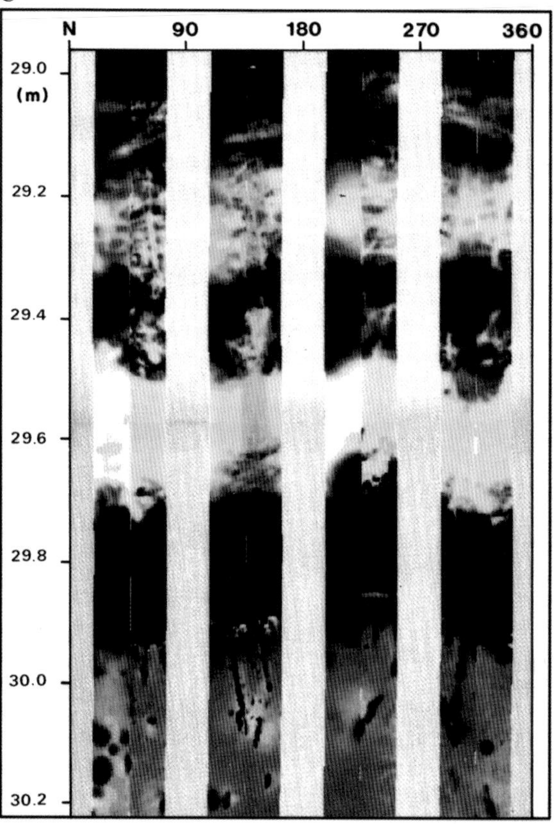

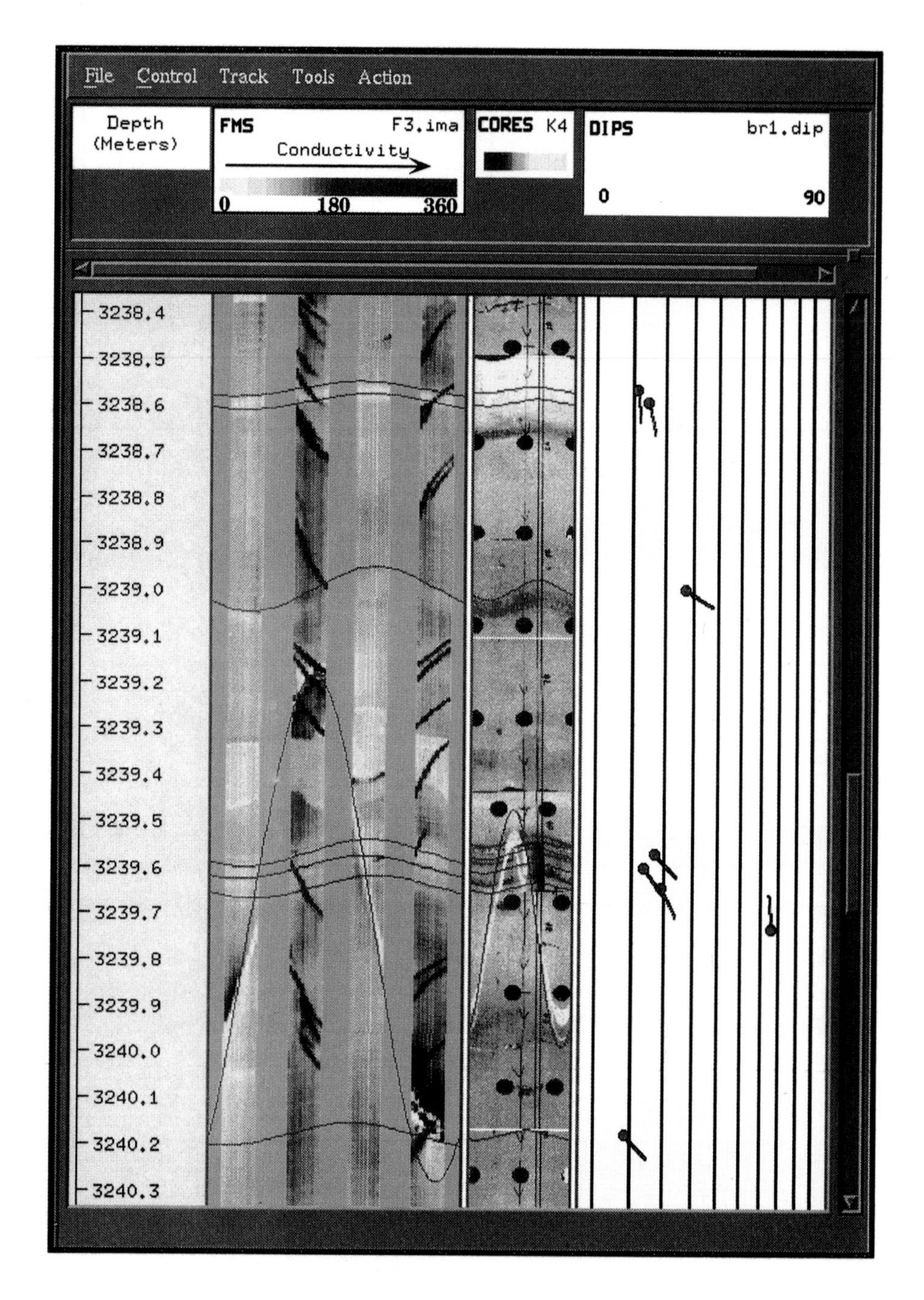

Plate 3
Monitor image produced by DIAMAGE*. This system enables cores to be photographed, the photos loaded and displayed on a screen beside corresponding FMI images and compared with dips calculated from the cored interval. Courtesy of Schlumberger.

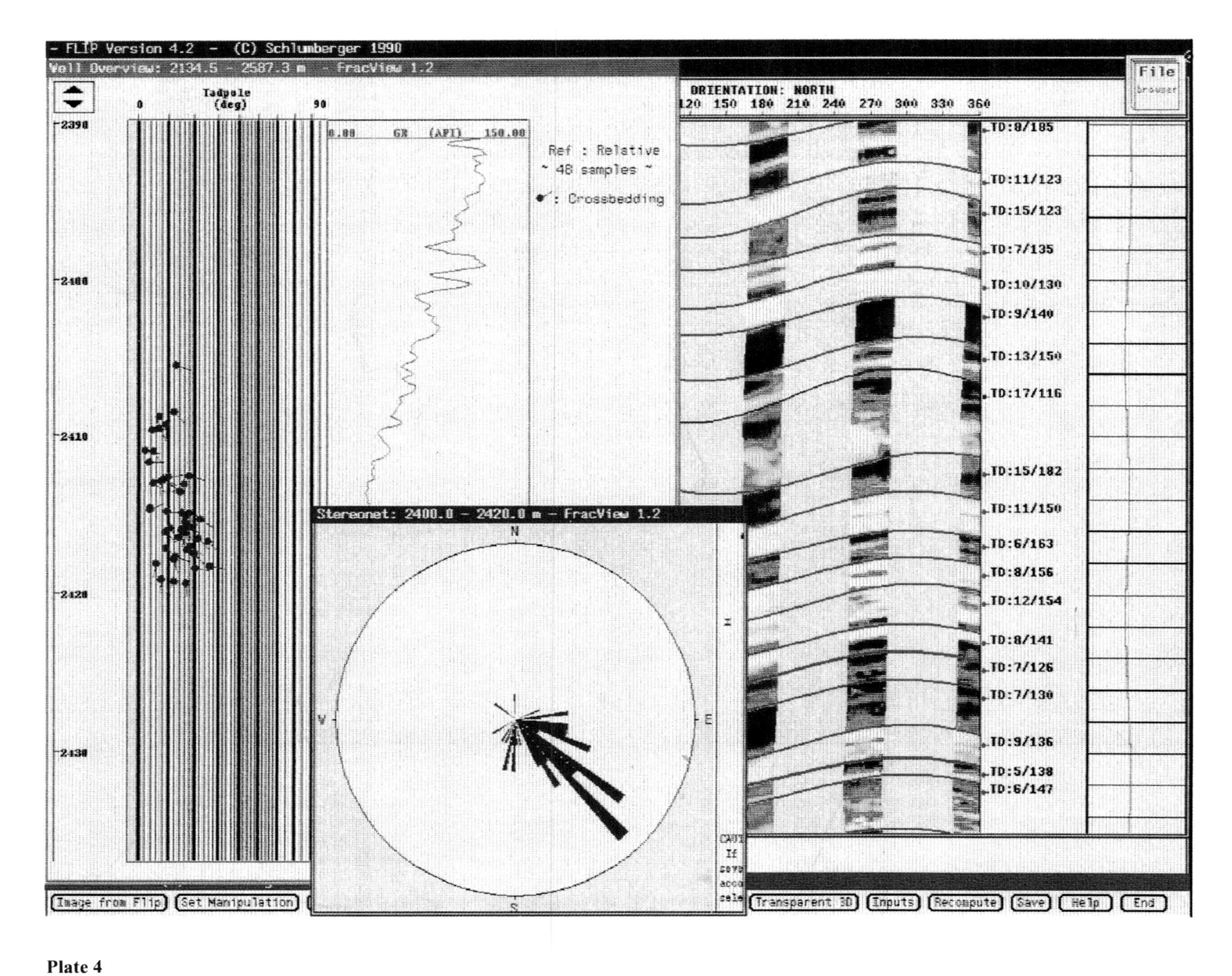

Plate 4
Work station display of borehole images, gamma calibration log, tadpole plot and dip azimuth. Courtesy of J. Roestenburg and Schlumberger.

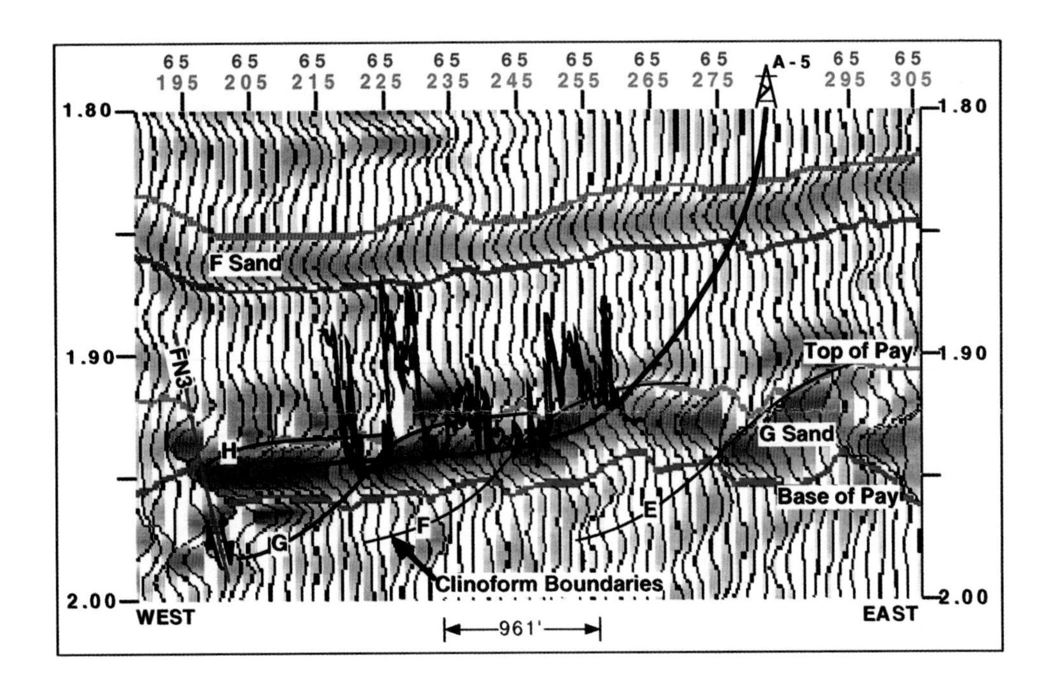

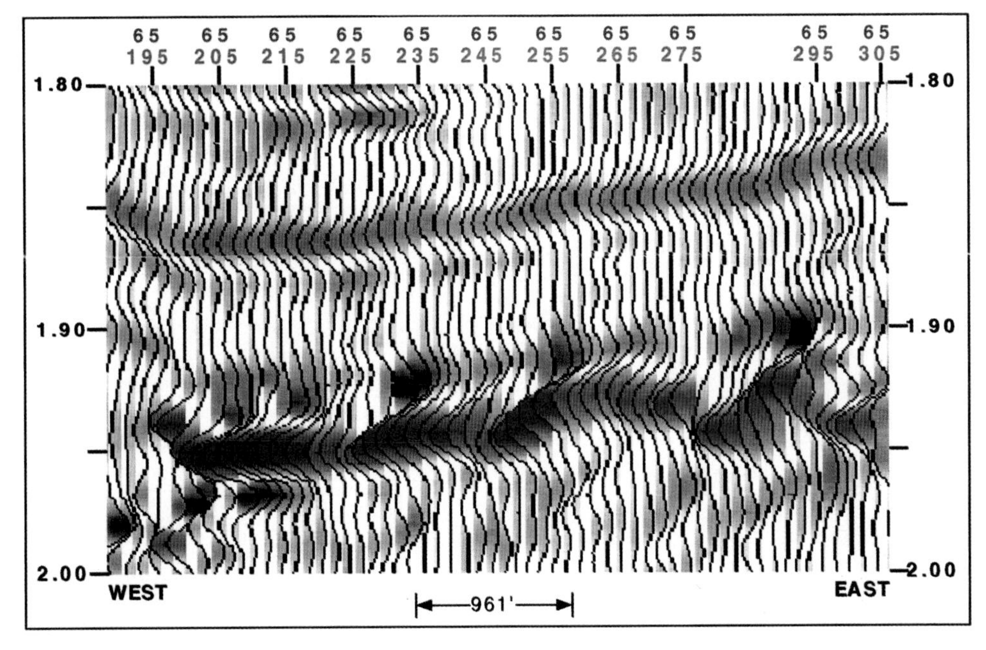

Plate 5

Seismic lines through BP's Amberjack Mississippi Canyon development, Gulf of Mexico. Upper: picked, interpreted and showing horizontal well path. Lower: unpicked. The prograding clinoforms of this delta consist of upward-coarsening increments some 120 m thick, in which mouth bar sands prograded out over slumped turbidite sands (see Mayall *et al.*, 1992, for a detailed account of the geology). These petroliferous sands are separated from one another by impermeable clinoform-defining shales. They were all penetrated by the single cunningly drilled horizontal well whose track is shown in the upper panel. Courtesy of BP Gulf of Mexico Mississippi Canyon Development, especially B. Cohn and W. H. Mills.

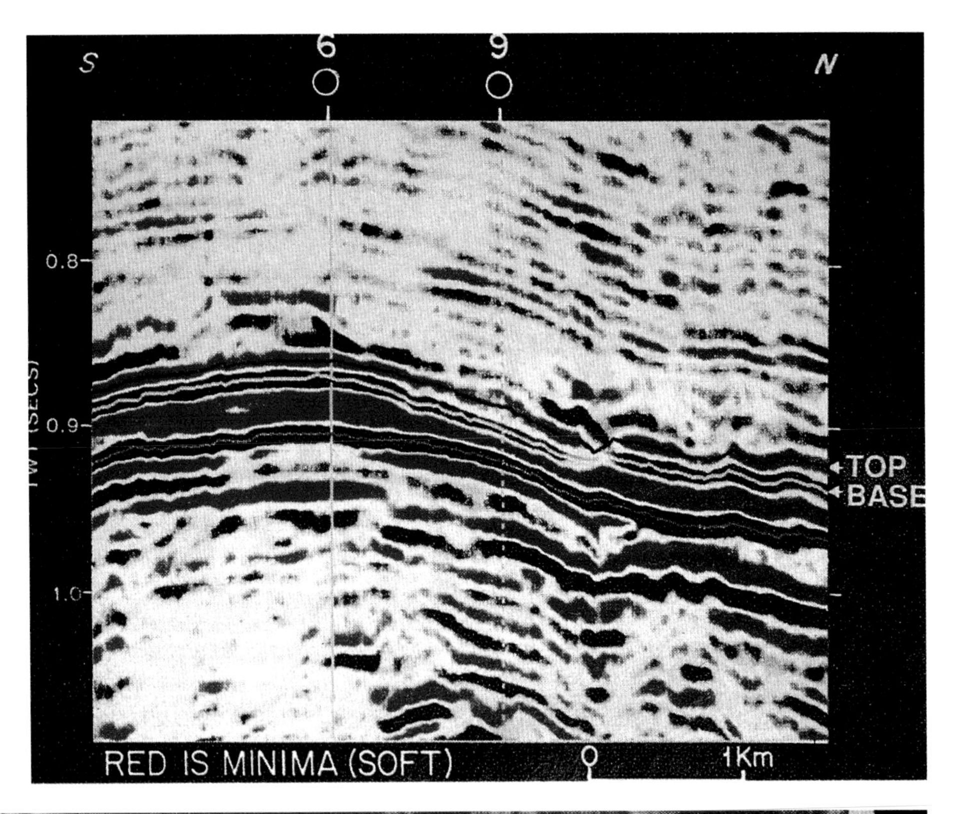

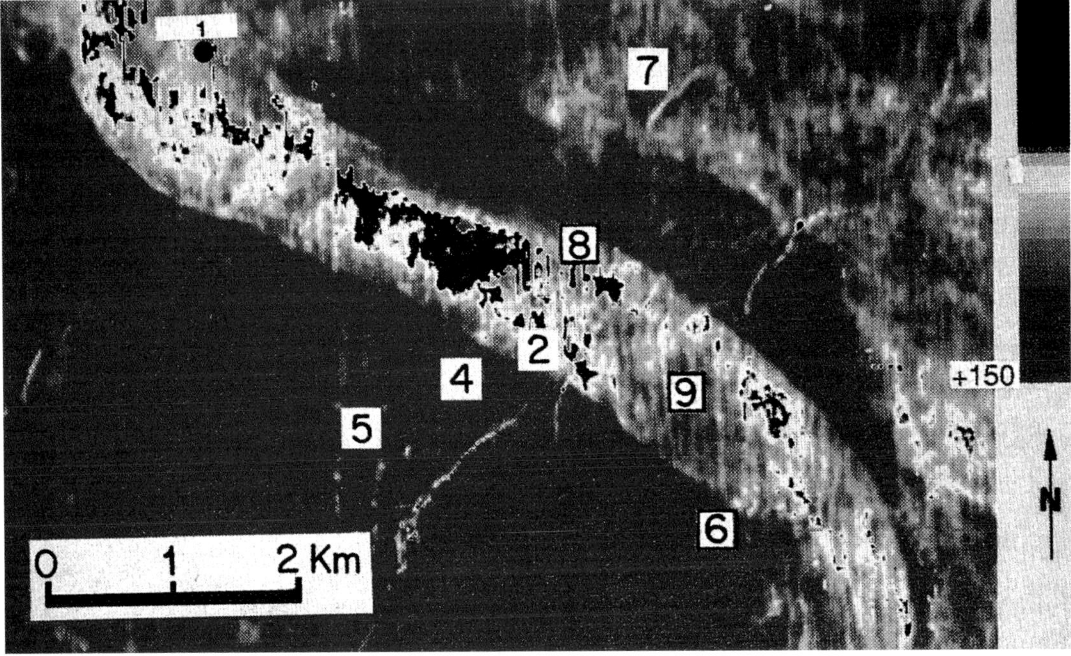

Plate 6
(This page and opposite.) Horizontal amplitude display map from the Balingian Province, Malaysia, showing a clearly defined channel, with crevasse splay lobe. From Rijks and Jauffred (1991).

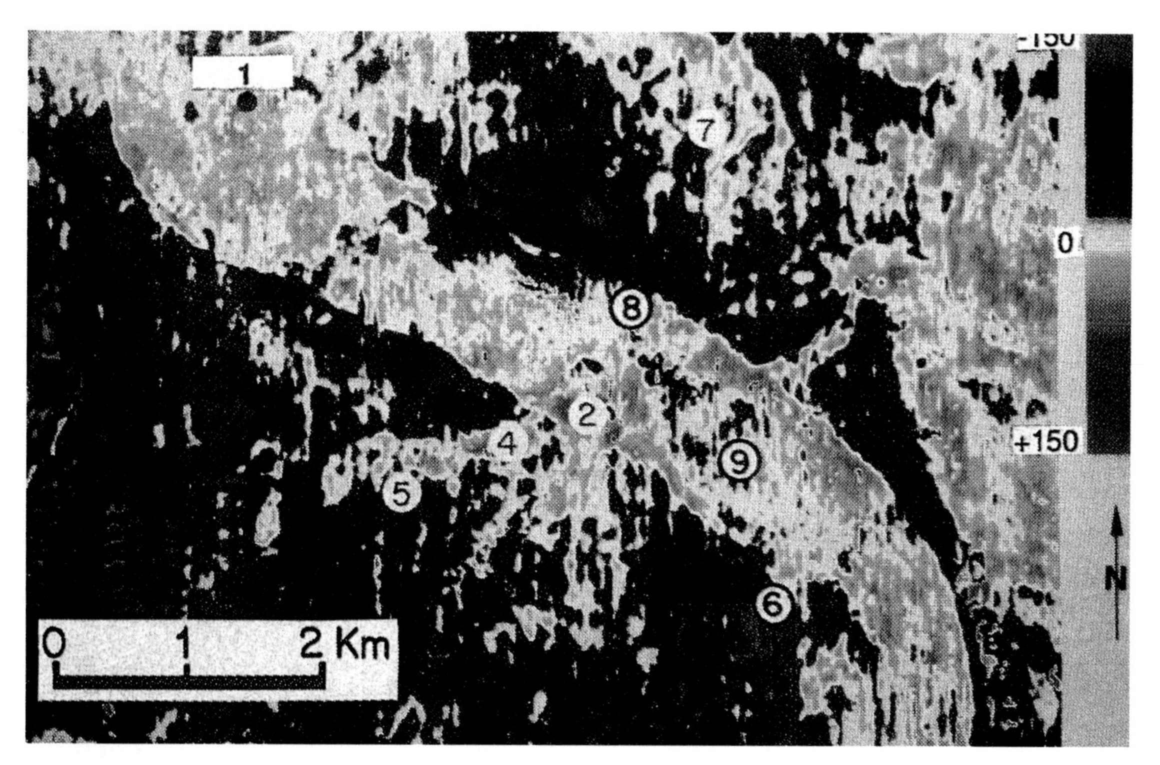
1
7
8
2
4
5
9
6
-150
0
+150
N
0 1 2 Km

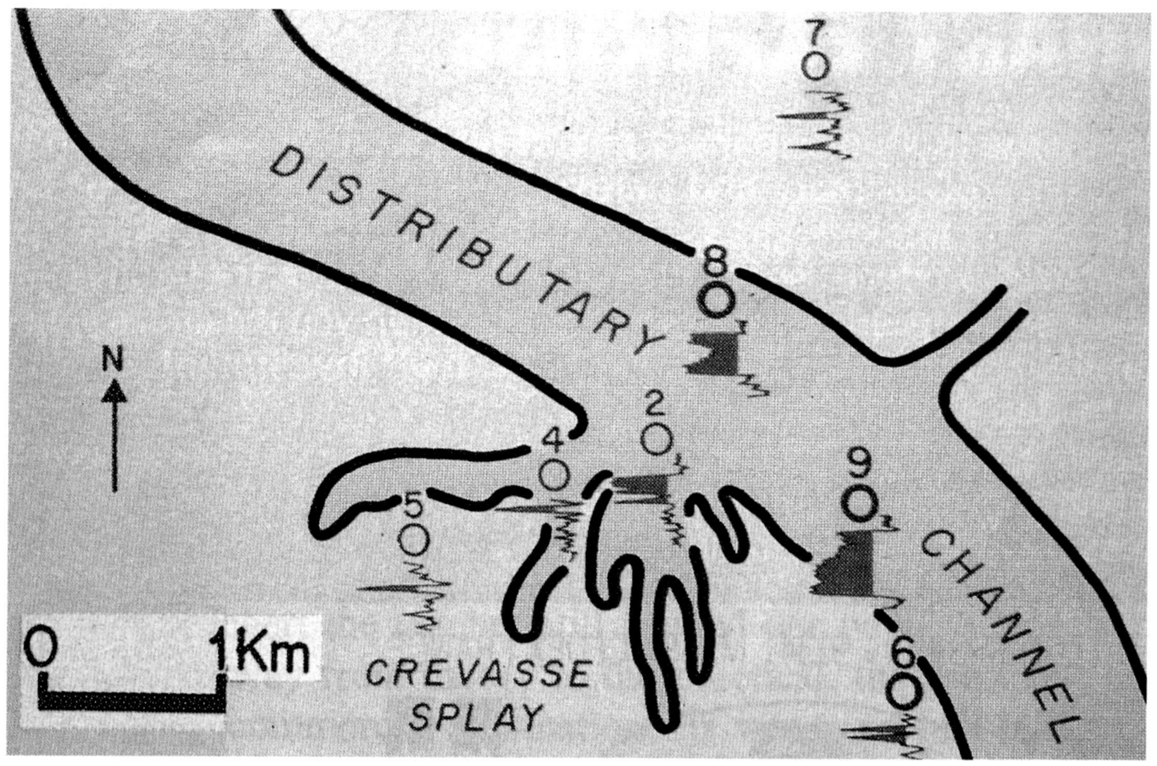
DISTRIBUTARY
CHANNEL
CREVASSE
SPLAY
7
8
2
4
5
9
6
N
0 1Km

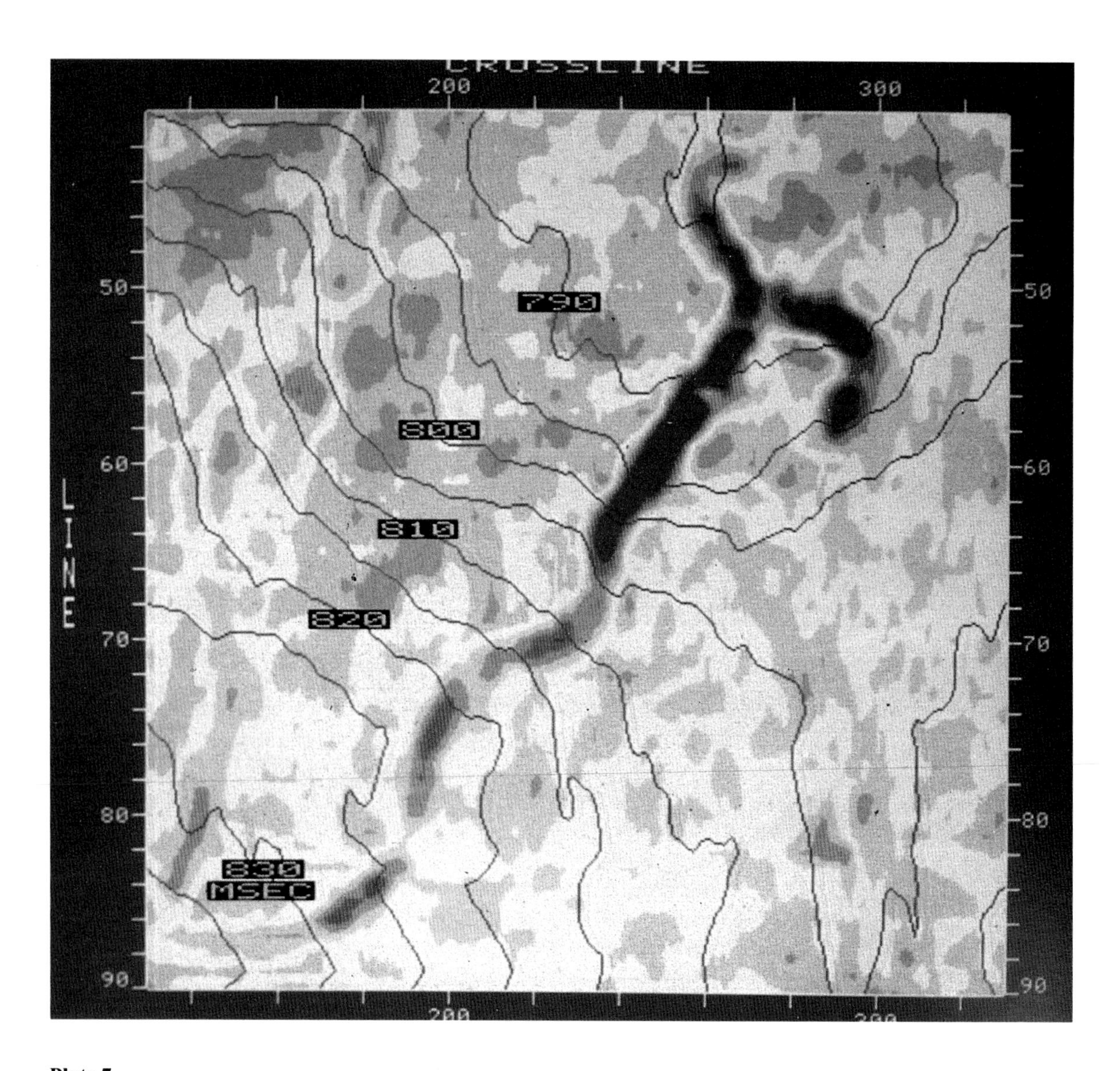

Plate 7
Horizon seiscrop section showing gas saturated delta distributary channel system, Gulf of Mexico. From Brown (1985).

Khreim Group (Ordovician-Silurian), Southern Desert of Jordan: interpretation

The Khreim Group consists of a series of upward-coarsening genetic increments, and as discussed at the beginning of this chapter, this is a diagnostic feature of deltaic deposits, though it is not necessarily restricted to them. The laminated shales were deposited in a low energy environment, and, as the graptolites in the lower part show, the waters were marine. The overlying flaser-bedded interlaminated sands and shales were deposited by currents of fluctuating velocity, in which episodes of ripple migration alternated with spells of negligible flow, during which the clay drapes were laid down. This, together with the bimodality of the ripple laminae, points to tidal conditions.

The abundant vertical burrows are a classic example of the *Scolithos* ichnofacies (refer back to Figure 2.15), and therefore suggest very shallow, probably intertidal conditions. This sub-facies would thus appear to have been deposited on tidal flats that built out over the deeper water marine muds.

The tidal flat sediments were cut through by the channels. These were infilled by sands deposited by unidirectionally migrating megaripples and ripples. There is, interestingly, no evidence of tidal currents within the channels. The unimodal cross-bedding dips suggesting that the river, aided perhaps by tidal ebb currents, was responsible for the sand deposition. The channels were infilled up to the level of the sequence boundary, and then abandoned. Each increment abandonment surface is clearly defined by extensive burrows, tracks and trails. Submergence, presumably relatively rapid, lead to renewed deposition of the deeper water marine muds. This pattern was repeated repeatedly.

Examination of Figure 6.8 shows that though some of the increments are laterally continuous, others are clearly very restricted. This suggests that the cyclicity is due in part, at least, to the autocyclic switching of delta distributary channels discussed earlier, though the more extensive cycles may be related to more widespread eustatic events. It is interesting to compare Figures 6.8 and 6.1.

The regionally diachronous nature of the sediments show how a major series of deltas gradually prograded out across the Saharan and Arabian shields towards the palaeo-Tethys Ocean. The overall regressive nature of the shoreline was intermittently interrupted by advances of the sea, of both local and regional extent.

Thus the Khreim Group of Jordan may serve as a typical case history of an ancient delta. Note, unlike many ancient deltaic deposits, coals are absent. Land plants had not yet evolved by the Silurian Period. The geometry of the Khreim Group channel sands deserves consideration. Not only are the equivalents of these beds potential petroleum reservoirs in the subsurface of

Arabia and the Sahara, but analogous geometries may be anticipated in deltaic petroleum reservoirs in rocks of different ages and on different continents. Such reservoirs provide challenges for development geologists and reservoir engineers alike, for it is essential to delineate the geometry of the channels, and to establish which ones are in communication with one another. This important topic will be considered at greater length in the following section on the economic aspects of deltaic deposits.

ECONOMIC SIGNIFICANCE OF DELTAIC DEPOSITS

Deltaic sediments are important sources of coal, oil, and gas. Peat formation is typical in the swamps and marshes of Recent deltaic alluvial plains. Detailed accounts of the sedimentology of coal deposits are given by Ward (1983), and Galloway and Hobday (1983), and of their possible significance as petroleum source rocks by Scott and Fleet (1994). In the effective exploitation of coal, it is obviously important to understand their depositional environment. Not all coals are deltaic; as already mentioned some occur in continental basins far from the sea (p. 127). Within the broad framework of a delta, coals can form in a number of different situations.

There are many different environments in which swamps grow. Therefore there are many different ways in which ancient coal beds may occur. Wanless, Baroffio and Trescott (1969) published a classic account of the relationship between depositional environments and coal bed geometry in the Pennsylvanian deltas of Illinois. Other examples will be found in Galloway and Hobday (1983).

Figures 6.10 and 6.11 illustrate some of the different types of coal bed geometry in plan and cross-section respectively. Sometimes, when a major delta is abandoned or when there is a major eustatic rise in sea-level, blanket deposits of peat may form (Figure 6.10a). On a smaller scale individual crevasse-splays may be covered by a bed of peat (Figure 6.10b). Repetition of this process can deposit coals that thin and split away from the channel margin (Figure 6.11b). Peat formation can also take place in abandoned fluvial tributary and deltaic distributary channels (Figure 6.10c); a minor variety of this situation is to be found in oxbow lakes (Figure 6.10d).

Blanket peats may form in flood basins on alluvial plains (Figure 6.11a). In such areas it is obviously very important to be able to map and predict such channel trends. This is not just because sandstone channels cut out valuable coal reserves, but also because they are sometimes highly permeable aquifers that can cause the flooding of underground mine workings. Lastly coal can occur in elongate bodies parallel to shorelines where peat grew in salt marshes and coastal lagoons (Figure 6.10f). This brief review of the relationship between coal bed geometry and depositional environment shows that facies analysis can aid the exploitation of coal deposits.

Deltas are often major hydrocarbon provinces for a number of reasons.

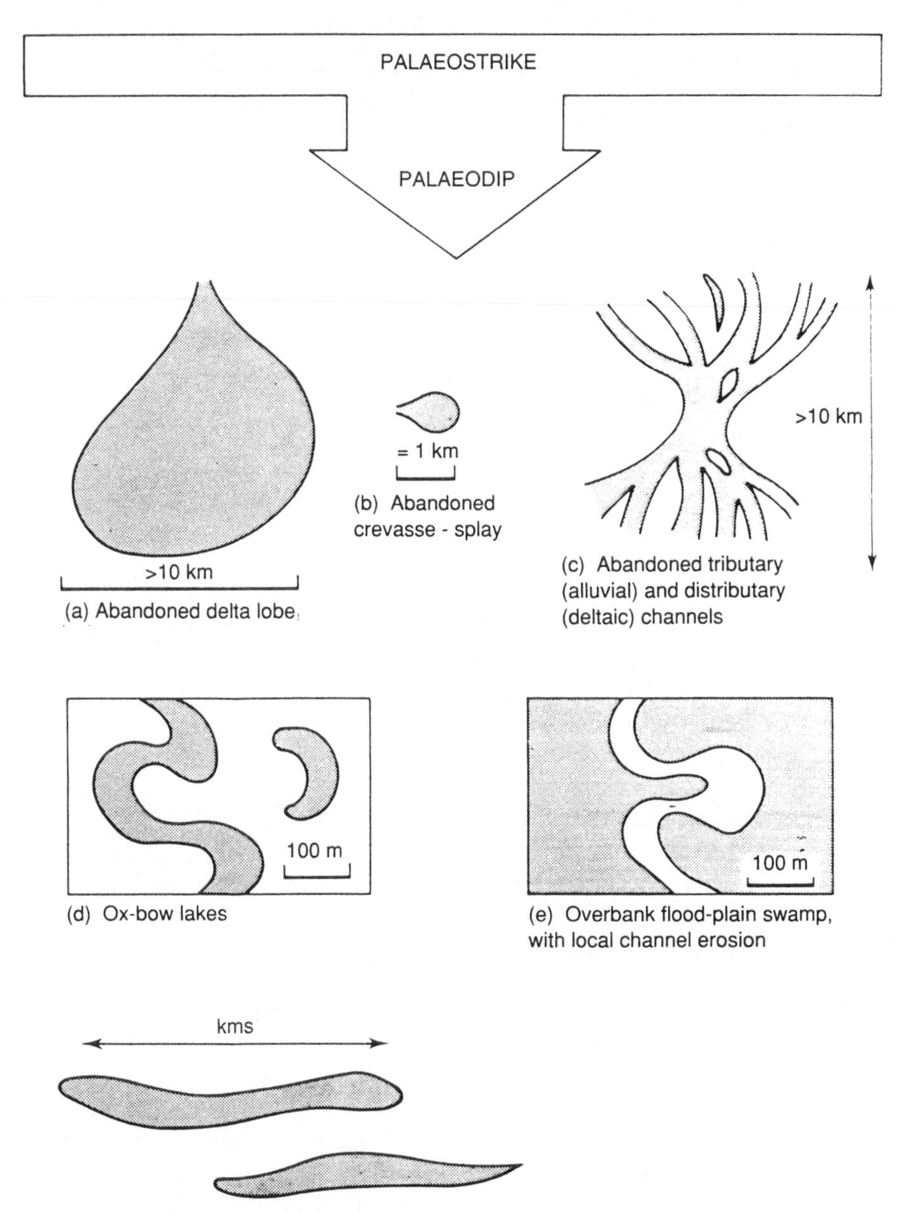

Figure 6.10 Cartoons to illustrate some of the ways in which coal beds may occur. Note that geometry is a function both of depositional environment and the effects of subsequent erosion.

The deltaic process is an excellent way of transporting sands (potential hydrocarbon reservoirs) far out into marine basins with organic rich muds (potential hydrocarbon source beds). Because deltas often form in areas of crustal instability structural deformation forms traps for migrating hydrocarbons. Deltas also form their own traps in growth faults, roll-over anticlines and associated stratigraphic traps (Selley, 1977, Coleman 1983).

Obviously it is important to understand the sedimentology of deltas so as

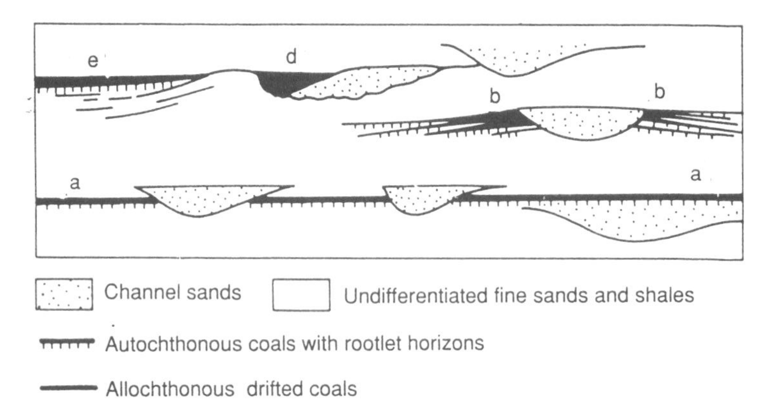

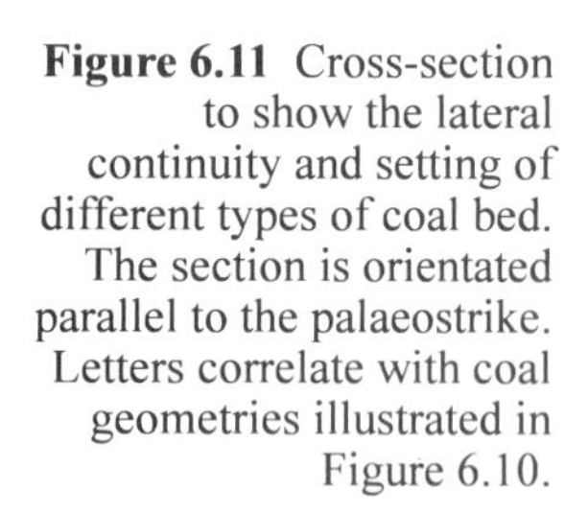

Figure 6.11 Cross-section to show the lateral continuity and setting of different types of coal bed. The section is orientated parallel to the palaeostrike. Letters correlate with coal geometries illustrated in Figure 6.10.

to predict the location and geometry of a hydrocarbon reservoir. Some of the modes of channel entrapment discussed for alluvial channels also apply to deltaic ones (refer back to p. 82 and Figure 3.12). Even where oil is structurally trapped in deltaic reservoirs environmental interpretation has an important part to play in both petroleum exploration and development (Whateley and Pickering, 1989).

Many oil fields in the northern North Sea produce from tilted truncated fault block traps. The main reservoir is the Middle Jurassic Brent Sandstone (Morton *et al.*, 1992). This was deposited by a delta system that prograded down the axis of the Viking Graben. Three major facies can be recognized: coarse pebbly non-marine sands and shales (fluvial), carbonaceous shales, sands and coals (deltaic), and clean sands that are occasionally shelly, glauconitic and burrowed (marine shoal). These are arranged in regressive: transgressive units (Figure 6.12).

There is a close correlation between facies and reservoir quality (porosity and permeability). Careful facies analysis has made it possible to map the progradation and destruction of the Brent delta. Palaeogeographic maps may thus be used to predict regional variations in reservoir quality (Figure 6.13).

Figure 6.14 illustrates the stratigraphy of the Brent sequence. The lower progradational phase of the delta deposited units with considerable lateral continuity. The upper part of the sequence is much more varied with local phases of transgression and truncation. During the development of fields there have been problems in differentiating deltaic channel sands from marine ones. The former are locally aligned down the delta slope, while the latter tend to parallel it (Figure 6.15).

Sand body geometry in the Brent delta is thus more complex than in the Khreim Group delta, where reworked shallow marine sands aligned parallel to the coast were absent.

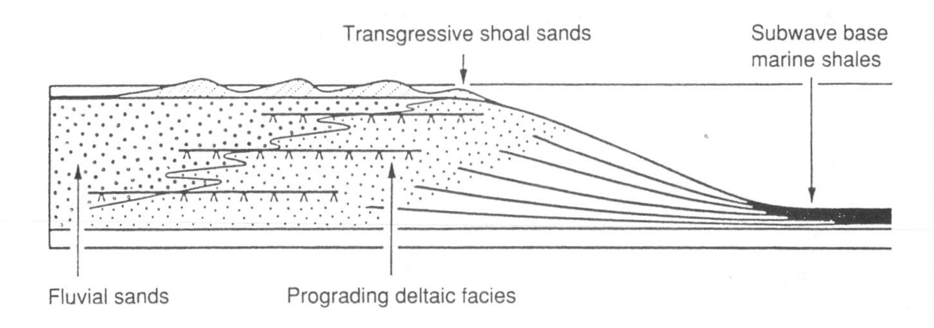

Figure 6.12 Geophantasmogram to show the relationship between the three facies of the Jurassic reservoir sands of the North Sea. From Selley, 1976a, by courtesy of the Geologists' Association.

SUB-SURFACE DIAGNOSIS OF DELTAIC DEPOSITS

Deltaic deposits may be easily recognized in the subsurface. The hard part is not to be able to identify a delta, but to unravel the complicated geometry of its sub-facies, particularly those of the delta plain. Subsurface diagnosis will now be considered according to the usual five parameters.

Geometry

The geometry of a delta may often be mapped seismically before a single well has been drilled. Major progradational foreset bedding is commonly seen in seismic sections through deltaic sediments (Figure 6.16).

Detailed studies may even be able to identify laterally stacked deltaic sand bodies (Figure 6.17).

3D seismic surveys can now sometimes map out radiating delta distributary channels (Figure 6.18), and even crevasse splays built out from distributary levees (Figure 6.19).

Coals have anomalously low seismic velocities so they may be identifiable when they are thick or shallow. Though many coals may lack continuity, those that form on delta abandonment surfaces will have considerable areal extent, and may help to delineate the upper surface of delta lobes.

Such excellent imaging is still normally only produced in shallow sediments, especially with gas-charged sands which are acoustically very slow. For deeper horizons, where 3D seismic is less effective, deltas may be mapped by means of palaeontology and wireline log correlation.

Lithology and sedimentary structures

Lithologically deltas are characterized by major wedges of terrigenous clastics, with widely varying grain-size and sand:shale ratio. At one

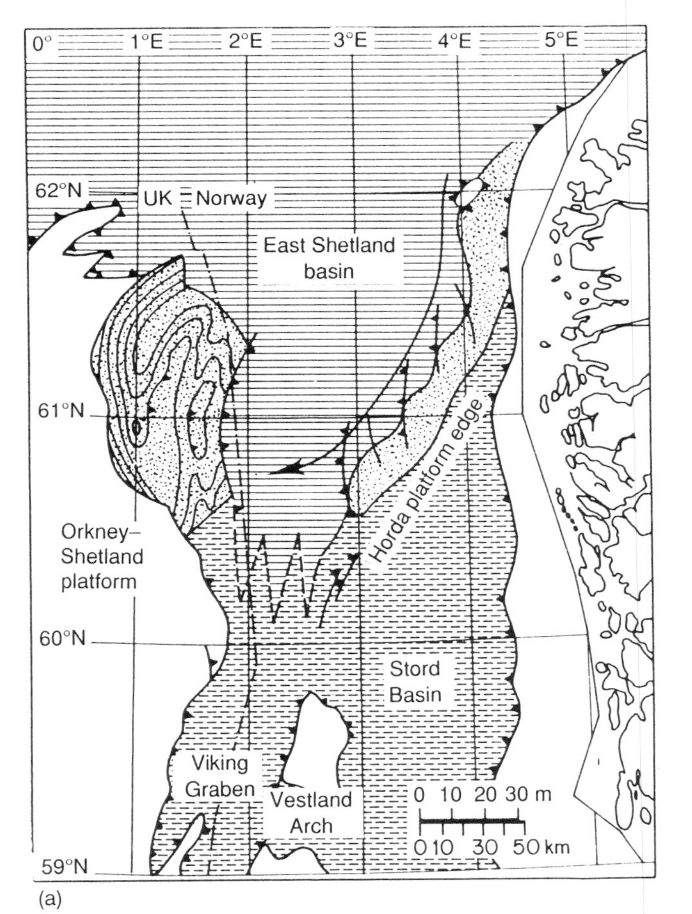

0°
1°E
2°E
3°E
4°E
5°E
62°N
61°N
60°N
59°N
UK
Norway
East Shetland basin
Orkney–Shetland platform
Horda platform edge
Stord Basin
Viking Graben
Vestland Arch
0 10 20 30 m
0 10 30 50 km
(a)

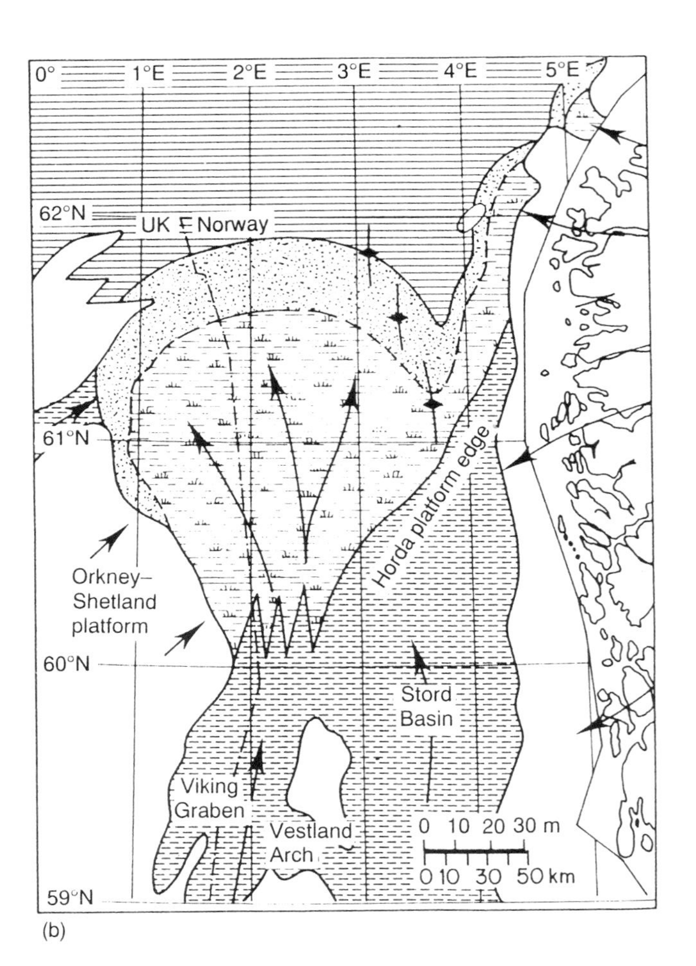

0°
1°E
2°E
3°E
4°E
5°E
62°N
61°N
60°N
59°N
UK
Norway
Orkney–Shetland platform
Horda platform edge
Stord Basin
Viking Graben
Vestland Arch
0 10 20 30 m
0 10 30 50 km
(b)

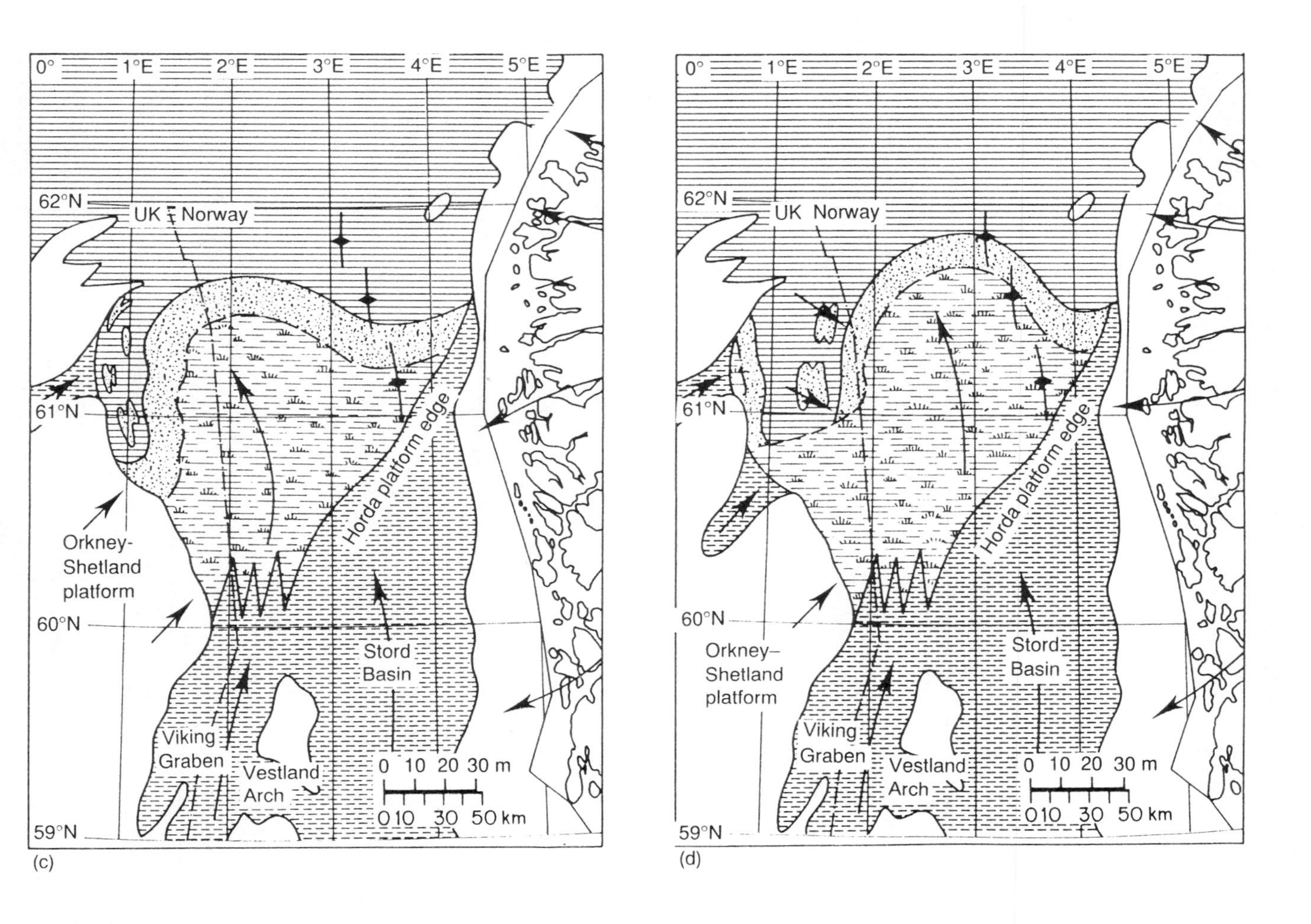

Figure 6.13 Successive palaeogeographic maps to illustrate the evolution of the Middle Jurassic Brent delta of the northern North Sea. From Eynon, 1981, by courtesy of the Institute of Petroleum.

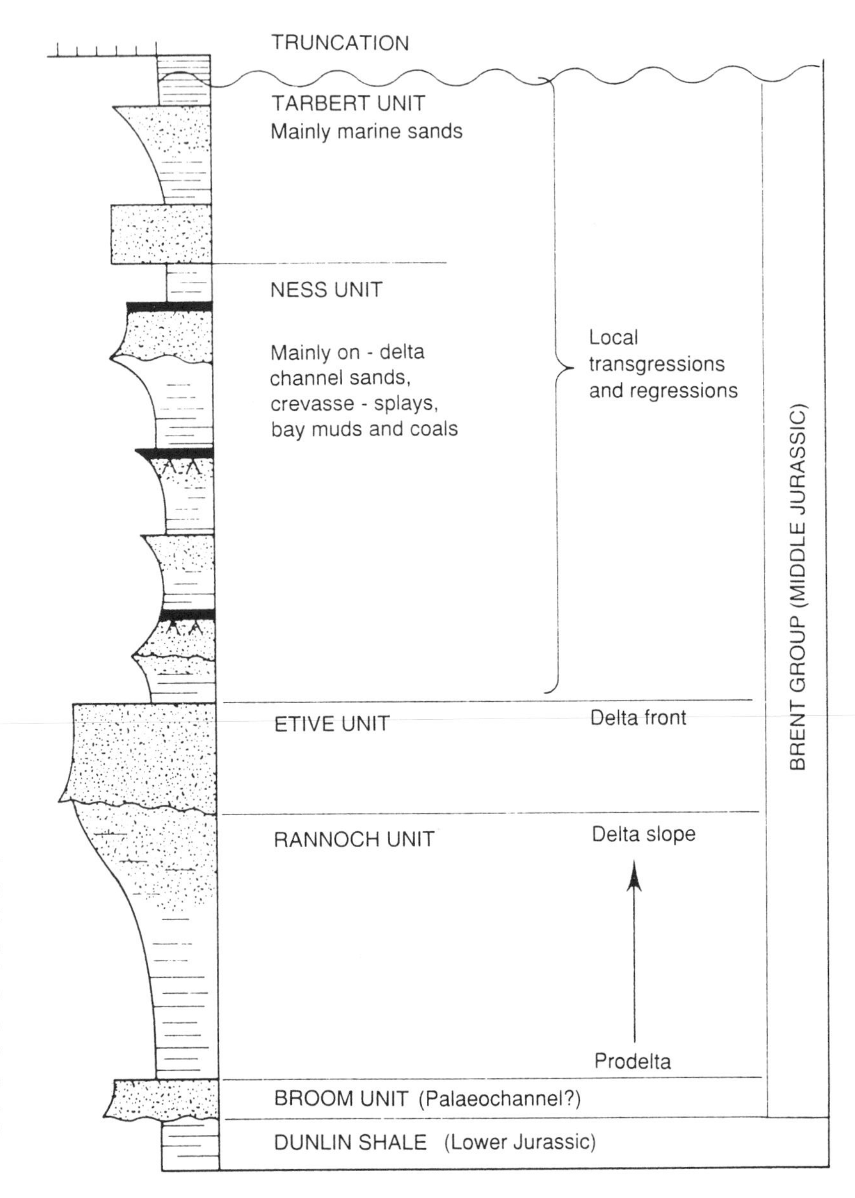

Figure 6.14 Summary stratigraphic section of the Middle Jurassic Brent Group of the Northern North Sea. Reservoir continuity is regular in the lower part, but becomes more complex towards the top due to the progradation and retreat of local subdelta lobes with transgressive marine sands.

extreme are fan deltas, whose sedimentology is often akin to that of fluvial fanglomerates (Colella and Prior, 1990). At the other extreme are muddy deltas, like the modern Mississippi, to which reference has already been made (p. 134). As already noted, coal or lignite beds are common, and disseminated carbonaceous detritus occurs in distributary channel sands, and

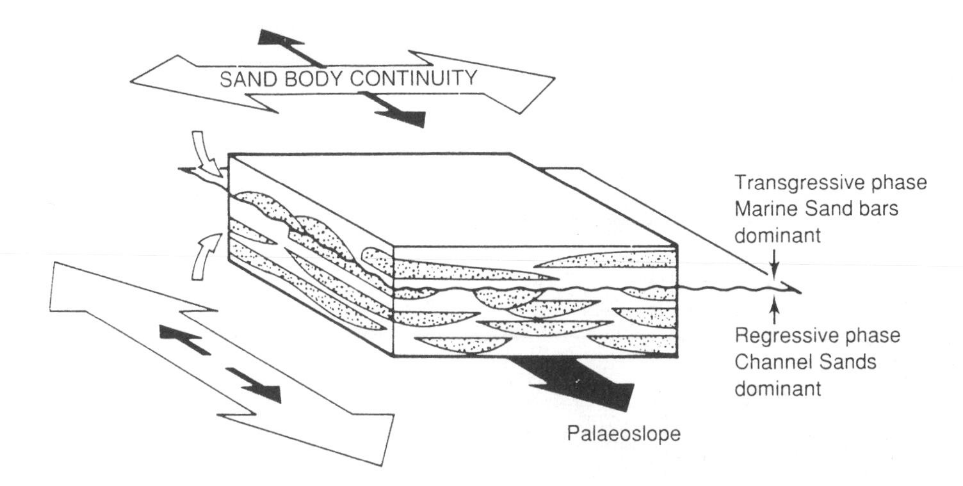

Figure 6.15 Cartoon to illustrate the relationship between reservoir continuity in a heterogenous delta complex with interbedded delta channel and marine shoreface sands.

other on-delta deposits. In the shoreface sands of wave dominated deltas, by contrast, carbonaceous detritus may be absent, and glauconite and/or shell debris may occur instead. A major deltaic lobe will show an overall upward-sanding sequence on wireline logs. The top set of a delta may show a wide range of log motifs.

Delta distributaries are normally clean sands, with sharp upper and lower contacts. They do not generally fine upward, as is the case with fluvial channels infilled by point bar sequences. A blocky log motif has been noted from ancient deltaic distributaries from the Gulf of Mexico to Indonesia. This shape may be explained as reflecting the sudden avulsion, and subsequent abandonment of a distributary channel.

A second characteristic motif of the delta plain is the upward-coarsening motif of the crevasse-splays. These prograde out from the levee into the interdistributary bays, generating an upward-coarsening motif. This is often capped by a bed of lignite or coal, indicating that the crevasse-splay built up to water level, before being abandoned and colonized by marsh vegetation. These crevasse-splay increments are normally only a few metres thick, in contrast to the main delta front prograde, that may be tens of metres or much more. The crevasse-splay motifs are often similar in scale to barrier beach sands (discussed in Chapter 7, p. 178). They can, however, readily be distinguished. Crevasse splay sands are commonly rich in carbonaceous detritus and mica flakes, reflecting their rapid deposition, and negligible reworking.

Carbonaceous detritus and mica are normally absent in the higher energy environment of a shoreface, where shell detritus and glauconite grains are much more likely to be found. Figure 6.20 shows the sequence of log motifs and sedimentary structures which may be anticipated in deltaic deposits. Figure 6.21 shows a borehole through Jurassic deltaic sediments of the North Sea, within which some of these motifs may be recognized.

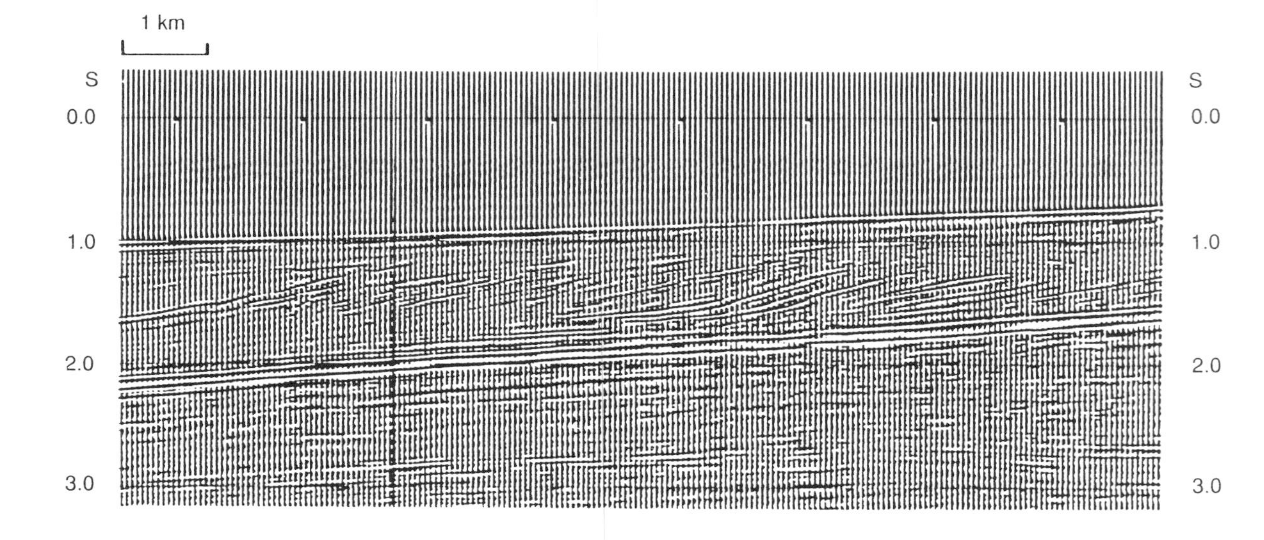

Figure 6.16 Seismic section through a delta showing progradation from topset through foreset to bottomset. From Fitch, 1976, Figure 26, by courtesy of Gebruder Borntraeger.

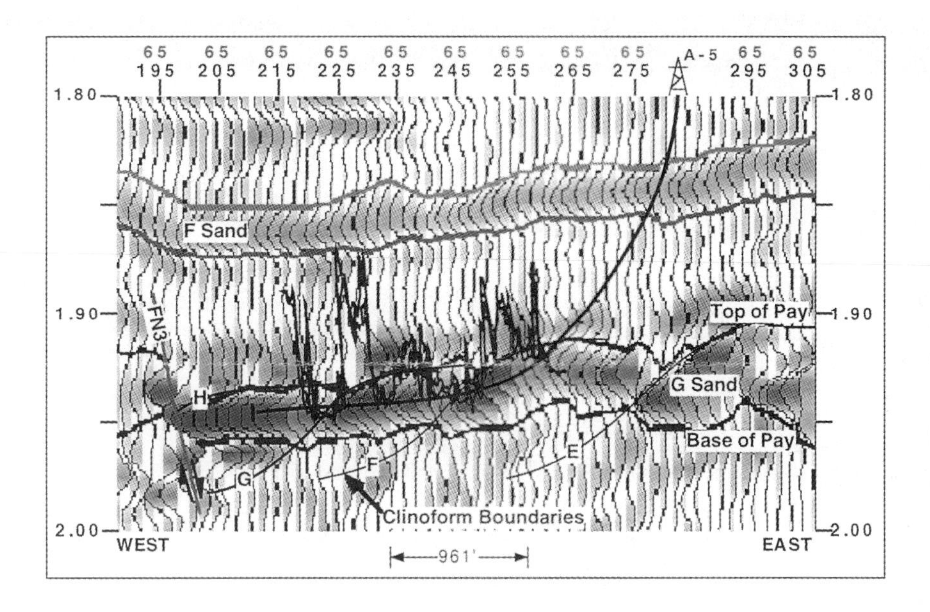

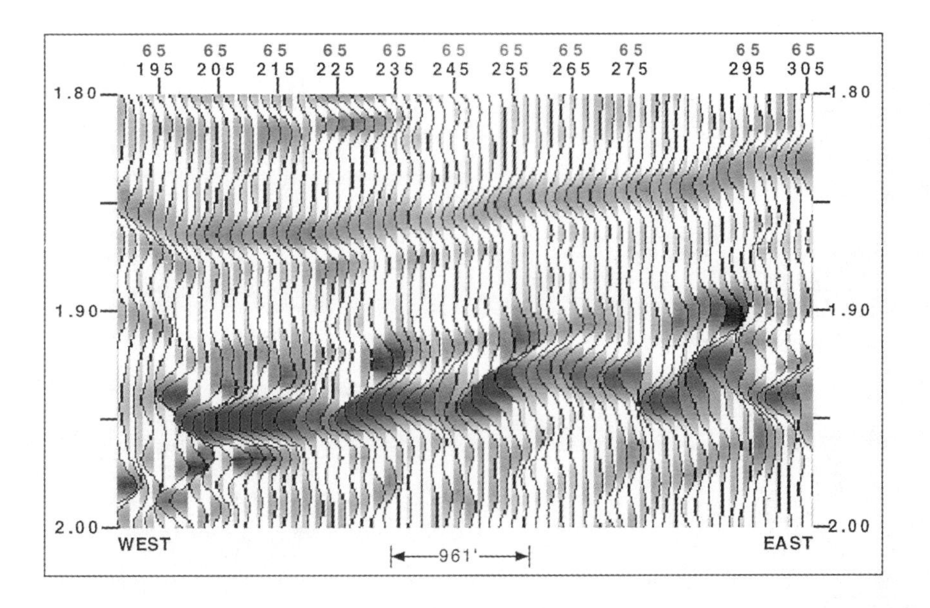

Figure 6.17 Seismic lines through BP's Amberjack Mississippi Canyon development, Gulf of Mexico. Upper: picked, interpreted and showing horizontal well path. Lower: unpicked. The prograding clinoforms of this delta consist of upward-coarsening increments some 120 m thick, in which mouth bar sands prograded out over slumped turbidite sands (see Mayall, *et al.*, 1992, for a detailed account of the geology). These petroliferous sands are separated from one another by impermeable clinoform-defining shales. They were all penetrated by the single cunningly drilled horizontal well whose track is shown in the upper panel. Courtesy of BP Gulf of Mexico Mississippi Canyon Development, especially B. Cohn and W. H. Mills (also reproduced as Plate 5).

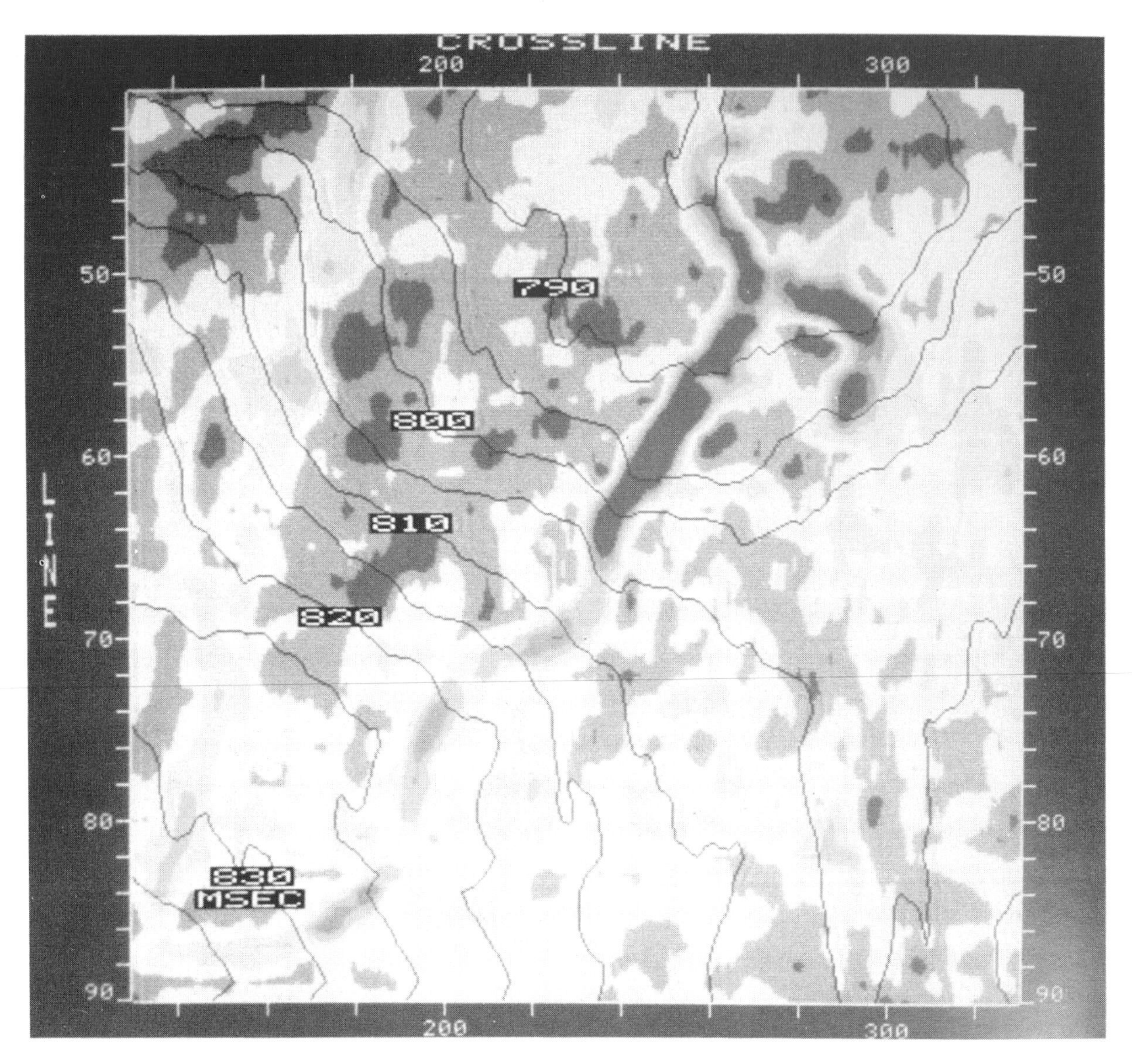

Figure 6.18 Horizon seiscrop section showing gas saturated delta distributary channel system. Gulf of Mexico. From Brown (1985) (also reproduced as Plate 7).

Palaeocurrents: dipmeter motifs

Dipmeter patterns in deltas are often complex but, if correctly interpreted, can make a valuable contribution to the unravelling of facies geometries. Two major types of motif are generally present, one in the channels, the other in the progrades. Delta distributary channels may show very complex dipmeter patterns. When the dips from one channel are plotted on a polar plot, however, a 90° bimodal pattern can appear. This may show one mode with high angle cross-bedding which dips down the channel axis, and another

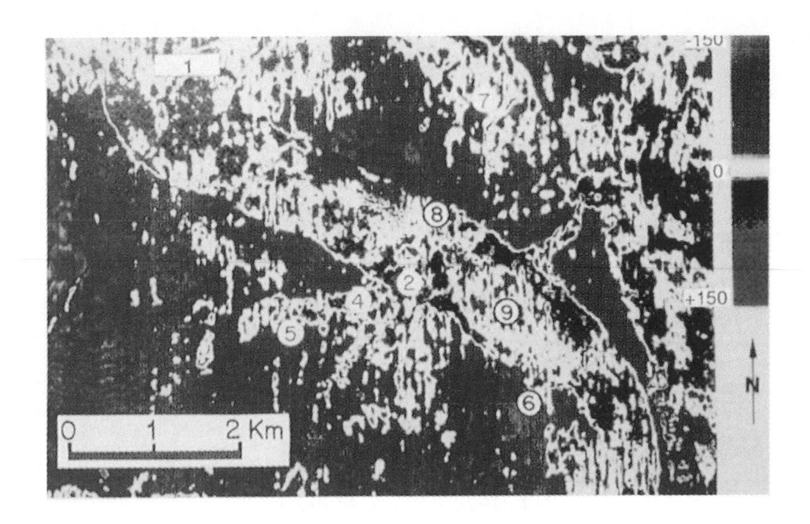

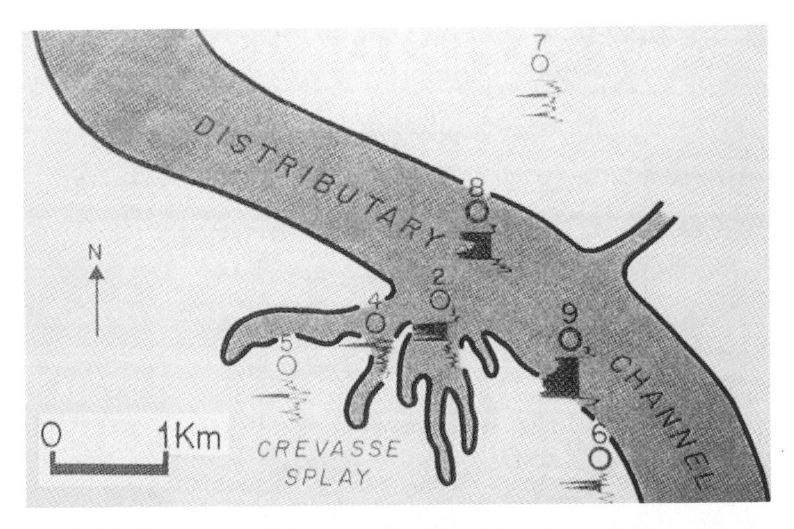

Figure 6.19 Horizontal amplitude display map from the Balingian Province, Malaysia, showing a clearly defined channel, with crevasse splay lobe. From Rijks and Jauffred (1991) (also reproduced as Plate 6).

mode which forms a red (upward declining) pattern. This reflects major point-bar foresets which dip into the channel axis (Figure 6.22).

This type of analysis of delta distributaries, which is generally also applicable to fluvial channels, is extremely important. From a single well it may be possible to ascertain both the channel trend, and the direction in which the axis lies. Information of this type is invaluable when looking for stratigraphically trapped oil in channel sands, as well as in unravelling the internal geometry of complex deltaic reservoirs. The foregoing presents the theory. It must be admitted that, in practice, it may be very difficult to produce a convincing

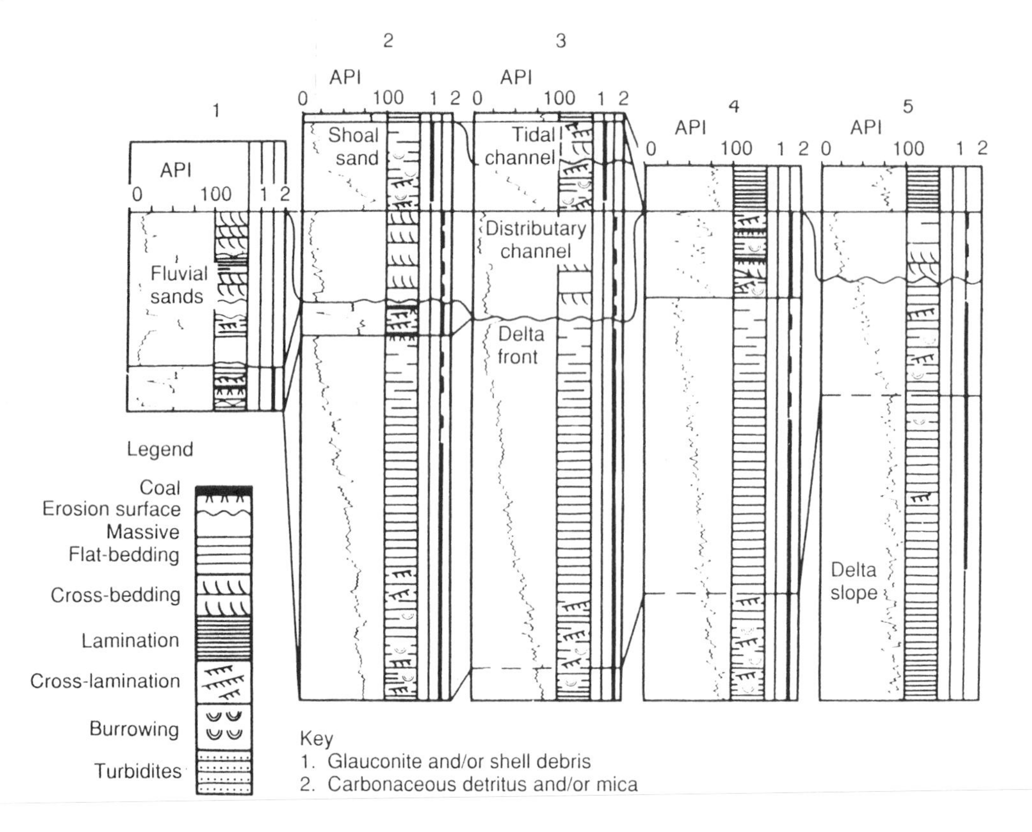

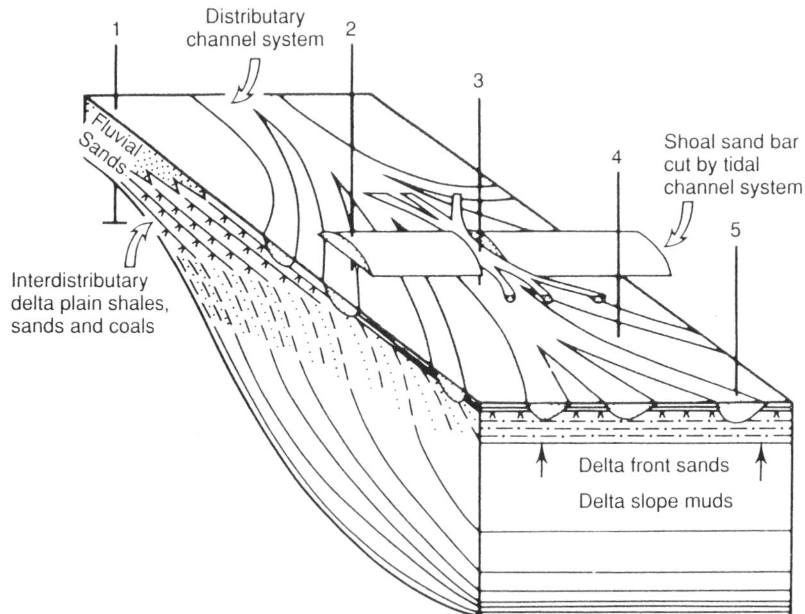

Figure 6.20 Cartoon to illustrate the sand bodies and log characteristics of deltaic sediments. When core material is available suites of sedimentary structures are often diagnostic. Even without core material, however, depositional environment may be detected from log motif (grain-size profile) and the distribution of glauconite, shell debris, mica and carbonaceous detritus.

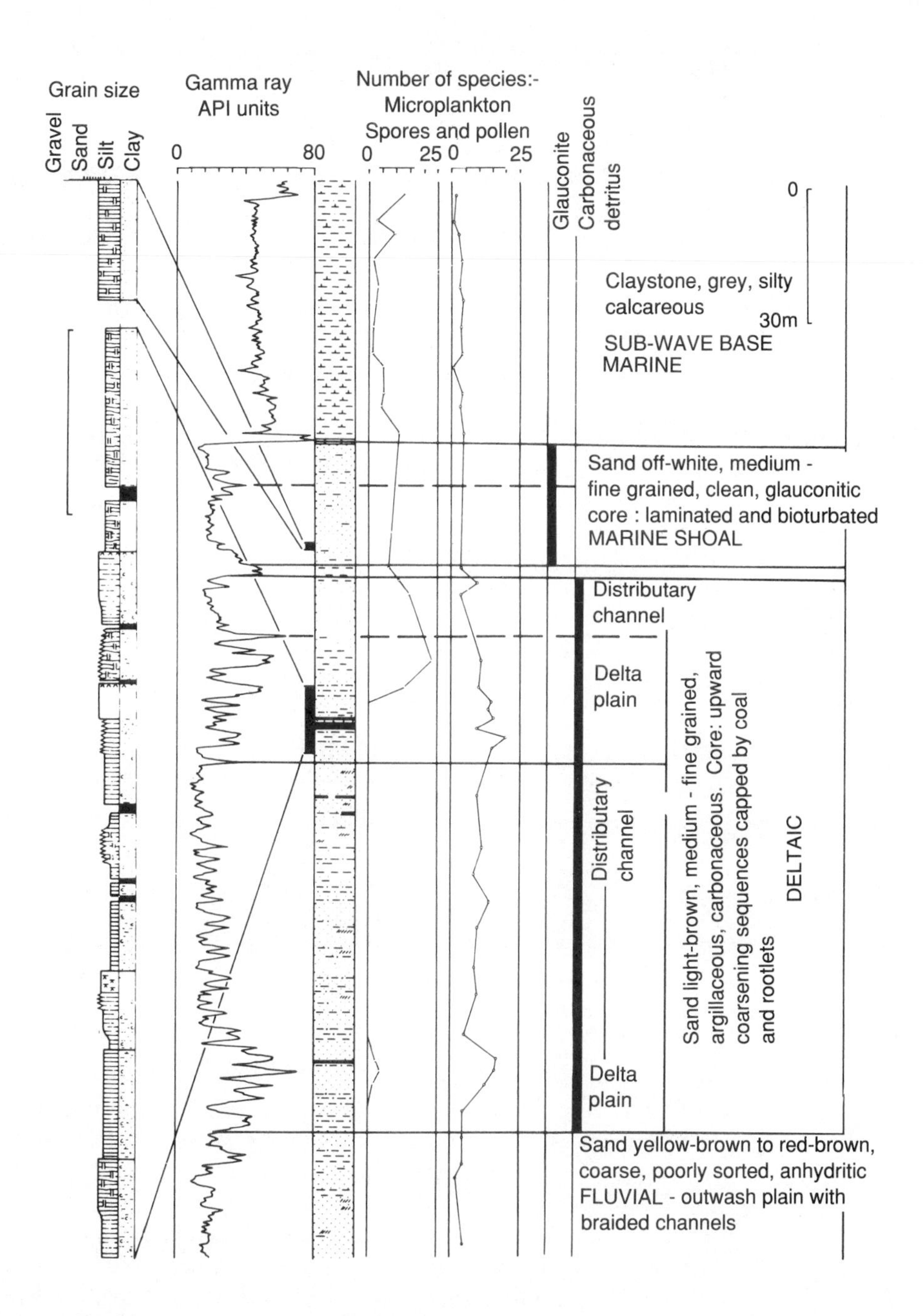

Figure 6.21 A borehole through Jurassic deltaic deposits beneath the North Sea. This shows how red sands and shales were deposited in a fluvial environment. These are succeeded by carbonaceous sands and shales with thin coals. Both the cores and the gamma log show upward-sanding progradational sequences. A subsequent marine transgression deposited glauconitic shoal sands on the abandoned delta. These were buried in turn beneath marine muds. From Selley (1976), Figure 4, by courtesy of the American Association of Petroleum Geologists.

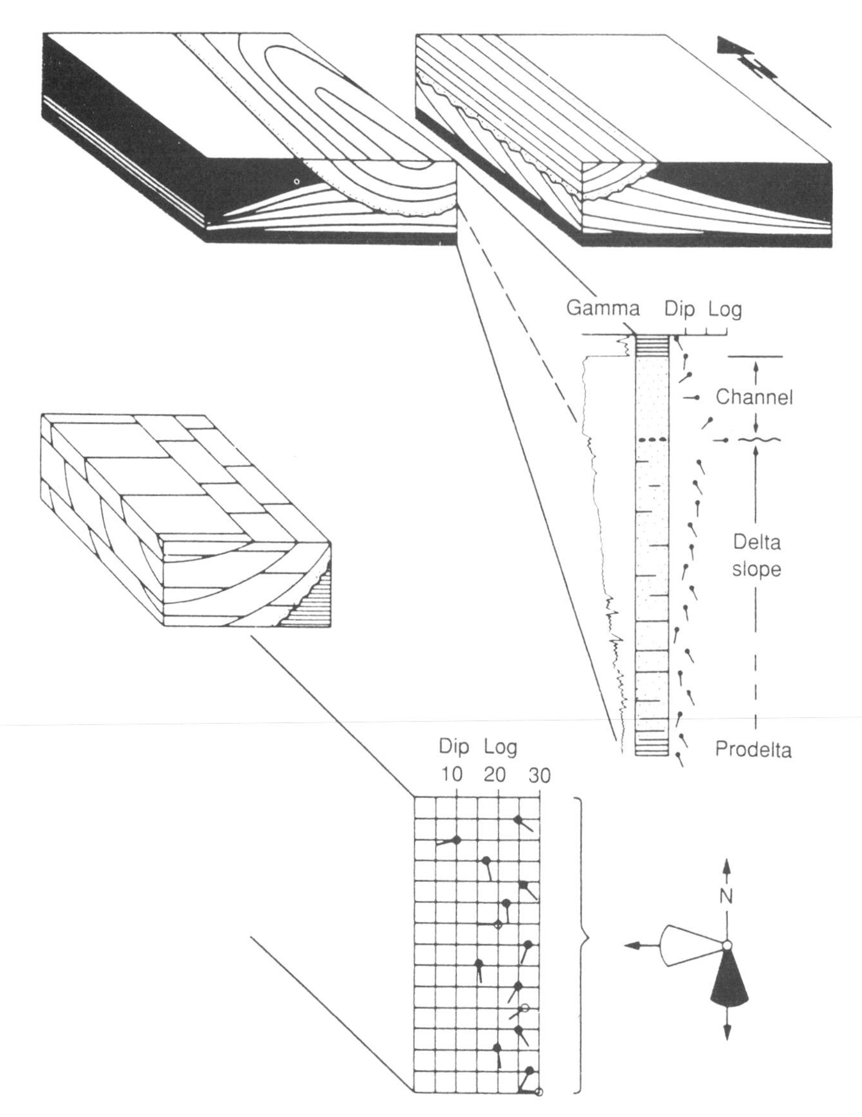

Figure 6.22 Illustration to show the dip motifs found in deltaic deposits. Upward increasing blue motif in the delta slope shows that this example prograded to the south. The upward-decreasing red pattern in the distributary channel reflects major bedding surfaces striking north-south along channel axis and dipping west into it. Close spaced dips (below) may indicate that the red readings dip west (open tadpoles); while steeper dips are due to southerly cross-beds dipping down channel (black tadpoles). From Selley (1977); reproduced by courtesy of Applied Science Publications.

interpretation of intra-channel dips. This is particularly true in homogenous sands, where there is little resistivity contrast for the imaging tool to detect. It is also true in oil-saturated sands, where, again, the resistivity logs have problems in detecting thin beds.

The second dipmeter motif of deltas is to be found in the upward-coarsening increments of the prograde. This is often characterized by an upward-increasing 'blue' motif (Figure 6.22). This motif is normally much more readily apparent than those of the channels. This is because the prograde consists of regularly interbedded sands and shales, with good resistivity contrast. Furthermore, the dips are of a single statistical population. They all

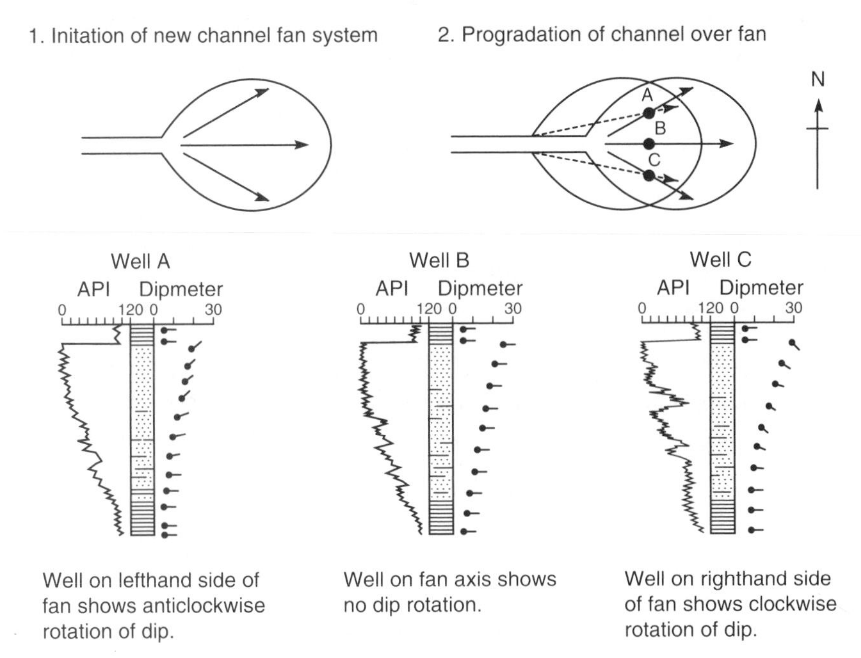

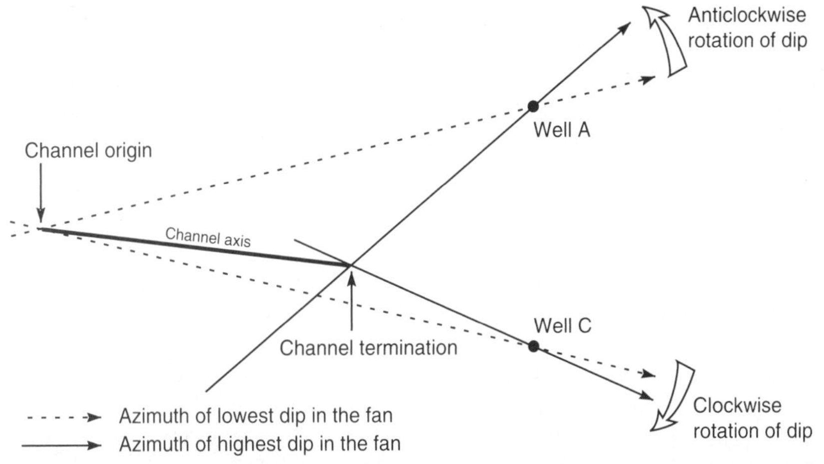

Figure 6.23 Upper: Diagrams to illustrate how depositional dips rotate in opposing directions on either side of a prograding delta lobe. Lower: Sketch to show how rotational progradation dips may be used to locate the optimum reservoir sand at the apex of a delta fan lobe (from Selley, 1989).

point in the direction of outbuilding. When the dips of two or more wells in one delta lobe are plotted back they may locate the apex of the lobe, which is the sandiest part, where the optimum reservoir conditions may be found.

There is a more sophisticated application of this principle. As a delta lobe progrades the depositional slopes rotate in a regular manner. The slope rotates anti-clockwise on the left flank (as viewed down current), and in a clockwise direction on the right flank of the lobe (Figure 6.23).

Consider two wells drilled on opposing sides of a delta lobe. If the dips at the base of the increment are plotted back they will intersect at the spot where fan progradation was initiated. If the dips at the top of the increment are plotted back, they will intersect at the spot where the fan was abandoned, and the distributary channel terminated. A line connecting the two intersections will define the crest of the fan, and most probably the axis of the distributary channel. The validity of this technique has been demonstrated in both ancient deltaic sediments, and also from bathymetric charts that record the progradation of a Recent Mississippi delta crevasse splay over the last 100 years (Selley, 1989).

Palaeontology

In simple terms a delta might be expected to be characterized by a regular alternation of marine biota in the prodelta, and non-marine biota on the delta plain. The truth is more complex. Because deltas bring so much sediment and fresh water in to the marine realm, sediment may be deposited on the pro-delta in what is, technically, an open marine environment. But the terrestrial sediment and fresh water may dilute and even totally exclude any marine faunal or floral influence. Thus fresh water pollen assemblages, together with land derived plant detritus, and astonished terrestrial vertebrates, may all be preserved in a marine environment uncontaminated by any marine fossils. This problem of dilution may even extend down in to deep water submarine fans, as discussed in Chapter 12.

SUMMARY

Deltas are very important depositional environments, so a summary may be appropriate. Deltas form where rivers shed their load at the edges of seas or lakes. The geometry of deltaic deposits reflects the relative ability of the river to deposit sediment, and of tide and wave processes to redistribute it. These processes will affect the nature and geometry of the various subfacies. Deltas are very important economically. They are important environments of peat formation and preservation, so most of the world's coal comes from ancient deltas. Similarly deltas are very important hosts to oil and gas. Not only do they contain their own source material, but they are commonly enveloped in organic-rich marine shale. Of their very nature, deltas contain sand reservoirs. They generate their own petroleum traps in many ways.

Subsurface diagnosis of deltas is easy, but the detailed interpretation of their various sub-facies is difficult. Once this was achieved by logs in general, and the dipmeter in particular. Now seismic surveys in general, and 3D surveys in particular, are directly imaging many deltaic reservoir sands. As remarked earlier, this success renders the main thrust of this book increasingly obsolete.

REFERENCES

The Ordovician-Silurian case history of the Sahara and Arabia was based on the author's own unpublished work, together with:

Al-Husseini, M. (1991a) Potential petroleum resources of the Palaeozoic rocks of Saudi Arabia. In *Proc. 13th World. Pet. Cong.* J. Wiley, Chichester, 3-13.
—— (1991b) Tectonic and depositional model of the Arabian and adjoining plates during the Silurian-Devonian, *Bull. Amer. Assoc. Petrol. Geol.*, **75**, 108–20.
Bellini, E. and Massa, D. (1980) A stratigraphic contribution to the Paleozoic of the southern basins of Libya. In M. J. Salem and M. T. Busrewil (eds), *The Geology of Libya*, Academic Press, London, Vol. 1, 3–56.
Mahmoud, M. D., Vaslet, D. and Husseini, M. I. (1992) The Lower Silurian Qalibah formation of Saudi Arabia: an important hydrocarbon source rock, *Bull. Amer. Assoc. Petrol. Geol.*, **76**, 1491–1506.
Powell, J. H., Moh'd, B. K. and Masri, A. (1994) Late Ordovician – Early Silurian glaciofluvial deposits preserved in palaeovalleys, in South Jordan, *Sed. Geol.*, **89**, 303–14.
Selley, R. C. (1970) Ichnology of Palaeozoic sandstones in the southern desert of Jordan, a study of trace fossils in their sedimentologic context. In T. P. Crimes and J. C. Harper (eds), *Trace Fossils*, Lpool. Geol. Soc., 477–88.
Tissot, B., Deroo, G. and Espitalie, J. (1975) Etude comparée de l'époque de formation et d'expulsion du pétrole dans diverses provinces géologiques, *Proc. 9th World Pet. Cong., Tokyo*, Applied Science Pubs, London, **2**, 159–69.
Turner, B. R. (1980) Palaeozoic sedimentology of the Southeastern part of Al Kufrah Basin, Libya, a model for oil exploration. In M. J. Salem and M. T. Busrewil (eds), *The Geology of Libya*, Academic Press, London, Vol. 2, 351–374.
—— (1991) Palaeozoic deltaic sedimentation in the southeastern part of Al Kufrah basin, Libya. In M. J. Salem, A. M. Sbeta and M.R. Bakbak (eds), *The Geology of Libya*, Vol. 6, Elsevier. Amsterdam, 3–56.

Other references cited in this chapter:

Brown, A. R. (1985) The role of horizontal seismic sections in stratigraphic interpetration. In O. R. Berg and D. G. Woolverton, (eds), *Seismic Stratigraphy II*, Amer. Assoc. Petrol. Geol. Mem., **39**, 37–48.
Colella, A. and Prior, D. (1990) *Coarse-grained Deltas*, Internat. Assn. Sedol. Sp. Pub., **10**.
Coleman, J. M. and Prior, D. B. (1982) Deltaic environments. In P. A. Scholle and D. Spearing (eds), *Sandstone Depositional Environments*, Amer. Assoc. Petrol. Geol., **31**, 139–78.
Elliott, T. (1986) Deltas. In H. G. Reading (ed.), *Sedimentary Environments and Facies*, 2nd edn, Blackwell, Oxford, 113–54.

Galloway, W. E. and Hobday, D. K. (1983) *Terrigenous Clastic Depositional Systems*, Springer-Verlag, Berlin.

Eynon, G. (1981) Basin development and sedimentation in the Middle Jurassic of the Northern North Sea. In G. D. Hobson and L. V. Illing (eds), *Petroleum Geology of the Continental Shelf of North West Europe*, Institute of Petroleum, London, 196–204.

Fisher, W. L., Brown, L. F., Scott, A. J. and McGowen, J. H. (1972) *Delta Systems in the Exploration for Oil and Gas,* Bureau Econ. Geol. Texas Univ.

Fitch, A. A. (1976) *Seismic Reflection Interpretation*, Gebruder Borntraeger, Stuttgart.

Herodotus (*c.* 430 BC) *The Histories of Herodotus of Halicarnassus*, Papyrus Publishing Co., Old Cairo; also 1962, translated by H. Carter, Oxford University Press.

Mayall, M. J., Yeilding, C. A., Pulham, A. J. and Sakurai, S. (1992) Facies in a shelf edge delta – an example from the subsurface of the Gulf of Mexico, Middle Pliocene, Mississippi Canyon, Block 109, *Amer. Assoc. Petrol. Geol. Bull*, **76**, 435–48.

Morton, A. C., Haszeldine, R. S., Giles, M. R. and Brown, S. (eds) (1992) *Geology of the Brent Group*, Geol. Soc. Lond. Sp. Pub., **61**.

Rijks, E. J. H. and Jauffred, J. C. E. M. (1991) Attribute extraction: an important application in any detailed 3-D interpretation study, *Geophysics: The Leading Edge of Exploration*, September, 11–19.

Scott, A. C. and Fleet, A. J. (eds) (1994) *Coal and Coal-bearing Strata as Oil-Prone Source Rocks*, Sp. Pub. Geol. Soc. Lond., **77**.

Selley, R. C. (1976a) The habitat of North Sea Oil, *Proc. Geol. Ass. London*, **87**, 359–88.

—— (1976b) Sub-surface environmental analysis of North Sea sediments., *Bull. Amer. Assoc. Petrol. Geol.*, **60**, 184–95.

—— (1977) Deltaic facies and petroleum geology. In G. D. Hobson (ed.), *Developments in Petroleum Geology*, Applied Science Publishers, London, 197–224.

—— (1989) Deltaic reservoir prediction from rotational dipmeter motifs. In M. K. G. Whateley and K. T. Pickering (eds), *Deltas: Sites and Traps for Fossil Fuels*, Geol. Soc. Lond. Sp. Pub., **41**, 89–95.

Wanless, H. R., Baroffio, J. R. and Trescott, P. C. (1969) Conditions of deposition of Pennsylvanian coal beds. In E. C. Dapples and M. E. Hopkins (Eds), *Environments of Coal Deposition*, Geol. Soc. Amer. Sp. Pub., **114**, 105–42.

Ward, C. R. (1983) *Coal Geology: Exploration, Mining, Preparation and Use*, Blackwells, Oxford.

Whateley, M. K. G. and Pickering, K. T. (eds) (1989) *Deltas: Sites and Traps for Fossil Fuels*, Geol. Soc. Lond. Sp. Pub., **41**.

7 Linear terrigenous shorelines

INTRODUCTION: RECENT LINEAR TERRIGENOUS SHORELINES

Deltas only form where rivers bring more sediment into the sea than can be re-worked by marine current. By their very nature, therefore, deltaic sequences indicate a regression of the shoreline. Where marine currents are strong enough to redistribute land-derived sediment, linear shorelines are formed with bars and beaches running parallel to the coast. Both deltas and linear shorelines deposit sediment in a wide range of sedimentary environments ranging from continental to marine. Studies of Recent sediments show that both linear and lobate shorelines can form upward-coarsening regressive sequences.

Because deltas and linear shorelines both deposit porous sands around marine basins they are important hydrocarbon reservoirs. Since their sand body geometries are quite different, however, it is important to be able to distinguish the two types. Studies of Recent terrigenous linear shorelines suggest that four major sedimentary environments can be recognized (e.g. Davis, 1978; Swift and Palmer, 1978 and McCubbin, 1982). These consist of two high-energy zones which alternate seawards with two low-energy zones. Basically from land to sea these are: fluviatile coastal plain, lagoonal and tidal flat complex, barrier island, and offshore marine shelf (Figure 7.1).

Considerable variation is found in the occurrence and distribution of these zones. This variation is largely dependent on the tidal range. Linear coasts have been classified by their tidal range in to Microtidal (< 2 m), Mesotidal (2–4 m) and Macrotidal (>4 m) (Barwis and Hayes, 1979). On Microtidal coasts barrier islands may be well developed, far from the shore, and cross-cut by few tidal channels. Mesotidal coasts have barriers close inshore, cut by extensive tidal channels. On Macrotidal coasts barrier islands may be absent, sand deposition taking place instead on tidal flats cross-cut by extensive tidal channels (Figure 7.2).

Each sub-environment of linear coasts deposits sub-facies which can be differentiated from one another by their geometry, lithology, sedimentary structures, palaeocurrent patterns and biota. These will now be summarized. Further data are given in the references previously mentioned. The fluviatile coastal plain will deposit alluvium similar to that described in Chapter 3.

Where the coastal plain has a gentle gradient the alluvium will be of the meanderlng river type. Where it is steep a braided outwash plain is generally developed. In the more normal former case fine-grained flood-plain sediments

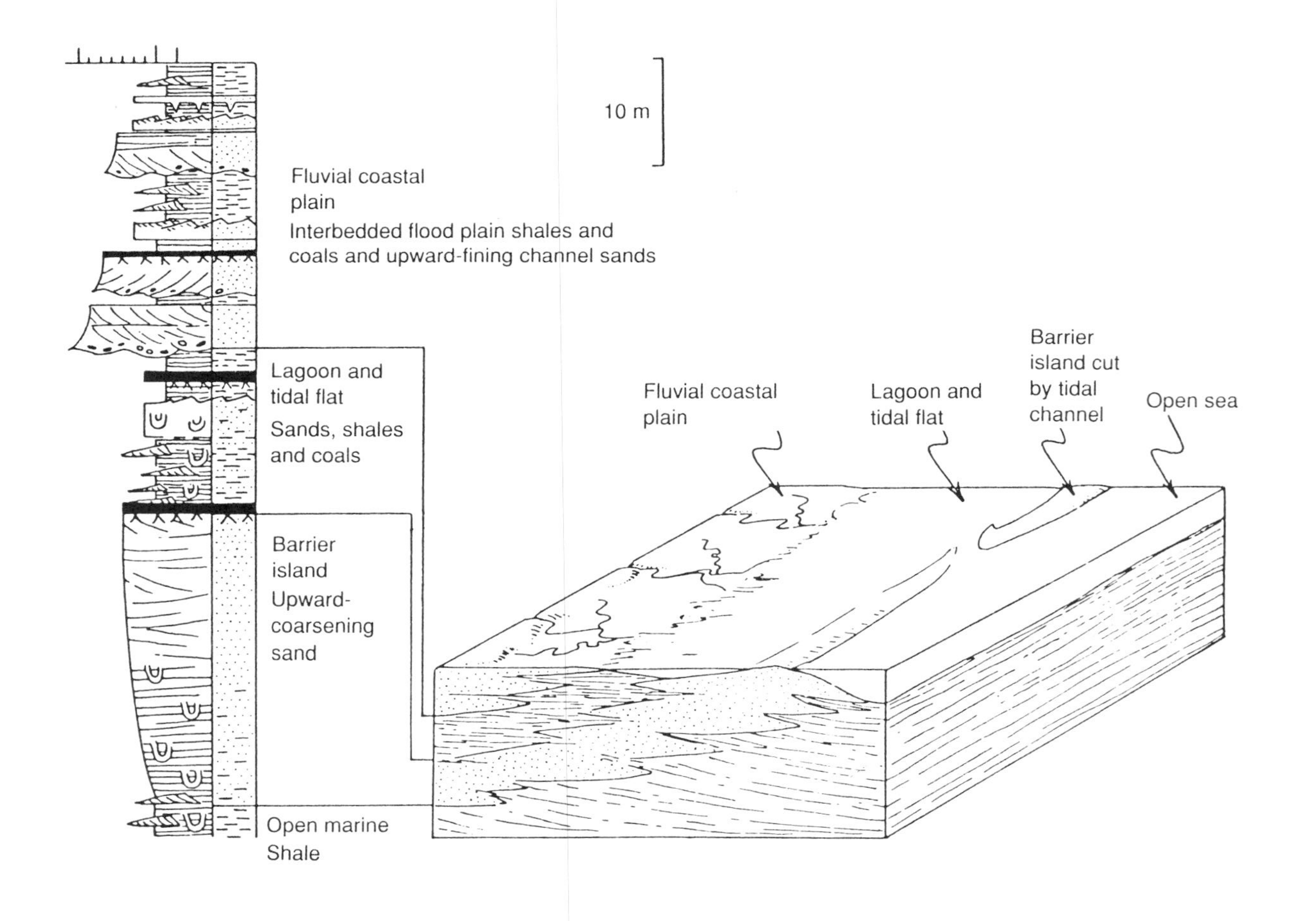

Figure 7.1 Geophantasmogram to show the environments, facies and sedimentary sequence produced by a prograding terrigenous coastline (from Selley, 1988).

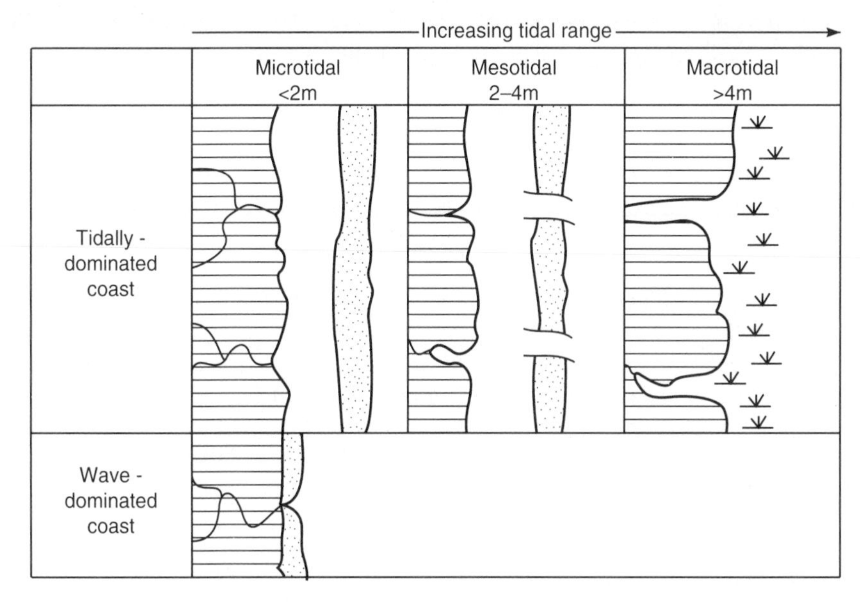

Figure 7.2 Sketches to show the way in which tidal range may control the geometry of linear coasts.

will dominate over upward-fining channel sand sequences. The biota is continental, with bones, wood, and other plant debris and freshwater invertebrates. The alluvial coastal plain passes transitionally seaward into tidal flats, and lagoons. Peat forms in the swamps.

Tidal flats deposit delicately interlaminated muds, silts, and very fine sands, often rippled and burrowed. They are cut by meandering gullies in which channel floor lag conglomerates are overlain obliquely by interlaminated fine sediment deposited on prograding point-bars (Van Straaten, 1959; Evans, 1965).

The deposits of lagoons are generally fine-grained too, but depending on their size and depth, may deposit sediment ranging from sand to mud. In regions of low sediment influx carbonate mud can be deposited. Evaporites may form in hypersaline lagoons. The fauna of lagoons is similarly variable depending on the salinity. It may range from freshwater, through brackish (with shell banks) to normal marine or, if restricted and of high salinity, a fauna may be absent.

Sedimentary structures of lagoons are similar to those of tidal flats, with delicately laminated muds and interlaminated and rippled sand silt and mud. Bioturbation is common. The lagoon is separated from the open sea by a barrier island complex. This is composed dominantly of well-sorted sand with a fragmented derived marine fauna. Studies of Recent barrier deposits (e.g. McCubbin, 1982) show that low angle seaward dipping bedding in the lower shoreface sand, passes up into trough and planar cross-bedding in the upper shoreface. These cross-beds dips generally landwards, though bipolar onshore: offshore dips have been recorded (Klein, 1967).

The barrier may be no more than an offshore bar exposed only at low tide, or it can form an island with eolian dunes on the crest. Intermittently along its length the barrier may be cut by tidal channels in which cross-bedded sands are deposited (Armstrong-Price, 1963; Hoyt and Henry, 1967). To landward the barrier may pass abruptly into the lagoon with the development of washover fans.

Alternatively a tidal flat may intervene. Barriers typically pass seawards with decreasing grainsize, via a transitional zone of interlaminated rippled, burrowed sand, silt, and clay, into an offshore zone where laminated marine mud is deposited below wave base. Thus as a barrier beach progrades seawards it deposits an upward-coarsening grain-size profile with a characteristic suite of sedimentary structures (Figure 7.1).

In summary, therefore, a Recent linear shoreline consists of two high-energy zones and two low-energy zones alternating with one another parallel to the coast. In some instances the barrier sand is forced landwards against the alluvial plain. A lagoonal tidal flat complex is then absent. This situation generally occurs on stormy coasts with low input of sediment from the land (Hoyt, 1968). The actual sedimentary sequence which is deposited from a linear clastic shoreline is a function both of sediment availability and of the rate of rise or fall of the land and sea.

All four facies described above will only be preserved where there is an abundant supply of land-derived sediment. When this happens and the shoreline is stationary all four sub-environments deposit sequences of the four subfacies side by side. Such static shorelines as these are rare in the geological column because they need a very delicate balance between sedimentation and rising sea level. The Frio sand (Oligocene) of the northwest Gulf of Mexico is an example of an ancient static clastic shoreline (Boyd and Dyer, 1966).

Where a high influx of sediment is accompanied by a regression of the shoreline (for which the former may be responsible) all four facies build out one above the other seaward. Such regressive sequences are similar to deltas, essentially coarsening upwards from marine shales at the base into continental sands at the top.

High sediment influx accompanied by a relative rise in sea level results in a complex transgressive sequence composed of a series of laterally stacked regressive increments. This curious stepwise arrangement of transgressive sections will be illustrated and explained in the following case history. Regressive and transgressive shorelines, with all four facies preserved, occur in the Upper Cretaceous rocks of the American Rocky Mountains which serves as the case history to now be described.

CRETACEOUS SHORELINES OF THE ROCKY MOUNTAINS, USA: DESCRIPTION

An easterly thinning clastic prism was deposited in a Cretaceous seaway which stretched across the midwest of North America from the Arctic to the Gulf of Mexico. Easterly thinning of this sequence is accompanied by a gradual decrease in grainsize. Coarse conglomerates derived from the rising Rocky

Mountains pass eastwards through sandstones to shales. Within these rocks it is possible to recognize three major sedimentary facies, which interfinger with one another laterally and are interbedded vertically (Figure 7.3).

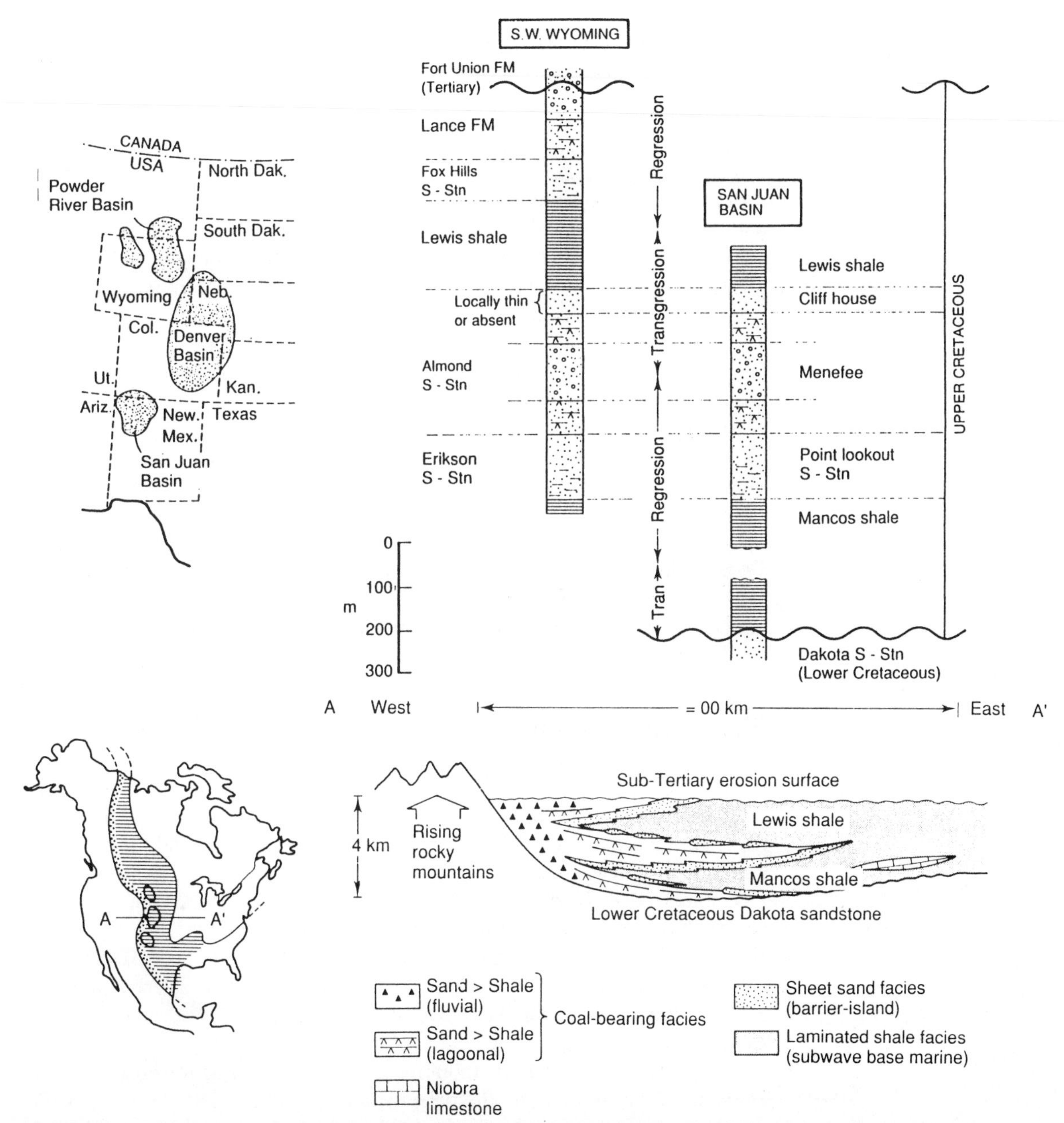

Figure 7.3 Maps, cross-section and stratigraphic columns to illustrate the distribution of coastal deposits laid down in the Cretaceous seaway of the Rocky Mountains. Note that formation names and thickness vary from place to place. Based on sources cited at the end of this chapter.

These may be listed from west to east as follows:

1. coal-bearing facies;
2. bioturbated sand facies;
3. laminated shale facies.

These three facies will now be described in turn.

Coal-bearing facies

Lithostratigraphically the coal-bearing facies includes the Menefee Formation of Colorado and New Mexico, and the Lewis Formation and parts of the Almond and Judith River formations of Wyoming and Montana.

This facies consists of wedges, up to 600 m thick in the west, which splits up into a number of easterly thinning tongues when traced away from the Rocky Mountains. In the west this facies is dominantly conglomeratic with minor amounts of coarse, poorly sorted sandstones and siltstones. Interbedded lavas and ashes are present, and the conglomerates and sands are often composed of volcanic detritus. These sediments show prominent channeling and cross-bedding.

Traced eastwards grainsize diminishes to medium and fine cross-bedded sands with channeled bases interbedded with shales. The shales are laminated and rippled. They contain a spectacular dinosaur fauna, and shells of the freshwater lamellibranch *Unio* and the brackish lamellibranchs *Corbula* and *Ostrea*. The oysters form reefs locally. The shales are generally dark in colour and often carbonaceous with fossil plant remains. The shales and sands are interbedded with coals, which are often present in very thick extensive units. These are commercially significant.

Bioturbated sand facies

Lithostratigraphically the bioturbated sand facies includes the Fox Hills Formation and the upper part of the Almond and the lower part of the Judith River formations of Wyoming and Montana. In Colorado and New Mexico sands of this facies are represented by the Cliff House and Point Lookout formations. This facies occurs in a range of geometries. At one extreme it occurs in thin isolated shoestrings aligned north–south along the depositional strike. These are sometimes laterally stacked to form sheets. Both vertically and laterally the bioturbated sand facies commonly separates the coal-bearing facies of the west from the laminated shales of the third facies to the east (Figure 7.3).

Individual sand sheets are about 30 m thick and can be traced for considerable distances both along the palaeoslope (west–east) and the palaeostrike (north–south). In Colorado and New Mexico the sheets contain local thickened benches which are laterally persistent along the palaeostrike for tens of kilometres. Isolated sand lenticles occur interbedded with laminated shales east of the main development of sheet sands. These have shoestring geometries with north–south trends. Examples include the Eagle sandstone of Montana, and the Bisti sand and Two Wells sand lentil of New Mexico.

Petrographically the bioturbated sand facies is regionally variable, ranging from glauconitic protoquartzite to feldspathic sand and sub-greywacke. The Gallup sand sheet in Arizona, Colorado, and New Mexico contains local heavy mineral placer deposits rich in ilmenite. Sorting is poor to moderate with considerable quantities of interstitial clay, though this is often diagenetic in origin. Texture, fauna, and sedimentary structures show a regular vertical arrangement within any one sand sheet (Figure 7.4).

Figure 7.4 Regressive Point Lookout barrier sand-sheet passing down transitionally into open marine Mancos shale. From Visher, 1965, Figure 5. Reproduced from the Bulletin of the American Association of Petroleum Geologists, by courtesy of the American Association of Petroleum Geologists.

In the case of a regressive sand this is as follows. The upper contact of the sand with shales of the coal facies is abrupt, and occasionally channelled and infilled with an oyster-bearing siltstone. The upper part of the sand sequence shows the coarsest grainsize and best sorting of the whole sheet. These fine sands are rarely cross-bedded and typically show low angle (5–15°) laterally persistent stratification with, generally, easterly dips. As its name implies, this facies is extensively bioturbated, with a diverse trace fossil assemblage.

Ophiomorpha is particularly characteristic. This burrow is comparable to those produced by the crustacean *Callianassa* on modern tidal and sub-tidal parts of beaches. The fine flat-bedded sands grade down into very fine silty sands. These are laminated, colour mottled, and sometimes burrowed. They contain the ammonites *Baculites* and *Discoscaphites*, together with the lamellibranchs *Inoceramus* and *Pholadomya*. This second unit of the bioturbated sand facies grades down into laminated siltstone. Using detailed wireline log correlations it has been shown that the sands are arranged in basinward prograding wedges. Key marker horizons, such as volcanic ash bands, demonstrate depositional topographies of up to 700 m.

Laminated shale facies

The third facies of the Cretaceous mid-west sediments includes the Lewis, Mancos, Bearpaw, and Pierre Shale formations. These are best developed to the east where they locally become calcareous and grade into chalky limestones (e.g. the Niobrara Formation). Traced westwards this facies thins and splits up into a number of tongues which interfinger with the bioturbated sand facies, forming the toesets of the progradational wedges of the sheet sands. Contacts are gradational with a siltstone sequence separating the shales from the very fine silty sands. In some areas the sheet sands are locally absent, and the laminated shales directly overlie the coal-bearing facies.

Lithologically this facies consists of grey laminated claystones with a fauna of ammonites and lamellibranchs similar to that recorded from the lower part of the sheet-sand facies, together with shark teeth.

CRETACEOUS SHORELINES OF THE ROCKY MOUNTAINS, USA: INTERPRETATION

The conglomerates and coarse cross-bedded channeled sands of the extreme west were clearly deposited in a high-energy environment by fast traction currents. The finer sediments with which they are interbedded eastwards indicate lower energy conditions. The lamellibranchs in the shales suggest deposition in waters whose salinity ranged from fresh to brackish. The coal beds indicate intermittent swamp conditions. Considered over all, therefore, the coal bearing facies seems to have been deposited on a piedmont alluvial plain which passed eastwards down slope into a region of swamps and brackish lagoons.

The bioturbated sand facies suggests that the lagoons were restricted to the east by a higher energy environment. The ammonites and lamellibranchs of the sands indicate that they were laid down in or close to a marine environment. The vertical sequence of sedimentary strutures, the bioturbation and the upward-increasing grainsize profiles are typical of Recent barrier beaches as discussed earlier.

The evidence points, therefore, to a barrier beach environment for this facies. The channels at the top of each sand sequence were probably cut by tidal currents flowing between the open sea to the east and the brackish lagoons to the west. The Rocky Mountain Cretaceous shoreline would appear to have been formed as a coalesced complex of wave-dominated Nile-type delta.

Alternatively the regionally extensive cyclicity of the formations may suggest that deposition took place by alternating deltaic regressive phases and barrier island transgressive phases. The laminated shales of the eastern part of the region were clearly laid down in low-energy conditions. Their fauna indicates a marine environment. It seems most probable, therefore, that the laminated shale facies originated below wave base on a marine shelf to the east of the barrier sands.

The preceding observations and interpretations show that these Cretaceous sediments provide a good example of a linear clastic shoreline. All four major environments found in Recent linear clastic coasts are represented in both regressive and transgressive phases of the shoreline. The coal-bearing facies represents the alluvial and lagoonal environments, the bioturbated sands indicate beaches and barriers islands, and the marine shale provides evidence of the open sea environment.

It is clear from the repeated vertical interbedding and lateral interfingering of all the facies that deposition took place synchronously in all the environments. Several lines of evidence point to fluctuations in the rate of advance and retreat of the shoreline. The local superposition of marine shales directly on the coal-bearing facies shows that sometimes the sea transgressed too fast for barrier sands to form. This may have been caused either by shortages of land-derived sediment, or by extremely rapid rises of sea level or subsidence of the land.

The isolated sand shoestrings within the marine shales also suggest that sometimes the sea advanced so fast that barriers were no sooner formed than they were submerged to form offshore bars and shoals. The location of these was sometimes controlled by palaeohighs on the sea floor. By contrast, the thick clean sand benches in the sheet-sand facies suggest that from time to time the shoreline was static. High barriers were then thrown up by the sea on which the sand was continually reworked and from which the clay was winnowed.

In conclusion these Cretaceous sediments provide a good example of a linear clastic shoreline in which rapid deposition allowed the sediments of all four environments to be deposited during both regressions and transgressions of the coast.

The Rocky Mountain Cretaceous sands have been studied intensely in the past, because they are petroliferous, with the marine shales acting as source beds, and the marine, and to a lesser extent, the channel sands, serving as petroleum reservoirs. More recently attention has been drawn to these sediments because they may be used to study modern sequence stratigraphic concepts. Research has centred around the cause or causes of the cyclicity. As observed in Chapter 1, coastlines may advance or retreat in response to global, eustatic sea level change, in response to local tectonic uplift and subsidence, and in response to fluctuating sediment supply. This may be due to tectonic, climatic or ecological changes.

When viewed in the broad sense the cyclicity of the Rocky Mountain shorelines appears to be symmetric, with marine muds passing up via barrier sands into lagoonal and fluvial sediments, and back, via shoreface sands into marine shales again (Figure 7.3). Detailed analysis of the data, however, shows that it is not that simple. Sedimentological sections and well logs show that the shoreface sands are progradational and upward-coarsening, not only for the regressive part of the cycle, as one might expect, but also for the transgressive part of the cycle, which one might not at first expect. This is true, whether the sand is in short supply (Figure 7.5a), and preserved as a series of isolated shoestrings, or sufficiently abundant as to be preserved as a blanket (Figure 7.5b).

These profiles suggest that transgressions occur in an episodic manner. A barrier island is breached, the sea floods landward to establish a new shoreface. If the sea is rising fast then the new shoreface will itself soon be swamped, ending up as an isolated shoestring sand, while a new strandline is established further landwards (Figure 7.5a).

If the sea level is static, or rising so slowly that sedimentation exceeds it, then the shoreline will prograde for a sufficient length of time for an upward-coarsening blanket sand to form, before it too is dramatically transgressed and buried beneath deep marine muds (Figure 7.7b). In other words, to misquote George Orwell, 'All sedimentation is regressive, some is more regressive than others.' Transgressions are represented in the stratigraphic record by sequence boundaries. These are regionally planar surfaces, though locally channeled, extensively burrowed, and overlain by lag gravels. When

(a)

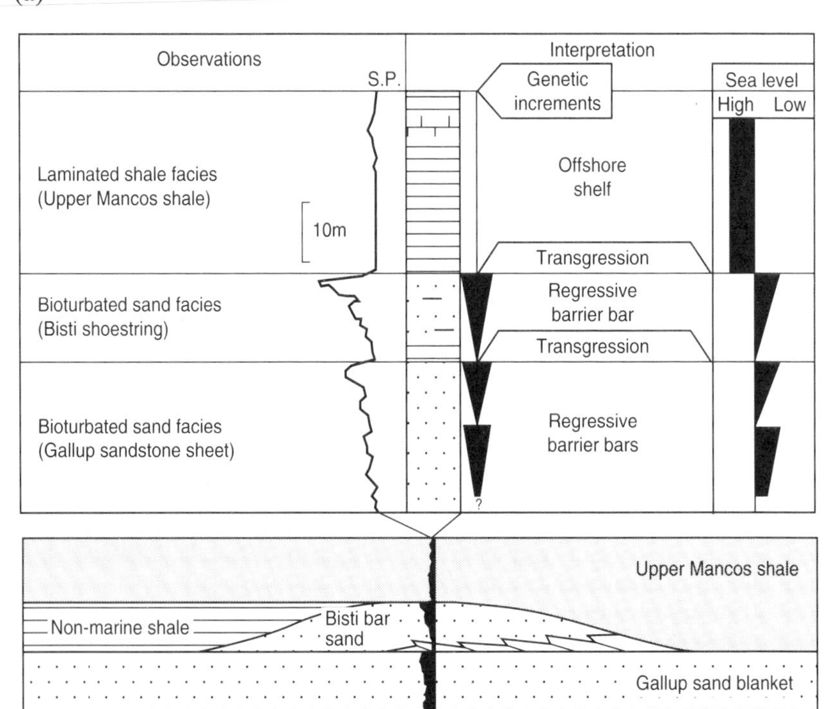

Figure 7.5 SP log motifs, showing grain size variations and sketch cross-sections: (a) the Bisti oil field, in an isolated shoestring barrier sand associated with an overall transgression of the Upper Mancos Shale over the Gallup Sand, New Mexico (based on data in Sabins, 1963 and 1972);

a sequence boundary is immediately overlain by deep water marine shales, inviting the appelation 'maximum flooding surface' in seismic sequence stratigraphic terminology.

The foregoing is the traditional view of the Rocky Mountain Cretaceous sediments. It is based firmly on the Concept of Walther's Law, outlined in Chapter 1. More recently it has been suggested that part of the sedimentary section is actually transgressive in origin. Based on a study of the Point Lookout Formation in Wyoming Devine (1991) argues that lagoonal and estuarine deposits formed behind barriers during times of rising sea level, and may therefore justifiably be termed 'transgressive'. Similarly Hart and Plint (1993), working on the coeval Cardium sands of Alberta, attribute the non-marine deposits to episodes of rising sea level, rather than to deposition behind a prograding barrier island sequence.

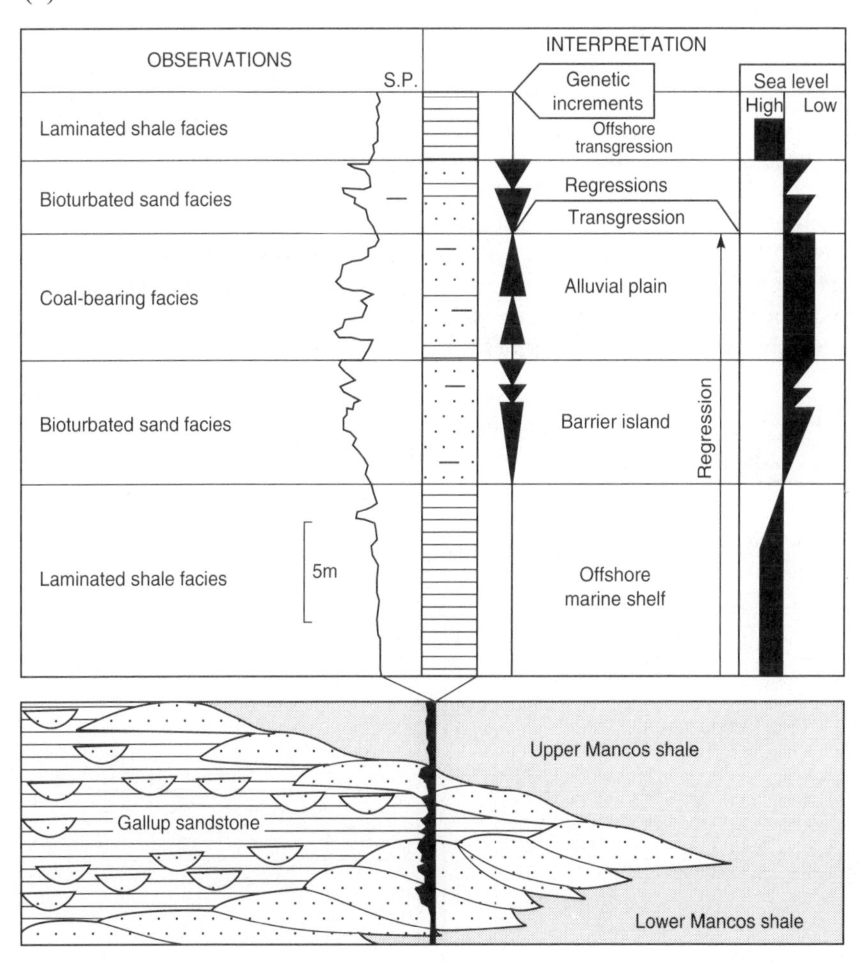

Figure 7.5 (b) cross-section through a superficially symmetric regressive: transgressive cycle in the Gallup Sandstone, New Mexico; examination of the SP log of a well drilled through this sequence reveals, however, that the transgressive sand is, in fact, made up of two upward-coarsening regressive increments (based on data in McCubbin, 1982).

GENERAL DISCUSSION OF LINEAR TERRIGENOUS SHORELINES

A regression of the sea from a low-lying land surface with no detritus may leave no mark in the geological record other than an unconformity. The actual shoreline may consist of a beach, or barrier beach and lagoon complex. As the sea retreats it will leave the old beach ridges stranded inland. These will be eroded and the sediment carried back to the sea to be re-deposited on the beach face. Net sedimentation above the unconformity is thus generally zero.

Transgressions are more complicated than regressions. Where the transgression is rapid, and sediment in short supply, deep marine shales may unconformably overly the old land surface (maximum flooding surface). When the advance of the sea is slower, and where there is sufficient input of sediment, then the four facies belts may be preserved in a mirror image of the regressive sequence. Detailed examination of a transgressive section, such as those of the Rocky Mountains just reviewed, generally shows that the reflection of the 'mirror' is imperfect in so far as the marine shoreface sands are concerned.

It is unusual for the barrier bar sands to grade up into the deeper water marine clays. It is more normal for them to show an upward-coarsening regressive sequence with a sharp contact with overlying marine shale. When this occurs it shows that the transgression is actually made up of a series of laterally stacked regressive increments. The sea apparently drowned a barrier beach and established a new shoreface further inland. The coast prograded seaward for a while depositing either an upward coarsening sequence or, on higher energy coasts, a sand which abruptly overlay a scoured pebble-covered sea floor. Renewed transgression drowned the second shoreface, and established a third one still further up the basin margin (Figure 7.6).

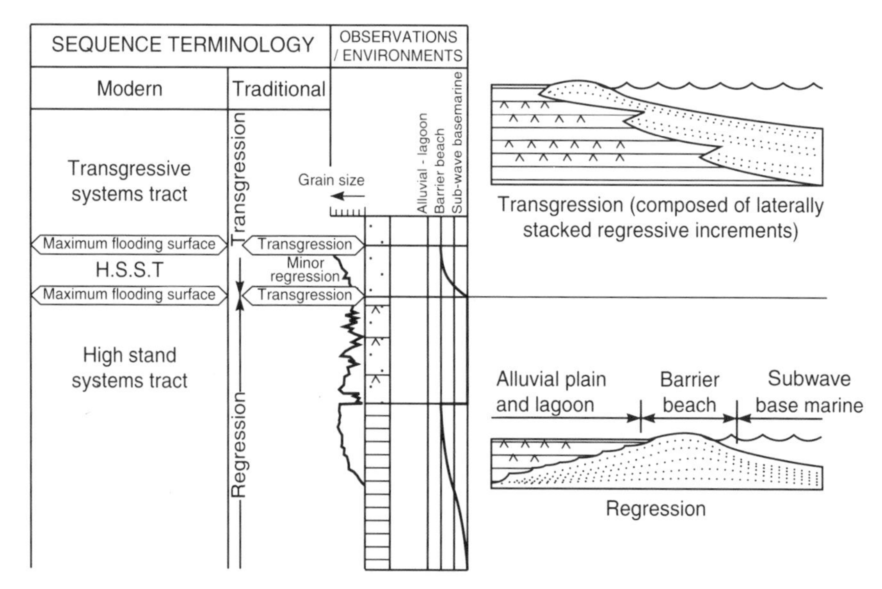

Figure 7.6 Sedimentary section and cartoons to show the differences between regressive and transgressive shoreline sequences using both traditional and modern sequence stratigraphic terminology. Note that the transgressive sands, though they may have a sheet geometry, are made up of a series of laterally stacked regressive increments. Each individual unit is an upward-coarsening prograding sand, though of limited duration and extent.

Where sediment supply is sufficient a blanket marine sand may thus be deposited. But detailed examination will show that this is made up of a series of laterally stacked sands with eroded tops and upward coarsening or sharp bases. When sediment supply is insufficient then a series of isolated shoestring sands may remain. Figure 7.7 attempts to illustrate the different types of sequence deposited in response to varying rates of sediment supply and of sea level change.

These processes are discussed in many of the papers previously referred to, but see especially McCubbin (1982).

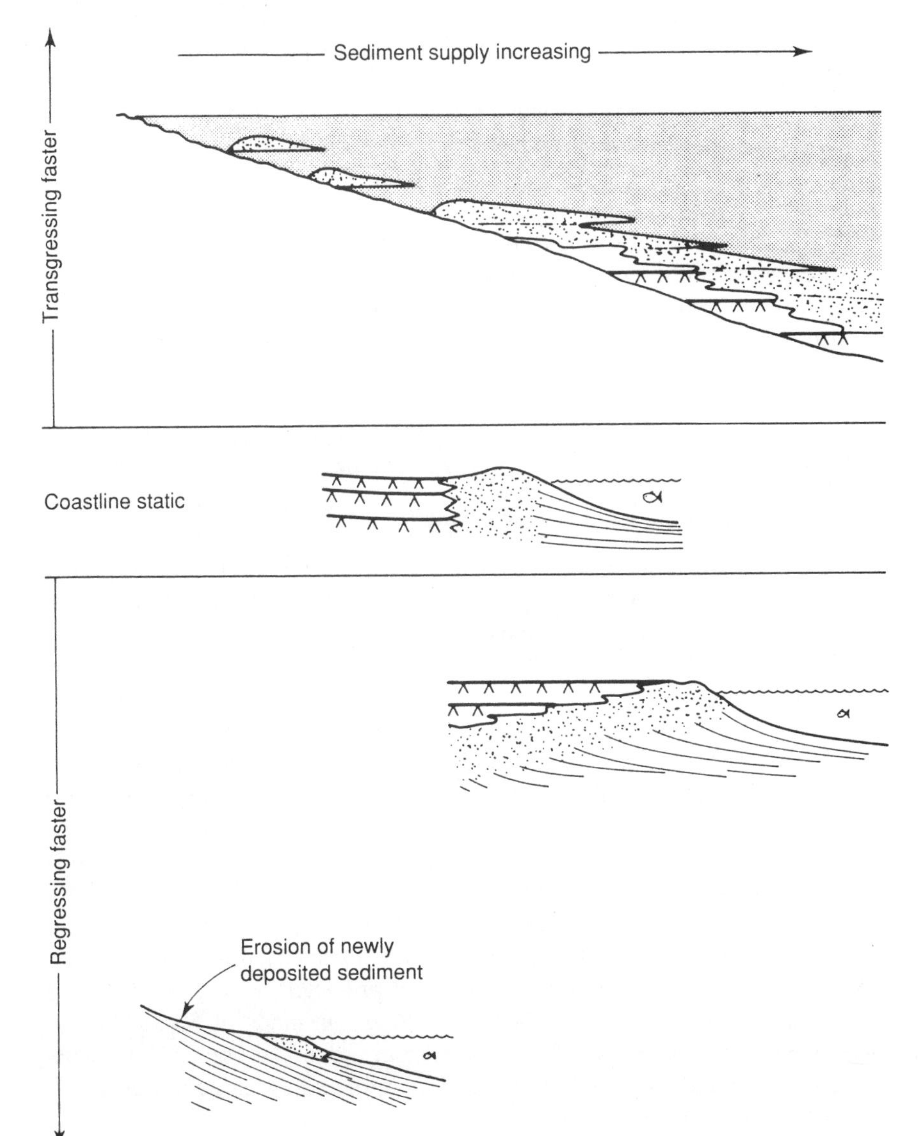

Figure 7.7 Cartoons to show how the geometry of linear shoreline facies are controlled by the rate of sediment input and the rate of rise or fall of sea level.

ECONOMIC SIGNIFICANCE OF LINEAR TERRIGENOUS SHORELINES

Beach and barrier sands are economically important as hosts to placer mineralization, and as petroleum reservoirs. These will now be considered in turn. The winnowing process on beaches can segregate and concentrate heavy minerals, notably gold, ilmenite, Zircon, cassiterite, rutile and garnet. The resultant placers concentrate along scoured sequence boundaries (unconformities) and in embayments that are open in the direction from which sediment is being transported by longshore drift. Detailed accounts of coastal placer deposits will be found in MacDonald (1983) and Force (1986).

Since linear shorelines deposit clean porous sands around marine basins they are often prolific oil and gas producers. The barrier sands obviously provide the best potential reservoirs with both the marine and the lagoonal shales being potential hydrocarbon source rocks. It is, therefore, very important to be able to recognize bar and beach sands and to make predictions of their geometry and trend. The Cretaceous sediments of the Rocky Mountain region contain good examples of linear shoreline sands. Some petroleum is trapped stratigraphically in channels that were incised and sand-infilled during regressions (low stands). But much of the reserves are found in the marine barrier sands.

These sands may be relatively easy to locate. It is not so easy though, to find parts with good reservoir properties. Where the sands are rich in volcanic detritus they often contain a diagenetic clay cement, which has significantly reduced the original porosity and permeability. Optimum reservoir characteristics are found where there is secondary fracture porosity, and where the sands are particularly well-sorted and clay-free due to extensive wave winnowing at the time of deposition.

Well-winnowed sands occur in two situations. They occur at the top of each sand sequence, where it is easy to find, and in the thick sand benches. Since the latter occur in narrow belts often only 3 or 4 kilometres wide, they are not always easy to locate, nor, once found, is their regional trend simple to predict. The Cretaceous shoreline did not extend in a straight line north to south from Canada to the Gulf of Mexico. Like Recent coasts, it had bays, capes, and spits, so that locally the thick barrier sand benches have trends varying from northwest to northeast.

Fortunately these strata are gently folded, and due to erosion the barrier sands crop out intermittently at the surface. In such situations the thick porous sands can be located and their trend predicted underground. Evans (1970) has shown how the Cretaceous Viking bar sands of Saskatchewan were deposited as a series of linear imbricately arranged bodies. The sands strike approximately parallel to the basin margin and the direction of imbrication is towards the basin centre.

The actual style of petroleum entrapment varies according to the extent of the sand. Obviously the blanket sands will contain oil in structural traps, whereas the shoestring sands will contain oil that is stratigraphically trapped. The Bisti field of the San Juan basin of New Mexico is a classic example of the latter (Figure 7.5a). In the Rocky Mountain basins it has been noted that sands deposited during regressive phases of the shoreline tend to make more effective traps, than do those associated with transgressions (Figure 7.8).

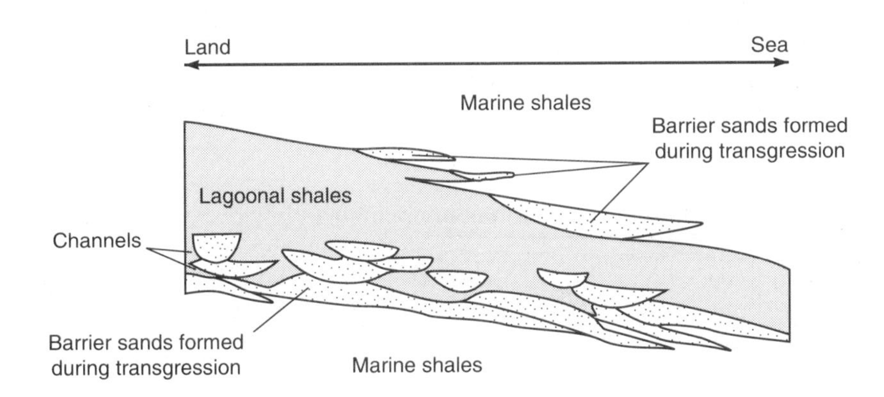

Figure 7.8 Sketch through a regressive:transgressive coastal cycle to show how regressive sands may lack effective updip seals, because they may be in fluid communication with sandfilled channels. By contrast barrier sands associated with transgressions may make more secure traps, because they are sealed up dip by lagoonal shales. Based on Mackenzie (1972).

It is argued that this is because the beach sands of the regressions pass up dip into sand-filled channels, and thus lack effective up dip seals to petroleum migration. The barrier sands associated with transgressions, by contrast, pass up dip into impermeable lagoonal and tidal flat shales (Mackenzie, 1972).

SUB-SURFACE DIAGNOSIS OF BARRIER SANDS

Because of their potential as hydrocarbon reservoirs great attention has been paid to the subsurface diagnosis of barrier sands (Davies *et al.*, 1971). Geometrically these deposits may have either shoestring or sheet geometries. Considered in detail the sheets are composed of a series of laterally stacked discrete beach increments, as shown by the Viking Cretaceous Sands of Canada already mentioned (Evans, 1970).

It has already been shown that prograding barrier bar sands have an upward-coarsening grainsize profile. They normally pass down gradationally into marine shales, and commonly have an abrupt top. Thus the upper surface of a barrier bar sand sometimes has sufficient acoustic impedance to generate a seismic reflector. This is of course the reverse situation to upward-fining channels where a reflector is to be anticipated at the base (compare with Figure 3.18).

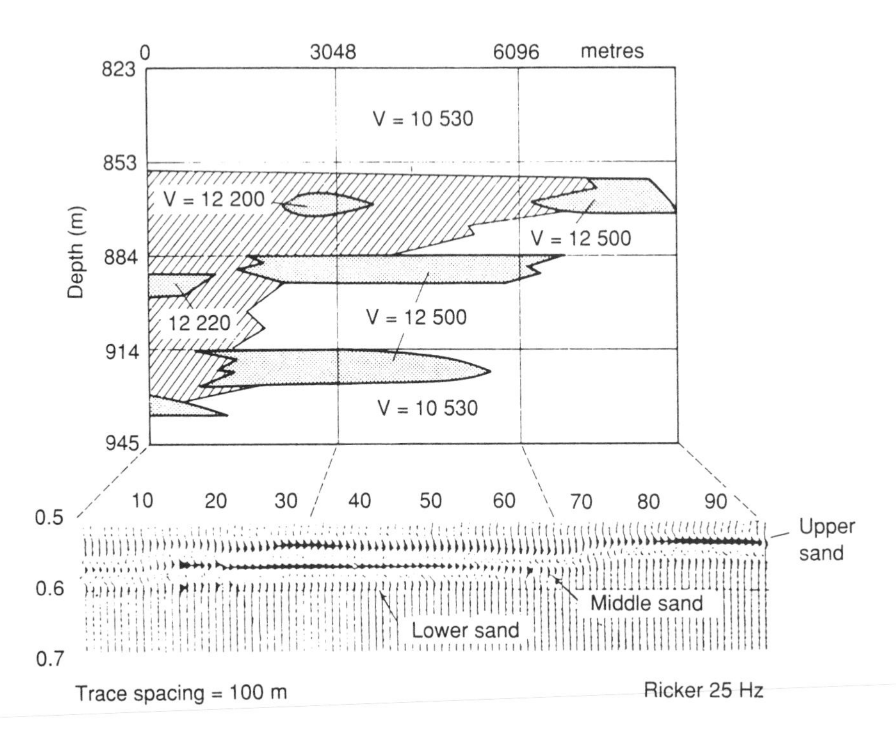

Figure 7.9 Geological model (upper) and seismic response (lower) for Cretaceous Belly River bar sands, Alberta. (From Meckel and Nath, 1977, by courtesy of the American Association of Petroleum Geologists.)

Figure 7.9 shows the seismic signature produced from modelling studies of Cretaceous Belly River bar sands from Alberta. These are a northward continuation of the Rocky Mountain Cretaceous sands of the USA discussed earlier in the chapter. It is sometimes possible to use seismic data to map individual sandbars and even the corrugated upper surface of barrier sand sheets. Thus environmental diagnosis may be made prior to drilling. More importantly reservoir geometry may be mapped without even considering depositional environments.

Lithologically marine sands can often be recognized in well samples by their textural and mineralogical maturity, by the presence of glauconite and shell debris, and by the absence of mica and carbonaceous detritus. Cores through barrier sand bars may show the suite of sedimentary structures associated with an upward-coarsening grain-size profile shown in Figure 7.1. Where cores are not available the grainsize motif may nonetheless be clearly shown by gamma or SP logs, and the structures, especially bioturbation, by borehole imaging tools.

The typical dip pattern for regressive barrier sands is an upward-increasing 'blue' motif. When plotted on a rose diagram a bimodal bipolar pattern may be seen. The dips of the blue motif represent low angle seaward dipping beds, while the opposed mode with higher dip is due to onshore directed foresets (Figure 7.10).

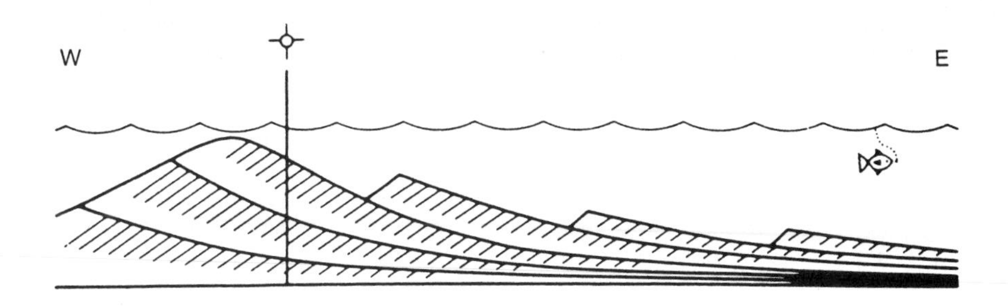

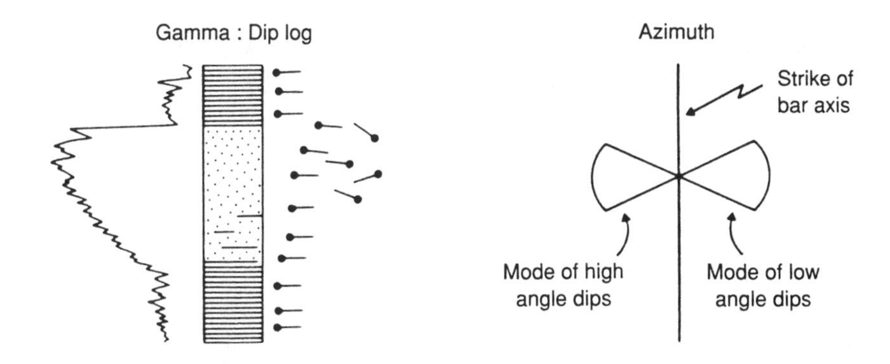

Figure 7.10 Gamma log and dipmeter motifs for barrier bar sand bodies.

Even if this bipolar pattern is indecipherable the direction of dip of the blue curve can be used to find the strike of the bar, and hence the direction in which the axis lies. This is extremely useful information to know, both in the quest for bar sand stratigraphic traps, and for regional palaeogeographic studies.

REFERENCES

The account of the Cretaceous shorelines of the Rocky Mountains was based on:

Asquith, D. O. (1970) Depositional topography and major marine environments, late Cretaceous, Wyoming, *Bull. Amer. Assoc. Petrol. Geol.*, **54**, 1184–1224.

Davies, D. K., Ethridge, F. G. and R. R. Berg (1971) Recognition of barrier environments, *Bull. Amer. Assoc. Petrol. Geol.*, **55**, 550–65.

De Graaff, F. R. Van (1972) Fluvial-deltaic facies of the Castlegate sandstone (Cretaceous), east-central Utah, *J. Sediment. Petrol.*, **42**, 558–71.

Devine, P. E. (1991) Transgressive origin of channeled estuarine deposits in the Point Lookout Sandstone, Northwestern New Mexico: A model for Upper Cretaceous, cyclic regressive parasequences of the U.S. Western interior, *Amer. Assoc. Petrol. Geol. Bull.*, **75**, 1039–63.

Evans, W. E. (1970) Imbricate linear sandstone bodies of Viking formation in Dodsland Hoosier area of southwestern Saskatchewan, Canada, *Bull. Amer. Assoc. Petrol. Geol.*, **54**, 469–86.
Hart, B. S. and Plint, A. G. (1993) Tectonic influence on Deposition and Erosion in a Ramp Setting: Upper Cretaceous Cardium Formation, Alberta Foreland Basin, *Amer. Assoc. Petrol. Geol. Bull.*, **77**, 2092–2107.
Hollenshead, C. T. and Pritchard, R. L. (1969) Geometry of producing Mesaverde Sandstones, San Juan basin. In J. A. Peterson and J. C. Osmond, *The Geometry of Sandstone Bodies*, Amer. Assoc. Petrol. Geol., 98–118.
Mackenzie, D. B. (1972) Primary stratigraphic traps in sandstone. In R. E. King (ed.), *Stratigraphic Oil and Gas Fields*, Amer. Assoc. Petrol. Geol., **16**, 47–63.
—— (1975) Tidal flat deposits in Lower Cretaceous Dakota Group near Denver, Colorado. In R. N. Ginsburg (ed.), *Tidal Deposits*, Springer-Verlag, New York, 117–25.
McCubbin, D. G. (1982) Barrier Island and Strand Plain Facies. In P. A. Scholle and D. Spearing (eds), *Sandstone Depositional Environments*, Amer. Assoc. Petrol. Geol., Tulsa, 247–80.
Martinsen, O. J., Martinsen, R. I. and Steidmann, J. A. (1993) Mesaverde group (Upper Cretaceous), southeastern Wyoming: allostratigraphy versus sequence stratigraphy in a tectonically active area, *Amer. Assoc. Petrol. Geol. Bull.*, **77**, 1351–73.
Miller, D. N., Barlow, J. A. and Haun, J. D. (1965) Stratigraphy and petroleum potential of latest Cretaceous rocks, Big Horn basin, Wyoming, *Bull. Amer. Assoc. Petrol. Geol.*, **49**, 227–85.
Sabins, F. F. (1963) Anatomy of a stratigraphic trap, *Amer. Assoc. Petrol. Bull.*, **47**, 193–228.
—— (1972) Comparison of Bisti and Horseshoe Canyon stratigraphic traps, San Juan Basin, New Mexico. In R. E. King (ed.), *Stratigraphic Oil and Gas Fields*, Amer. Assoc. Petrol. Geol., **16**, 610–22.
Weimer, R. J. (1961) Spatial dimensions of Upper Cretaceous Sandstones, Rocky Mountain Area. In J. A. Peterson and J. C. Osmond (eds), *The Geometry of Sandstone Bodies*, Amer. Assoc. Petrol. Geol., 82–97.
—— (1966) Patrick Draw Field, Wyoming, *Bull. Amer. Assoc. Petrol. Geol.*, **50**, 2150–75.
—— and Haun, J. D. (1960) Cretaceous stratigraphy, Rocky Mountain Region, USA, *Int. Geol. Congr. Norden*, pt 12, 178–84.
—— (1988) Record of relative sea level changes, Cretaceous of Western interior, USA, *Soc. Econ. Pal. Min. Sp. Pub.*, **42**, 266–88.
—— (1992) Developments in sequence stratigraphy: foreland and cratonic basins, *Amer. Assoc. Petrol. Geol. Bull.*, **76**, 965–82.
—— and Land, C. B. (1975) Maestrichtian deltaic and interdeltaic sedimentation in the Rocky Mountain region of the United States. In W. G. E. Caldwell (ed.), *The Cretaceous System in the Western Interior of North America*, Geol. Assn. Canada, Sp. Pub., **13**, 633–66.

Other references cited in this chapter:

Armstrong-Price, W. (1963) Patterns of flow and channelling in tidal inlets, *J. Sediment. Petrol*, **33**, 279–90.
Barwis, J. H and Hayes, M. O. (1979) Regional patterns of modern barrier island and tidal inlet deposits as applied to paleoennvironmental studies. In J. C. Ferm and J. C. Horne (eds), *Carboniferous Depositional Environments of the Appalachian Region*, University of South Carolina, 472–98.

Boyd, D. R. and Dyer, B. F. (1966) Frio Barrier bar system of South Texas, *Bull. Amer. Assoc. Petrol. Geol.*, **50**, 170–8.

Davies, R. A. (ed.) (1978) *Coastal Sedimentary Environments*, Springer-Verlag, Berlin.

Davis, R. A. and Ethington, R. L. (eds) (1976) *Beach and Nearshore Sedimentation*, Soc. Econ. Paleont. Miner, Sp. Pub., **24**, Tulsa.

Evans, G. (1965) Intertidal flat sediments and their environments of deposition in the Wash, *Quart Jl. Geol. Soc. Lond.*, **121**, 209–45.

Force, E. R. (1986) Descriptive model of shoreline placer Ti. In D. P. Cox and D. A. Singer (eds), *Mineral Deposit Models*, USGS Bull., 1693.

Ginsberg, R. N. (ed.) (1975) *Tidal Deposits*, Springer-Verlag, New York.

Gregory, J. L. (1966) A Lower Oligocene delta in the subsurface of south-eastern Texas. In M. L. Shirley and J. A. Ragsdale (eds), *Deltas*, Houston Geol. Soc., 231–8.

Hoyt, J. H. (1968) Genesis of sedimentary deposits along coasts of submergence, *Report 23rd Internat. Geol. Cong. Prague.*

—— and Henry, V. J. (1967) Influence of island migration on barrier island sedimentation, *Bull. Geol. Soc. Amer.*, **78**, 77–86.

Klein, G. de V. (1967) Paleocurrent analysis in relation to modern sediment dispersal patterns, *Bull. Amer. Assoc. Petrol. Geol.*, **51**, 366–82.

Macdonald, E. H. (1983) *Alluvial Mining*, Chapman & Hall, London.

Meckel, L. D. and Nath, A. K. (1978) Geological considerations for stratigraphic modelling and interpretation. In C. E. Payton (ed.), *Seismic Stratigraphy - Applications to Hydrocarbon Exploration*, Amer. Assoc. Petrol. Geol., **26**, 417–38.

Rainwater, E. H. (1966) The geologic importance of deltas. In M. L. Shirley and J. A. Ragsdale (eds), *Deltas*, Houston Geol. Soc., 1–16.

Swift, D. J. R. and Palmer, H. D. (eds) (1978) *Coastal Sedimentation*, Dowden, Hutchinson & Ross, Stroudsburg.

Van Straaten, L. M. J. U. (1959) Minor structures of some recent littoral and neritic sediments, *Geol. Mijnb.*, **21**, 197–216.

8 Mixed terrigenous: carbonate shorelines

INTRODUCTION

Mixed terrigenous:carbonate shorelines are defined as those in which carbonate deposition occurs so close to the land that it contributes, not just to the open sea sediments, but also to the shoreline deposits themselves (see p. 132). These conditions can be brought about by three factors acting singly or in concert. Low input of terrigenous sediment to the shoreline may be due to low runoff or, if the hinterland is low-lying, low sediment availability. Third, if the shoreline itself has a very gentle seaward gradient it will have an extremely broad tidal zone and an extremely wide development of the facies belts paralleling the shore. In such instances, terrigenous sediment may be dumped around river mouths in estuarine tidal flats. There may be insufficient current action to re-work these deposits and carry sand out to the barrier zone. In this case, onshore currents may pile up bars of carbonate sand far to seaward of the river mouths.

Because of the tendency of sedimentologists to specialize in either siliciclastics or carbonates mixed deposits have tended to be ignored (Doyle and Roberts,1988). Modern mixed coastlines are well-described, however, from the Great Barrier Reef of Australia, Indonesia, the Gulf of Mexico and the adjacent Caribbean (Roberts, 1987). Ancient examples are also well known, particularly in Miocene sediments of the Mediterranean littoral (e.g. Martini *et al.*, 1992). A particularly good example of a Miocene mixed terrigenous: carbonate shoreline occurs in the Sirte basin, Libya. All three of the factors outlined above seem to have operated in this instance. The Libyan Miocene case history will now be described. It is well documented, and easily accessible, and has been used for the field study of a terrigenous:carbonate shoreline by students and oil company staff for many years.

A LIBYAN MIOCENE SHORELINE

The Sirte basin developed in the Late Cretaceous on the North African margin of the Tethys ocean. It is essentially a failed rift basin, initiated by the break-up of Africa and what is now South America. By the Miocene Period the embayment had largely been infilled by carbonates and shales. Through Late Tertiary time a gradual regression of the sea was interrupted by minor marine transgressions. One of these in Early Miocene time (Aquitanian-Burdigalian) deposited the shoreline sediments now to be described. These have been studied in the region of Marada Oasis and the Jebel Zelten (Figures 8.1 and 8.2).

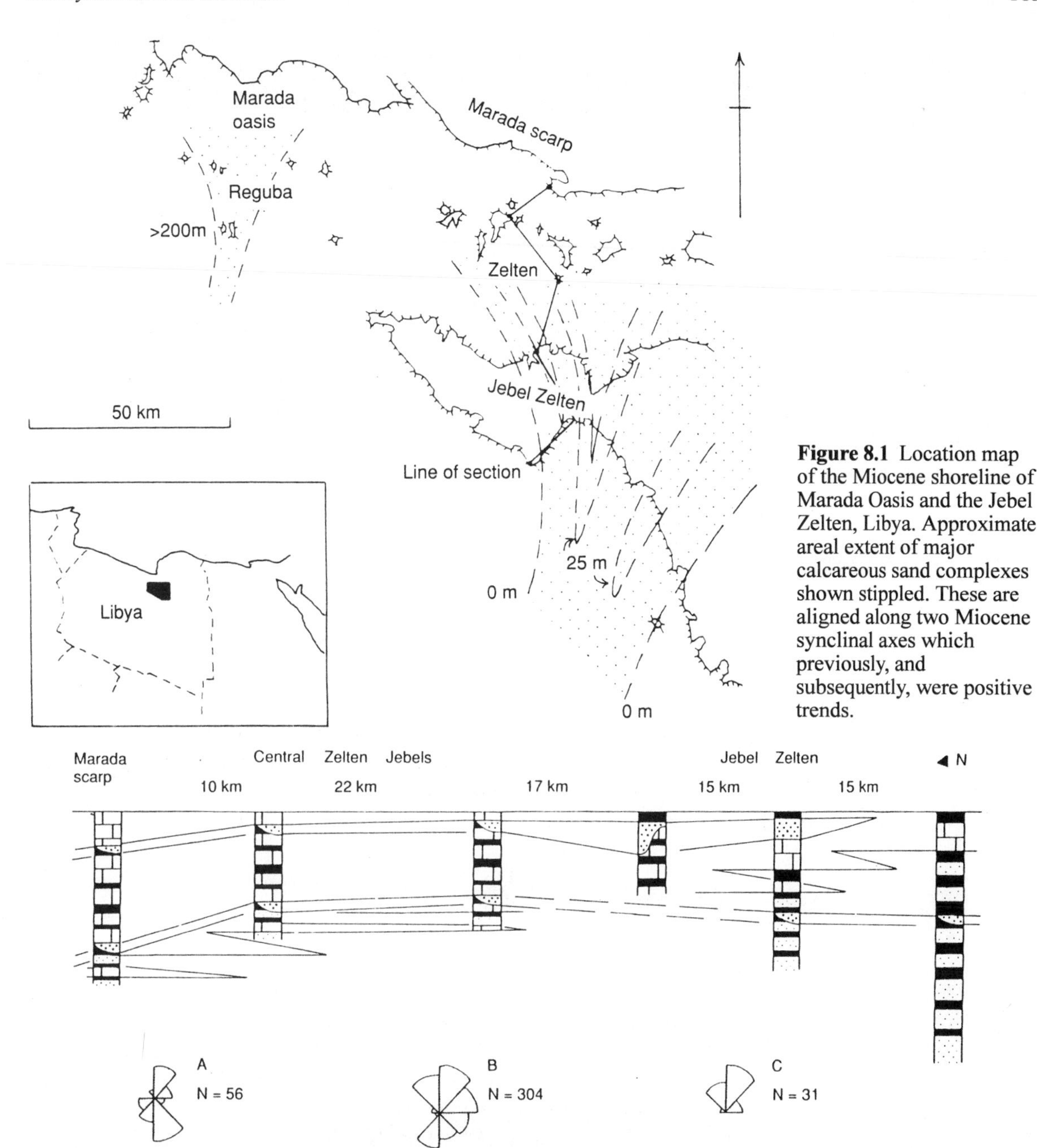

Figure 8.1 Location map of the Miocene shoreline of Marada Oasis and the Jebel Zelten, Libya. Approximate areal extent of major calcareous sand complexes shown stippled. These are aligned along two Miocene synclinal axes which previously, and subsequently, were positive trends.

Figure 8.2 Diagrammatic measured sections indicating facies changes across the Miocene shoreline between the Marada scarp and the Jebel Zelten. Datum is post-Marada erosion surface. Right hand section 180 m thick. Key: open marine and barrier limestone: blocks; lagoonal and intertidal shales and sands: black; fluviatile sands: sparse stipple; estuarine calcareous sandstones: dense stipple. Grain size and frequency of cross-bedding in limestones increases over the crest of the anticline in the central Zelten jebels. Azimuth diagrams show crossbedding dip orientations North vertical. Left: limestones show bipolar pattern possibly due to tidal currents; southerly directed major mode indicates dominance of onshore transport. Centre: calcareous sandstones show bipolar pattern with major mode directed offshore, probably due to deposition by tidal currents in estuarine channels. Right: unimodal seawards orientation of cross-bedding shown by fluviatile sands.

In the western part of the region, an unconformity between Miocene and Oligocene sediments crops out at the surface. This seems to have been subjected to intensive sub-aerial weathering. The Oligocene limestones immediately beneath the unconformity are extensively oxidized, replaced by gypsum, and penetrated by the roots of trees which grew on the pre-Miocene land surface. The succeeding Miocene shoreline deposits, some 200 m thick, are overlain by a later Miocene formation. The contact with this is unconformable in the Jebel Zelten, but becomes transitional when traced northwards towards the centre of the basin.

This contact has a present-day northerly dip of about 1:1000. Used as a datum, this surface shows that the underlying Miocene sediments were previously folded into a gentle east–west trending anticline with a culmination over the Zelten oil field. The Miocene shoreline deposits are composed of a diversity of repeatedly interbedded facies which may be listed as follows:

Facies	
1. Skeletal limestone facies	Predominate in north (seawards)
2. Laminated shale facies	
3. Interlaminated sand and shale facies	
4. Cross-bedded sand and shale facies	Predominate in south (landwards)
5. Calcareous sandstone channel facies	North-south radiating shoestring complexes

Each facies will now be described and interpreted in turn.

SKELETAL LIMESTONE FACIES

Description

The skeletal limestones compose nearly the complete 200 m section exposed in the north. Traced southwards they die out due to interfingering with shales and sands. Petrographically these are poorly sorted medium- and coarse-grained packstones, composed of a framework of bioclastic debris and pellets with interstitial micrite and sparite. They have a high intergranular porosity.

Finer calcarenites, calcilutites, and calcirudites are also present. There are considerable regional variations in the grainsize of this facies, coarser calcarenites being concentrated over the crest of the Zelten anticline, while finer carbonate sands and muds are present to north and south. The coarser-grained limestones are flat-bedded or cross-bedded in isolated sets about a metre thick. In plan view it can be seen that both tabular planar and trough foresets are present, often on a vast scale. Individual planar foresets can be traced along strike for over 100 m. Troughs are up to 50 m wide (Figure 8.3).

Figure 8.3 Wind-eroded surface showing large-scale cross-stratification in offshore bar detrital limestone facies. From Selley, 1968, Plate 21(a), by courtesy of the Geological Society of London.

The orientation of these structures indicates deposition from alternately onshore (south) and offshore (north) flowing currents, though with a dominance of onshore transport. The finer limestones are generally massive, and often bioturbated. At many points limestone beds are cross-cut by large channels, several metres deep. These are infilled by calcarenites similar to those in which they are cut. The channel margins and floors are often lined with re-worked, bored limestone cobbles and boulders up to a metre in diameter. The limestones are largely composed of a diverse biota in all stages of preservation from entire and obviously *in situ*, to highly comminuted. This includes calcarous algae, bryozoa, corals, lamellibranchs (such as oysters and scallops), gastropods, echinoids, foraminifera (including miliolids and peneroplids), and a diverse suite of trace fossils (including *Ophiomorpha*).

Interpretation

Detailed comparison of the fossils with Recent forms indicate that they grew in a variety of habitats ranging in depth from 0–50 m, from fully marine to slightly brackish salinities and from low- to high-energy bottom conditions. The coarse bioclastic cross-bedded limestones were probably deposited in turbulent conditions. Comparison with Recent carbonate sands suggests that they originated from migrating offshore bars and shoals. Recent examples of these deposit foresets on their steep slopes and sub-horizontal bedding on their gentle backslopes (see McKee and Sterrett, 1961).

The troughs and channels cross-cutting the shoals may be analogous to those which tidal currents cut through Recent carbonate barriers (e.g. Jindrich, 1969). This interpretation is supported by their bipolar palaeocurrents (Figure 8.2). Steep channel margins and intraformational limestone conglomerates testify to penecontemporaneous diagenesis comparable to the formation of 'beachrock' by the sub-aerial exposure of Recent carbonate beaches. Considered over all, therefore, it seems most probable that the coarser limestones were deposited by shoreward migrating offshore bars which, intermittently exposed above sea level, became barrier islands. The finer-grained, more generally massive and burrowed calcarenites point to deposition in less turbulent conditions largely below wave base, and perhaps in depths as great as that suggested by the fossils (Figure 8.4).

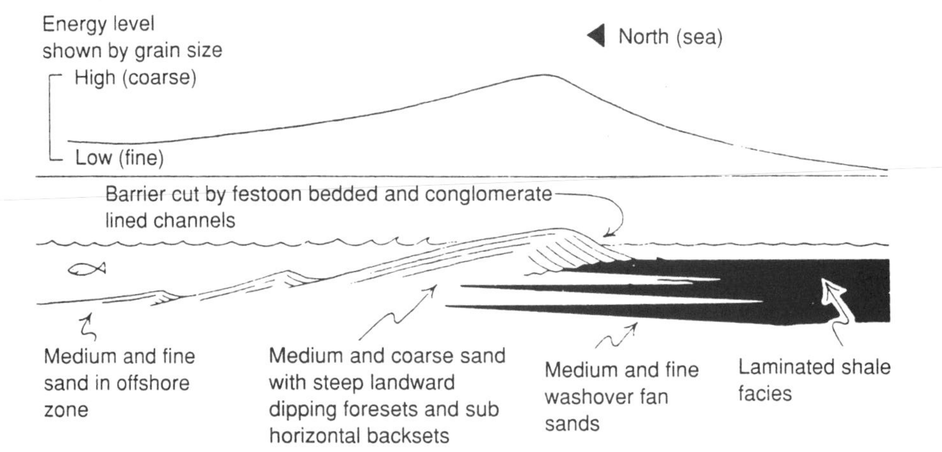

Figure 8.4 Illustration of the presumed origin of offshore bars in the skeletal limestone facies. From Selley, 1968, Figure 6, by courtesy of the Geological Society of London.

LAMINATED SHALE FACIES

Description

Beds of shale 1 or 2 m thick occur across the whole region, interbedded with limestones in the north and sands in the south. They are grey or green in colour, often highly calcareous, and contain thin white calcilutite bands. Internally, the shales are laminated throughout and occasionally rippled and burrowed. Within the shales, and separating them from limestones beneath, it is common to find oyster reefs. These are often up to a metre thick and composed of the *in situ* shells of long thin oysters, oriented vertically with their umbos pointed downwards (Figure 8.5). Apart from the oysters, this facies contains rare bryozoa and calcareous algae and plant debris.

Figure 8.5 Oyster shell bank with valves in vertical growth position. From Selley, 1968, Plate 21(b), by courtesy of the Geological Society of London.

Interpretation

The lamination and fine grainsize of this facies indicate deposition of clay out of suspension in a low-energy environment. The fossils suggest a range of salinity from normal marine to brackish. These conditions could be fulfilled either in relatively deep water below wave base, or in shallow water sheltered from the open sea. The brackish element of the fauna and the way in which the shales interfinger seaward with carbonate shoals suggest that the latter alternative is the correct one. The laminated shale facies may, therefore, be attributed to a lagoonal environment.

INTERLAMINATED SAND AND SHALE FACIES

Description

The third facies of this Miocene shoreline occurs in lenticular units two or three metres thick, interbedded with all the other facies. It is erratically distributed through the region, being most common in the Jebel Zelten and lower parts of the section exposed to the north. It consists of sands and interlaminated sand and shales, with rare thin lignites and rootlet beds. The sands are fine-grained, well-sorted, and argillaceous. Internally they are massive, laminated, or micro cross-laminated. Associated with the sands are delicately interlaminated very fine sands, silts, and clays. These are typically rippled and highly burrowed with vertical, often U-shaped, sand-filled tubes attributable to the ichnogenus *Diplocraterion* (Figure 8.6).

Figure 8.6 Interlaminated very fine sand and shale with *Diplocraterion* burrows. From Selley, 1968, Plate 22(b), by courtesy of the Geological Society of London.

These sediments are cross-cut by curvaceous channels up to 20 m deep and 80 m wide. The channels are infilled by sediments showing a regular sequence of sedimentary structures and a vertical decrease of grainsize (Figure 8.7).

The scoured channel floor is overlain by a conglomerate which grades up into several metres of cross-bedded calcareous sand. This is overlain by interlaminated rippled and burrowed very fine sand, silt, and clay beds which dip obliquely off the channel walls and are succeeded transitionally by horizontally laminated shale (Figure 8.8).

Interpretation

The over-all fine grainsize of this facies indicates deposition in a relatively low energy environment. Ripples of sand and interlamination of sand, silt, and clay point to sedimentation from gentle currents of pulsating velocity which alternatively caused sand ripples to migrate and then halted to allow clay to settle out of suspension. The resultant bedding type and the associated intensive burrowing are closely comparable to Recent tidal flat deposits such as those described from the North Sea coasts (Van Straaten, 1954; Evans, 1965). Likewise, the morphology and sediments of the channels are

Figure 8.7 Fining upward tidal-creek sequences. (1) basal erosion surface; (2) channel-lag conglomerate; (3) cross-bedded coarse-medium sandstone of channel bar; (4) cross-laminated fine sand of point bar; (5) laminated siltstone channel fill. From Selley, 1968, Plate 23, by courtesy of the Geological Society of London.

analogous to those formed by the tidal gullies which drain Recent mud flats (e.g. Van Straaten, 1954, Figure 4). Further evidence of the shallow environment of this facies is provided by the lignites and rootlet horizons. These may have originated in salt marshes on the landward side of the tidal flats. Rare desiccation cracks also testify to the shallow origin of this facies.

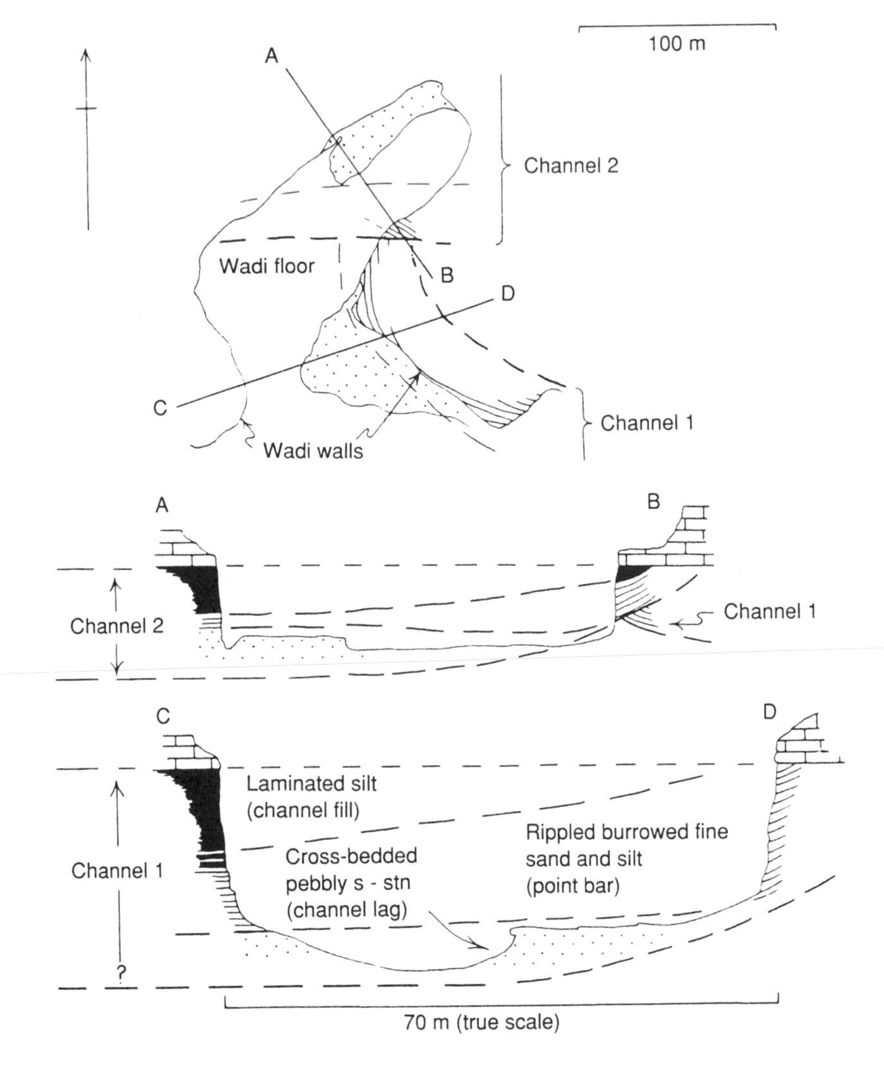

Figure 8.8 Plan and sections from dissected wadi system, showing upward-fining, but largely shale infilled channels, interpreted as tidal gullies draining tidal flats. From Selley, 1968, Figure 10, by courtesy of the Geological Society of London.

CROSS-BEDDED SAND AND SHALE FACIES

Description

This is the most southerly of the four facies belts which are aligned subparallel to the Sirte basin shore. It is best developed in the south scarp of the Jebel Zelten and also occurs, interbedded with the previously described facies, at the base of jebels exposed northwards as far as the Marada escarpment. This facies is divisible into three interbedded sub-facies. Two-thirds of it is composed of poorly sorted pale yellow unconsolidated sands. These range in grainsize from coarse to fine. The coarser sands occur in sequences 5–6 m thick with erosional channelled bases, often veneered by thin quartz pebble conglomerates. Internally they are both tabular planar and trough cross-bedded with set heights of 30–40 cm arranged in vertically grouped cosets. The orientations of foresets and trough axes indicate deposition from unidirectional northerly flowing currents (Figure 8.2). The finer sands are argillaceous, massive, and, sometimes, flat-bedded.

The second sub-facies, interbedded with the first, consists of laminated shales which occur both as sheets and infilling abandoned channels. They can be distinguished from the laminated shale facies to the north since they are not calcareous, and they swell and fall apart in water. This phenomenon suggests that they are largely composed of montmorillonite.

The third sub-facies consists of thin sequences, only a few centimetres thick, of lignite, sphaerosideritic limestone, and ferruginized sand pierced throughout by rootlets. The cross-bedded sand and shale facies contains an abundant, diverse, and well-preserved vertebrate fauna. This includes bones of terrestrial mammals such as ancestral elephants, camels, giraffes, antelopes, and carnivores, together with aquatic forms such as crocodiles, turtles and fish. These bones occur within the sands together with transported tree trunks. The interbedded shales contain plant debris and the continental gastropod *Hydrobia*.

Interpretation

This facies differs from the three previously described in the dominance of terrigenous sediment and continental fossils. These facts, together with the sedimentary structures and northerly directed palaeocurrents, suggest deposition in a fluviatile environment (see Chapter 3 for the diagnostic features of alluvium).

Accordingly, the sand sub-facies can be attributed to deposition within river channels, while the shales and fine sands probably originated on levees and floodplains. The channel-fill shales are abandoned ox-bow lake deposits. The thin limestones, ferruginous layers, lignites, and rootlet beds represent old soil horizons which perhaps formed on the levees and low-lying flood basins between channels. The large amounts of shale imply that the alluvium was due to sinuous meandering rivers rather than braided ones. The vertebrate fauna suggests a savannah climate. In conclusion, it can be seen that the cross-bedded sand and shale facies originated in a low-lying alluvial coastal plain which, seawards, merged imperceptibly into the tidal flats to the north.

CALCAREOUS SANDSTONE CHANNEL FACIES

Description

The four facies belts paralleling the Sirte basin just described are locally cross-cut by northerly trending calcareous sandstone channels of the fifth and last facies. These can arbitrarily be sub-divided into two types: small isolated channels of sandy limestone, and large radiating channel complexes of calcareous sandstone. Channels of the first type are concentrated near the base of the southern scarp of the Jebel Zelten, where they are interbedded with the fluviatile deposits. These also occur at the base of jebels around the Nasser (formerly Zelten) oil field and can be traced at the same level as far north as the Marada scarp. Channels of this type are about 10 m deep and 300 m wide. In some areas they are well-exposed where they have been exhumed from the softer sands and shales with which they are interbedded (Figure 8.9).

Figure 8.9 Meandering estuarine channel of resistant sandy limestone, exhumed from soft fluvial sands and shales, now crops out as a long sinuous flat-topped jebel. From Selley, 1968, Plate 24(a), by courtesy of the Geological Society of London.

These channels are composed of poorly sorted medium and coarse-grained sandy limestones with fragments of marine shells mixed with quartz and micrite matrix. They are floored by intraformational bored limestone pebble conglomerates. The sands are cross-bedded in a wide variety of scales and types, and, less commonly, massive, flat-bedded, or rippled. Burrowing is present, especially in the finer sediment towards the top of each channel which is often intensively burrowed and sometimes shot through with plant rootlets. The major channels occur at two points and, due to the good exposure, can be isopached (Figure 8.1). At Reguba this facies is about 200 m thick. Near the crest of the south scarp of the Jebel Zelten is a sheet of calcareous sandstone, lenticular in an east-west direction, about 25 km wide and 30 m deep.

On the north scarp of the Jebel Zelten this has split up into a series of discrete channels which can be traced north to the upper part of the sections in jebels around the Nasser oil field. One or two sandy channels occur at the same level in the Marada scarp. Petrographically this facies consists of coarse and very coarse, sometimes pebbly, calcareous sandstones. These are arranged in a series of coalesced channels infilled with various kinds of large-scale trough and planar cross bedding. These sometimes show penecontemporaneous deformation attributable to quicksand movement. Apart from bioturbation, this facies contains no fossils *in situ*. There are fragments of marine shellfish, bones, teeth and wood. Palaeocurrents determined from cross-bedding are bipolar with the major mode pointing northwards (Figure 8.2). Around the Jebel Zelten palaeocurrents plotted at outcrop show a regionally radiating pattern.

Interpretation

Clearly this facies was deposited from fast-flowing currents confined to channels. The palaeocurrents and the mixture of marine carbonate sediment and shells with terrigenous sand, bones, and wood indicate to and fro current movement. It seems highly probable, therefore, that this facies originated in estuaries subject to strong tidal currents. The location of the two major channel complexes at Reguba and Jebel Zelten may not be a matter of chance. They both trend along north–south pre-Miocene palaeohighs which host several major oil fields. This suggests that the negative movement occurred along these two trends in Miocene time, favouring the development of estuaries where they cross-cut the shoreline.

GENERAL DISCUSSION OF THE MIOCENE SHORELINE OF THE SIRTE BASIN

This case history provides a good example of mixed carbonate:terrigenous shoreline. From north to south relatively deep water carbonates pass up slope into coarser cross-bedded shell sands, deposited on shoals and bars. These inter-finger southwards with fine-grained terrigenous muds and sands of lagoonal and tidal flat origin. In turn these pass landwards into an alluvial coastal plain facies. Locally, this shoreline was interrupted by two major estuaries which supplied coarse sand to the shoreline (Figure 8.10).

Currents were seldom strong enough to carry this detritus out to the bar zone. Subtle syn-sedimentary movement seem to have controlled the distribution of facies. The concentration of coarse grainsize and cross-bedding in the limestones over the east–west trending crest of the Zelten anticline suggests that it may have been a Miocene palaeohigh on which the barrier carbonates were deposited. Likewise the two major estuaries seem to have been

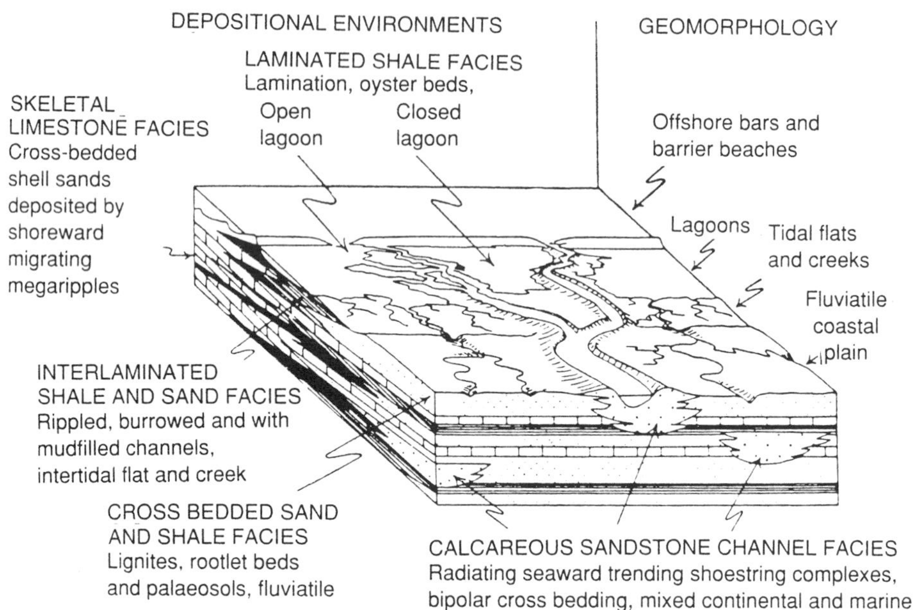

Figure 8.10 Block diagram illustrating the supposed origin of the Miocene shoreline of the Sirte basin, Libya. From Selley, 1968, Figure 17, by courtesy of the Geological Society of London.

located along northerly subsiding axes. Palaeocurrent analysis shows that the carbonate sands were deposited by predominantly up slope shoreward-flowing currents, thus coming to rest in waters shallower than that in which they formed. In contrast, quartz sand was carried off the Sahara shield to be deposited in the alluvial plain and tidal flats. Only the finest fraction was transported as far as the lagoons. Mixing of the land-derived quartz sand and marine carbonate detritus occurred only in the estuarine channels, perhaps due to tidal currents.

GENERAL DISCUSSION AND ECONOMIC ASPECTS

Shorelines where barrier carbonates are juxtaposed with continental terrigenous facies are transitional between clastic shores where carbonate sedimentation, if present, is restricted to the offshore zone, and carbonate shores where terrigenous sediment is negligible. These are described in the previous and subsequent chapters respectively. Apart from the Mediterranean Miocene cases noted in this chapter, other mixed shorelines occur in the Permian rocks of West Texas. Here shoal calcarenites pass landwards through lagoonal deposits into continental red beds and evaporites. These are overlain by reef limestones with similar shoreward facies changes.

Indeed, the combinations of conditions which favour the occurrence of mixed carbonate:clastic shores (i.e. low influx of terrigenous sediment and aridity) are particularly favourable for reef growth, as exemplified by the present-day Great Barrier Reef of Australia. Perhaps reefs are more characteristic of mixed shorelines than are carbonate sand banks. Other ancient examples are known from the Lower Cretaceous of northen Spain (Garcia-Mondejar and Fernandez-Mendiola, 1993) and from Triassic sediments of the Alps (Bechstadt and Schweizer, 1991).

Mixed terrigenous:carbonate shorelines are of considerable economic significance. This need not be discussed here. The importance of alluvial deposits has already been described on pp. 81 and 144. The economic aspects of carbonate bars and reefs are discussed on pp. 220 and 241 respectively.

REFERENCES

The account of the Libyan Miocene shoreline was based on the author's own fieldwork and:

Doust, H. (1968) *Palaeoenvironmental Studies in the Miocene (Libya, Australia)*, Vol. I, unpublished Ph.D thesis, University of London.

El-Hawat, A. S. (1991) Carbonate-terrigenous cyclic sedimentation and palaeogeography of the Marada Formation (Middle Miocene), Sirt Basin. In M. J. Salem, A. M. Sbeta and M. R. Bakbak (eds), *The Geology of Libya*, Vol. II, Elsevier, Amsterdam, 427–48.

Savage, R. J. G. and Hamilton, W.R. (1973) Introduction to the Miocene mammalian faunas of the Jabal Zelten, Libya, *Bull. Br. Mus. (Nat. Hist.), Geol.*, **22**, 515–27.

Savage, R. J. G. and White, M. E. (1965) Two mammal faunas from the early Tertiary of Central Libya, *Proc. Geol. Soc. Lond.*, **1623**, 89–91.

Selley, R. C. (1966) The Miocene rocks of Marada and the Jebel Zelten: a study of shoreline sedimentation, *Petrol. Explor. Soc. Libya.*

—— (1967) Paleocurrents and sediment transport in the Sirte basin, Libya, *J. Geol.*, **75**, 215–23.

—— (1968) Facies profile and other new methods of graphic data presentation: application in a quantitative study of Libyan Tertiary shoreline deposits, *J. Sediment. Petrol.*, **38**, 363–72.

—— (1968) Nearshore marine and continental sediments of the Sirte basin, Libya, *Quart. Jl. Geol. Soc. Lond.*, **124**, 419–60.

—— (1972) Structural control of Miocene sedimentation in the Sirte basin. In C. Gray (ed.), *The Geology of Libya*, The University of Libya, 99–106.

Other references listed in this chapter were:

Bechstadt, T. and Schweizer, T. (1991) The carbonate-clastic cycles of the East Alpine Raibl group: result of third order sea-level fluctuations in the Carnian, *Sed. Geol.*, **70**, 241–70.

Doyle, L. J. and Roberts, H. H. (eds) (1988) *Carbonate – Clastic Transitions*, Developments in Sedimentology No. 42, Elsevier, Amsterdam.

Evans, G. (1965) Intertidal flat sediments and their environments of deposition in the Wash, *Quart. Jl. Geol. Soc. Lond.*, **121**, 209–45.

Garcia-Mondejar, J. and Fernandez-Mediola, P. A. (1993) Sequence stratigraphy and systems tracts of a mixed carbonate and siliciclastic platform-basin setting: the

Albian of Lunada and Soba, Northern Spain, *Bull. Amer. Assoc. Petrol. Geol.*, **77**, 245–75.

Jindrich, V. (1969) Recent carbonate sedimentation by tidal channels in the Lower Florida keys, *J. Sediment Petrol.*, **39**, 531–53.

McKee, E. D. and Sterrett, T. S. (1961) Laboratory experiments on form and structure of longshore bars and beaches. In J. A. Peterson and J. C. Osmond (eds), *Geometry of Sandstone Bodies*, Amer. Assoc. Petrol. Geol., 13–28.

Martini, I. P., Oggiano, G. and Mazzei, R. (1992) Siliciclastic-carbonate sequences of Miocene grabens of northern Sardinia, western Mediterranean Sea, *Sed. Geol.*, **76**, 63–78.

Roberts, H. H. (1987) Modern carbonate-siliciclastic transitions: humid and arid tropical examples, *Sed. Geol.*, **50**, 25–65.

Van Straaten, L. M. J. U. (1953) Composition and structure of Recent marine sediments in the Netherlands, *Leid. Geol. Meded.*, **19**, 1–110.

9 Terrigenous shelf sediments

INTRODUCTION: SHELF SEDIMENTARY PROCESSES

The continental shelf extends from low tide to the 200 m isobath. This water depth normally marks the boundary between the gently sloping continental shelf, and the continental slope. This continues, with a steeper dip of some 4°, down to the continental rise. The continental shelf edge marks the aproximate limit of continental crust.

It is difficult to use modern shelf seas as analogues for their ancient counter parts for two reasons. First, at the present time, the earth lacks the vast sub-horizontal shelves that existed in earlier times. For example, it is possible to trace a remarkable uniform Palaeozoic stratigraphy in shallow marine formations across much of modern North America. Similarly it is possible to trace another uniform stratigraphy in shallow marine Mesozoic formations across much of Arabia. These are examples of sediments deposited on broad shelves, commonly referred to as 'epeiric seas', the like of which are absent today. This reflects the fact that the earth is now in an unstable and exciting phase of its history, in which the vast continental plates of the past have been rifted and drifted apart. Thus most present-day shelves are seldom more than 200 km across.

There is a second difference between present-day shelves and many ancient ones. Modern shelves have been subaerially exposed several times during glacially induced drops in sea level throughout the Pleistocene Ice Age. Much of the modern sediments of marine shelves, though now reworked by marine currents, were actually first transported out on to the shelves by fluvio-glacial processes. Thus these modern reworked relict sediments are not suitable analogues for many ancient shelf deposits (Emery, 1968; McManus, 1975). One or two modern shelves are known, however, in which the sediment is graded and in equilibrium with the present-day current regime. The Bering shelf has been cited as one such example (Sharma, 1972). For further accounts of modern marine shelves see Bouma *et al.* (1982) and McCave (1985).

The deposits of modern shelves come from several sources. Some detritus is being brought in by fluvial processes, some is relict from glacial low stands, and some forms on the shelf at the present time. This authigenic sediment includes relatively minor amounts of glauconite and phosphate, but often very major amounts of carbonate material. Carbonate continental shelf deposits will be dealt with in a separate chapter, which follows. This chapter will be solely concerned with terrigenous siliciclastic shelf sediments.

There is a range of processes that transport and deposit sediment on continental shelves. These vary from place to place, and with water depth. The main processes are tidally generated currents, wind generated currents, including those due to storms, and density flows. The relative importance of tidal currents is largely related to tidal range. The higher the range, normally the more tidal flow and hence higher current velocity. On the continental shelf of Northwest Europe the tidal range is as much as 6 m .

The resultant powerful tidal currents transport and deposit sand in huge tidal current sand ridges (also termed sand waves or megaripples). These are tens of metres high, several km wide, and tens of km in length. They are composed of clean quartz sand, with variable amounts of skeletal sand and glauconite. Internally these sand bodies are cross-bedded, with a wide range of scales and directions, reflecting the varied intensity and orientation of the currents from which they were deposited (Stride, 1982).

Wind may also generate currents on continental shelves. These currents will vary in intensity and direction correlative with the intensity and direction of the wind. Their importance will also be related to the strength of tidal currents. Storms, accompanied by high wind velocities and low barometric pressure, may generate localized areas of high water level on a continental shelf. When the storm abates, or passes inland, the water drains off the shelf, and in so doing deposits a very characteristic sedimentary unit known as a 'tempestite' (Ager, 1973). This is a graded bed with a scoured, often erosional, base.

Internally 'tempestites' contain curious curvaceous cross-stratification. This is termed 'hummocky cross-stratification' (Harms, 1975). HCS, as it is colloquially termed, is quite unlike the regular cosets of normal traction deposited cross-bedding. Each tempestite reflects the erosion and deposition of a single storm event, as the excess water drains off the continental shelf back in to the ocean. Storm deposits are commonly reworked by tidal, or wave generated currents on the shelf. They are thus preserved in the deeper water of the outer shelf in a zone between fair weather and stormy weather wave base.

The last process to operate on continental shelves is the density flow. Density flows are discussed more fully in Chapter 12. Suffice to say for now that turbidity currents, a particular type of density flow, may transport sediment on to and beyond continental shelves. The resultant deposits, termed turbidites, are normally reworked on the shelf by storm, wind or tidal currents. Turbidites are thus only preserved in quieter deeper water. Figure 9.1 illustrates the processes and products of continental shelves.

CASE HISTORY: THE CAMBRO-ORDOVICIAN SANDSTONES OF THE SAHARA AND ARABIA

As noted in Chapter 3 geologists have long remarked on the remarkable uniformity of Lower Palaeozoic stratigraphy from the Atlantic coast of North

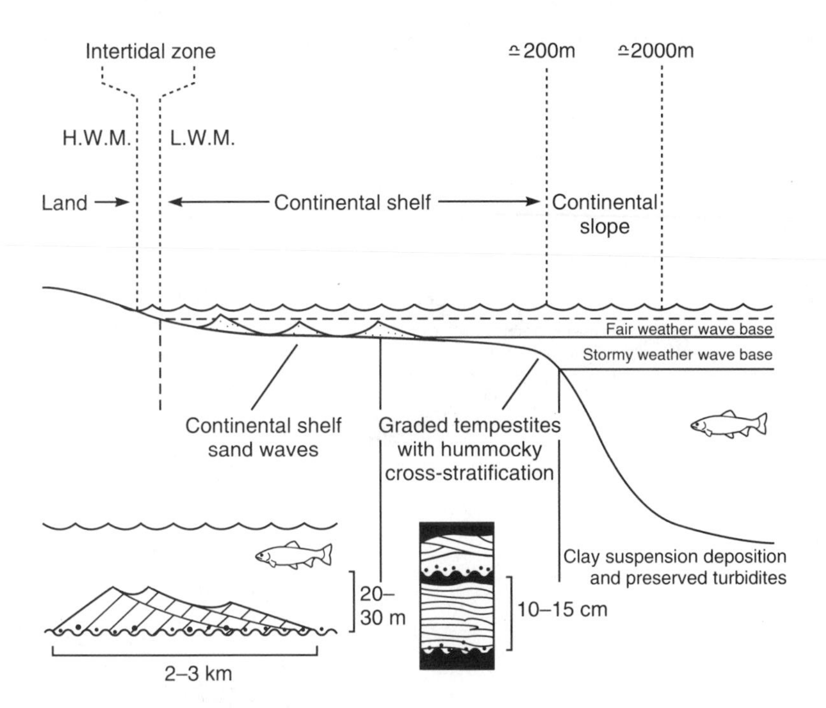

Figure 9.1 Cartoon to show the processes and products of continental shelf sedimentation. This figure is applicable to both terrigenous and carbonate shelves.

Africa east to the shores of the large gulf between Arabia and Iran. Laterally extensive blankets of sandstone and shale crop out around the PreCambrian shields of the Sahara and Arabia, and can be traced intermittently northward and eastward towards the Mediterranean and Gulf coasts respectively (Figure 9.2).

These formations show a remarkably uniform vertical sequence of facies. The sequence normally commences with coarse pebbly channeled sands (attributable to fluvial environments). These pass up into better sorted finer sands, commonly bioturbated (attributable, as will now be shown, to shallow marine environments). This facies passes up in to graptolitic shales and turbidites, in to prograding deltaic, and fluvial sands (Figure 9.3).

The first facies was described as a braided outwash plain in Chapter 3. The third facies was described and attributed to a deltaic environment in Chapter 6. The second facies will now be described and diagnosed as a shallow marine continental shelf deposit.

Non-pebbly sheet sand facies: description

The braided alluvial pebbly channel sand facies described in Chapter 3 is overlain by the non-pebbly sheet sand facies. This includes the Um Sahm Formation of Jordan, and the Haouaz Formation of Libya. These formations

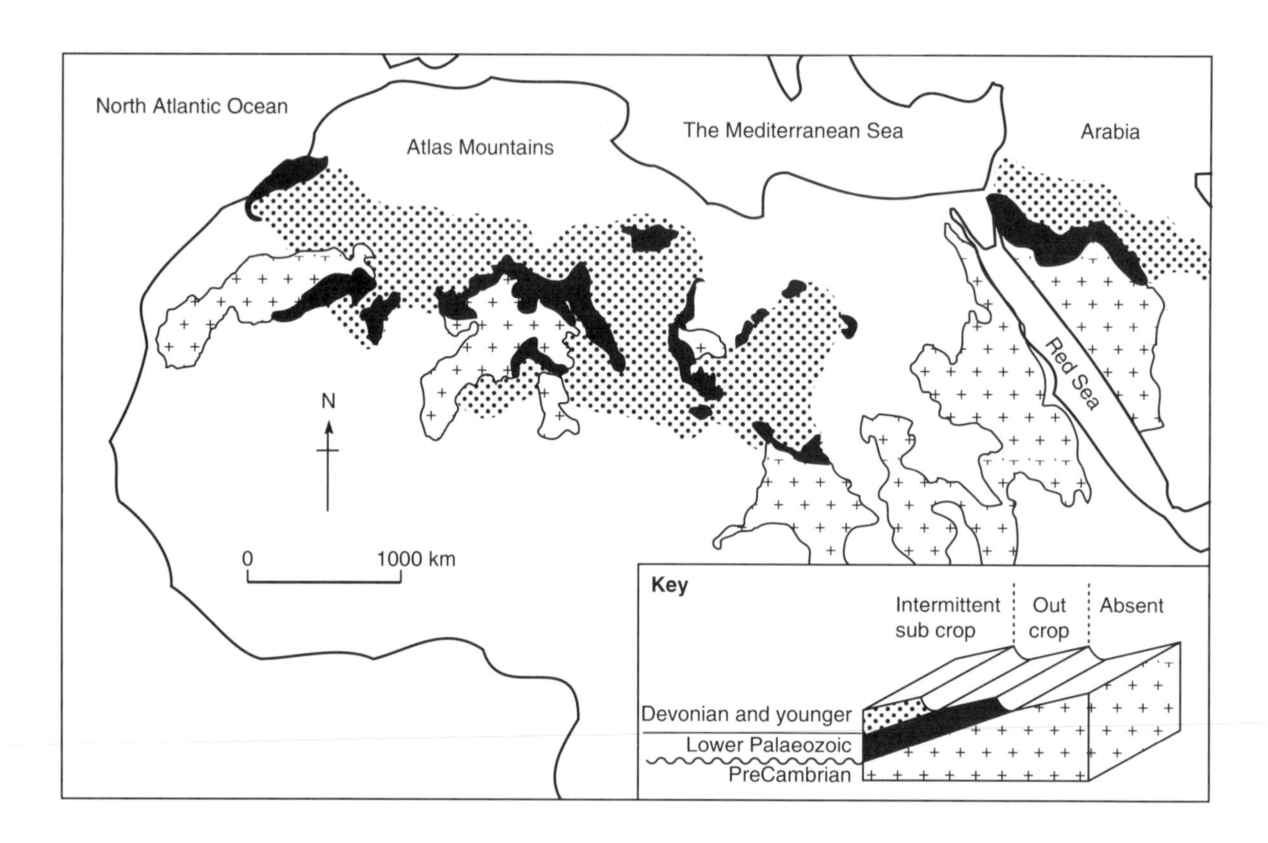

Figure 9.2 Distribution of Lower Palaeozoic sediments in North Africa and Arabia.

locally exceed 250 m in thickness. The contact with the underlying fluvial sands is abrupt in many areas, commonly planar, and devoid of channeling. But the two facies are locally interbedded. At the top the non-pebbly sheet sand facies is abruptly overlain by the graptolitic shales of the Hanadir and Tannezuft formations of Arabia and Libya respectively. Locally, however, there are some tongues of graptolitic shale within these sands, such as the Melez Chogranne shale of Libya.

This facies is composed largely of well-sorted medium and fine grained proto-quartzites. Coarse sandstones are sometimes present towards the base, with occasional scattered granules and rare quartz pebbles up to 1 cm long.

The commonest sedimentary structure in this facies is cross-bedding. This occurs in tabular planar cosets up to 3 m high with sets 5–15 cm high (Figure 9.4).

Sometimes, however, individual sets up 2 m height occur. Troughs are rare. Foresets are homogeneous and accretionary, reflecting the good sorting of this facies. Heterogeneous and avalanche foresets, such as occur in the poorly sorted pebbly sands beneath, are seldom present. There are very few body fossils in these sandstones. There are, however, occasional trace fossils. The most

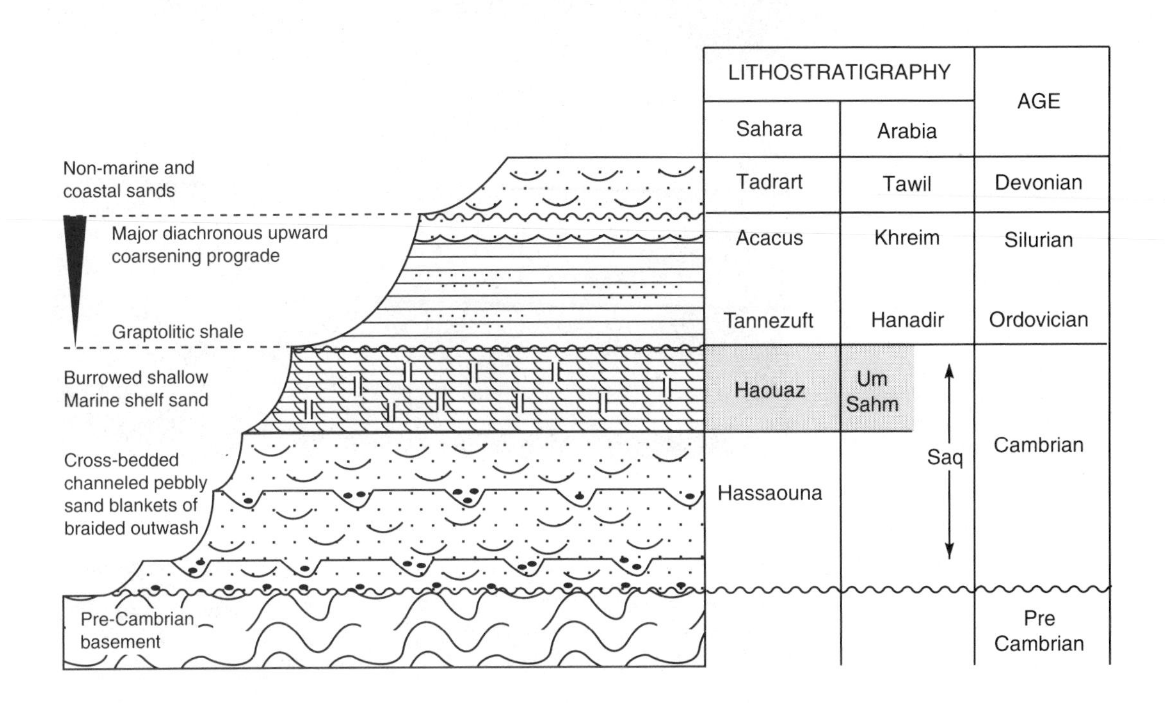

Figure 9.3 Stratigraphic section of the Lower Palaeozoic succession of North Africa and Arabia. This case history is concerned with the Um Sahm and Haouaz sandstones, interpeted as examples of continental shelf environments.

characteristic type is the vertical burrow that is known in the Sahara as *Tigillites*, in Arabia as *Sabellarifex* (though the Bedu of the Howetat tribe know them as 'Angel's tears'), and elsewhere in the world as *Scolithos* or *Monocraterion*. These burrows are sometimes so abundant that they destroy any original sedimentary structures that may have once existed (Figure 9.5).

The sandstones are interbedded with rare grey argillaceous micaceous siltstones with thin very fine sandstone layers. These units are each between 1–3 m thick and have sheet geometries, in contrast to the abandoned channel silts of the facies beneath.The siltstone sheets are generally laminated throughout with occasional thin beds of very fine sand and isolated sand ripples. These sandstones are rippled throughout. Microcross-laminated cosets are generally absent, the sands being composed of congeries of rippled lenses separated by argillaceous laminae and clay drapes. The bases of the siltstone sheets are generally transitional, their tops are abrupt, rarely erosional (Figure 9.6).

Cross-bed dip directions in the sandstones are generally unimodal, and indicate deposition from currents flowing down the local depositional slope. That is to say normally in a northeasterly direction in Arabia, and generally northerly in the Sahara.

Figure 9.4 Coset of tabular planar cross-bedding in Um Sahm Formation, Jordan.

Figure 9.5 Sandstone with extensive vertical burrows termed *Tigillites*, indicative of shallow marine conditions. Haouaz Formation, Jebel Eghei, southern Libya.

Non-pebbly sheet sand facies: interpretation

The abundance and orientation of cross-bedding in this facies points to deposition from unidirectional lower flow regime traction currents. The fine grainsize however shows these currents to have had significantly lower velocities than those which deposited the braided alluvial pebbly channel sands beneath. The predominance of tabular planar cross-beds, and the absence of channeling indicates that these were openflow currents that were

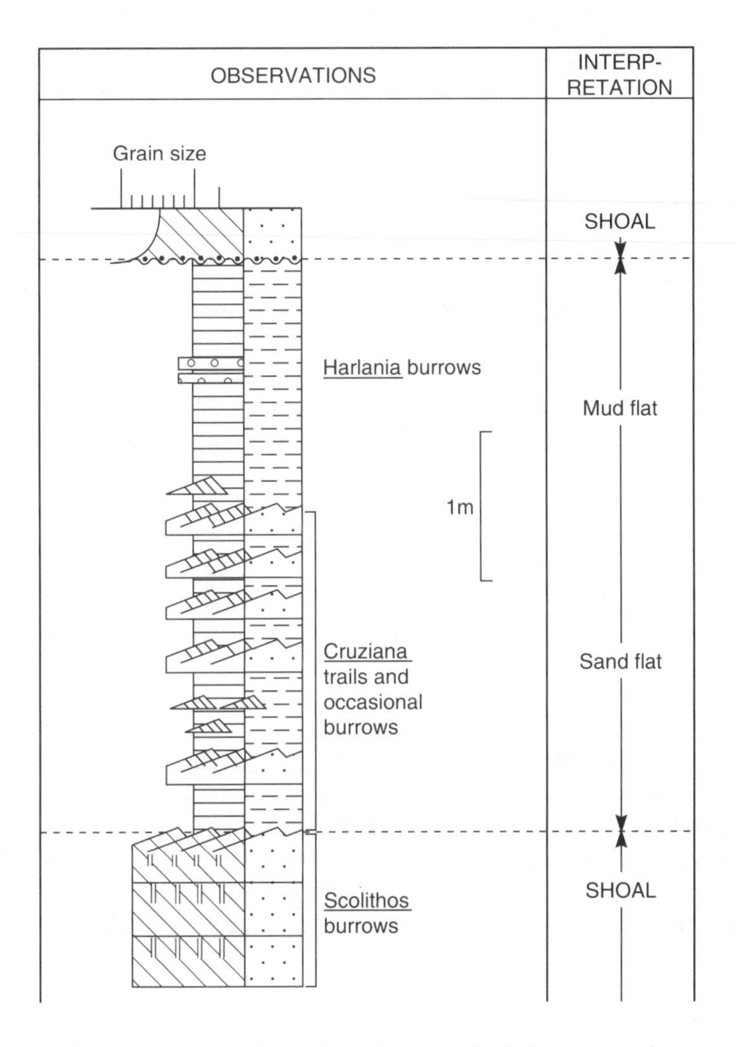

Figure 9.6
Sedimentological log through shale member in the upper part of the Um Sahm Sandstone Formation, Jordan, showing interpretation as an intertidal flat sequence within continental shelf shoal sands.

not confined by channel banks. The vertical burrows characterize the *Scolithos* ichnofacies which is diagnostic of shallow marine conditions (refer back to pp. 51–52).

The laminated shale units indicate sporadic lower energy conditions when suspended sediment settled out. The associated rippled very fine sands show that gentle traction currents sometimes occurred, while the absence of cross-laminated cosets and the presence of clay drapes on ripple crests suggests that these currents pulsated gently. Such conditions are more likely to be found in tidal realms rather than in the more regular flows of river channels. A shallow tidal environment is also suggested by the suite of trace fossils. *Skolithos*, *Cruziana* and *Harlania* are all characteristic of shallow marine deposits.

The weight of the evidence suggests therefore that the Cambro-Ordovician non-pebbly sheet sand facies of the Sahara and Arabia originated in a marine shelf environment. The cross-bedded sands were probably deposited from the slip off faces of migrating megaripples and bars similar to those described from modern shelf sand waves. The bioturbated silts and very fine sands suggest intermittent regressive phases when shallower tidal flat deposits were laid down.

ECONOMIC ASPECTS OF TERRIGENOUS SHELF SANDS

Shelf deposits are of great economic significance. Many of the world's major oil fields occur in shallow marine sands. Economically important continental shelf sand formations are especially characteristic of the Lower Palaeozoic in many parts of the world. These sands are often hypermature texturally and mineralogically, reflecting prolonged and repeated episodes of weathering, erosion, transportation and deposition on vast stable continental shelves. These sands were deposited with excellent primary porosity, and, because of their mineralogical maturity, have often undergone little subsequent diagenesis.

These formations thus serve as very important aquifers, petroleum reservoirs and sources of glass sand. North American examples include the Simpson Group (Lower Ordovician), of Oklahoma, and the Clinton Sands (Silurian) of the Cincinnati arch. The Cambro-Ordovician sands of the Sahara, and Arabia, also serve as petroleum reservoirs and aquifers. It is particularly important to be able to differentiate the braided alluvial sands from the shallow marine ones. This is because in the former the shales occur as abandoned channel shoestrings, which are not permeability barriers, whereas in the latter, the tidal flat shales have blanket geometries. They may therefore serve as permeability barriers to petroleum migration and ground water flow.

SUB-SURFACE DIAGNOSIS OF TERRIGENOUS SHELF DEPOSITS

Whereas it may be easy to diagnose that a particular sand formation is of shallow marine origin, it may actually be very difficult to differentiate shelf sands from barrier island or beach sands. All may be composed of clean quartzose sand, with varying amounts of glauconite and carbonate detritus. The detailed geometry of the sand, which is essential to differentiate these three environments, may no longer be detectable, because prograding beaches and barrier islands may migrate across a continental shelf to deposit blanket sands similar in character to those of the open shelf itself. Indeed, a range of resultant geometries may be preserved that extend from isolated shoestrings via laterally stacked shoestrings, to blankets with corrugated ridges, where megaripple palaeotopography is preserved, and ultimately to planar blankets (Figure 9.7).

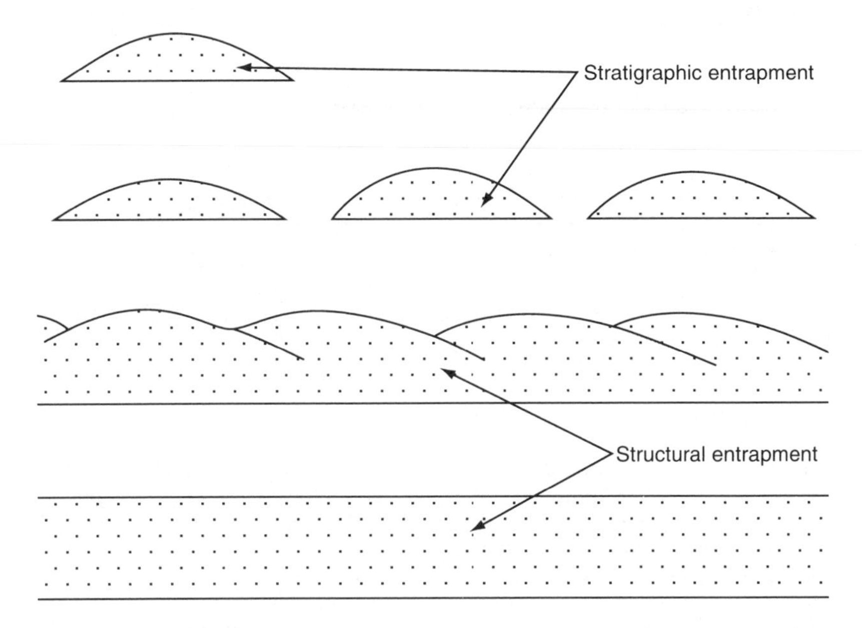

Figure 9.7 Cross sections to illustrate the varied geometry of shallow marine sands. Note how laterally extensive blanket formations may be deposited by the lateral migration across a continental shelf of offshore sand ridges, of nearshore barrier bars and of beaches. Geometry is not diagnostic of the specific shallow marine environment. What is important, however, is that all of these environments tend to deposit sands that are elongated along the shoreline, and hence parallel to the basin margin. Note how the continuity of the sand bodies dictates their style of petroleum entrapment.

The progradational nature of many shallow marine sands generates discrete laterally stacked sands that are separated from one another by thin shales that may serve as permeability barriers. These present considerable petroleum production problems. Where reservoir pressure data are absent or unreliable it may not be easy to correlate the impermeable shale zones from well to well (Figure 9.8).

This problem has been encountered in many shallow marine sands, notably in the Cretaceous Viking sands of Alberta, Canada (Evans, 1970).

Seismic data may be used to map the diverse geometries of shallow marine sands. As remarked above, these geometries cannot be used to distinguish continental shelf sands from barrier and beach ones. In simple economic terms, however, the distinction is academic, since the object of the exercise is to map reservoir geometry. Shallow marine sands tend to be aligned parallel to the coast, and hence to the basin margin, irrespective of their particular depositional environment.

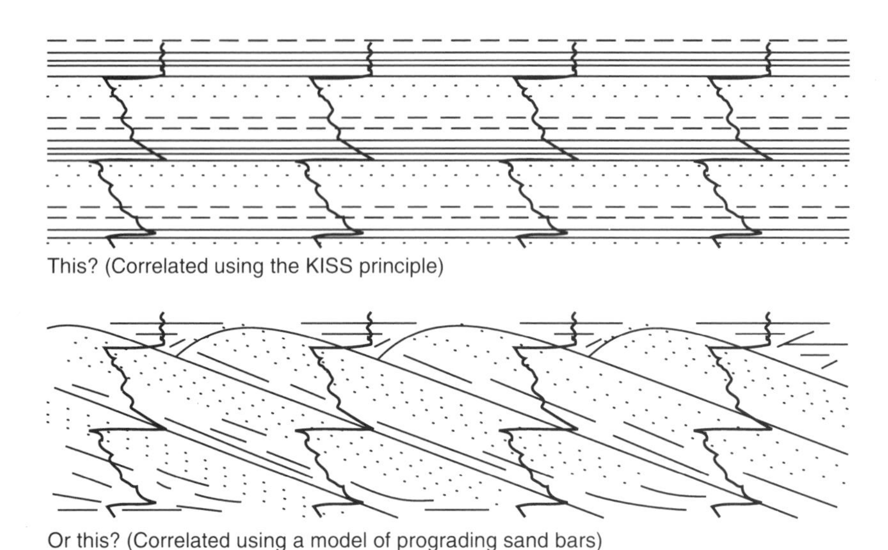

Figure 9.8 Hand-crafted gamma log cross sections to show the problems encountered when trying to correlate shallow marine reservoir sands. The upper correlation is constructed using the KISS principle (Keep It Simple, Stupid). If this is used as a basis for petroleum production, it may soon appear that water injected in to the lower sand in well D increases petroleum production in the upper sand in well C, and so forth. The lower model, based on appreciation of the progradational nature of shallow marine sands, may prove more useful. Real examples of this problem are well known. One instance is cited in the text.

Lithologically, again, it may be easy to identify a sand as of shallow marine origin, on the basis of its textural maturity, the presence of shell fragments, and occasional grains of glauconite. But it may be impossible to differentiate open shelf sands from those of barrier islands and beaches by lithology alone. Vertical grainsize profile is likewise of little use. Barrier islands, beaches and tidal sand waves may all prograde out over muds deposited in deeper quieter water. All three types of sand may thus exhibit upward-coarsening grainsize profiles in cores, and on gamma and SP logs.

Where continental shelf sand waves migrate over scoured equilibrial surfaces, as in the modern North Sea, however, the grainsize profile may be of a uniformally clean sand marked by abrupt upper and lower contacts. The sharp base marks where the sand overlies a scoured surface, often with a lag gravel. The sharp top marks where the sand body was blanketed by deeper water mud when sea level rose. Examples of this grainsize motif are recognizable in older sands of the North Sea basin (Figure 9.9).

Similarly sedimentary structures can be used to identify a sand as shallow marine, but not to differentiate an open shelf sands from inshore barriers and beaches. All three types of sand body contain cross-bedding, flat-bedding and cross-lamination, to varying degrees. Studies of both modern and ancient shelf sands have shown a diversity of cross-bedding directions. Unimodal

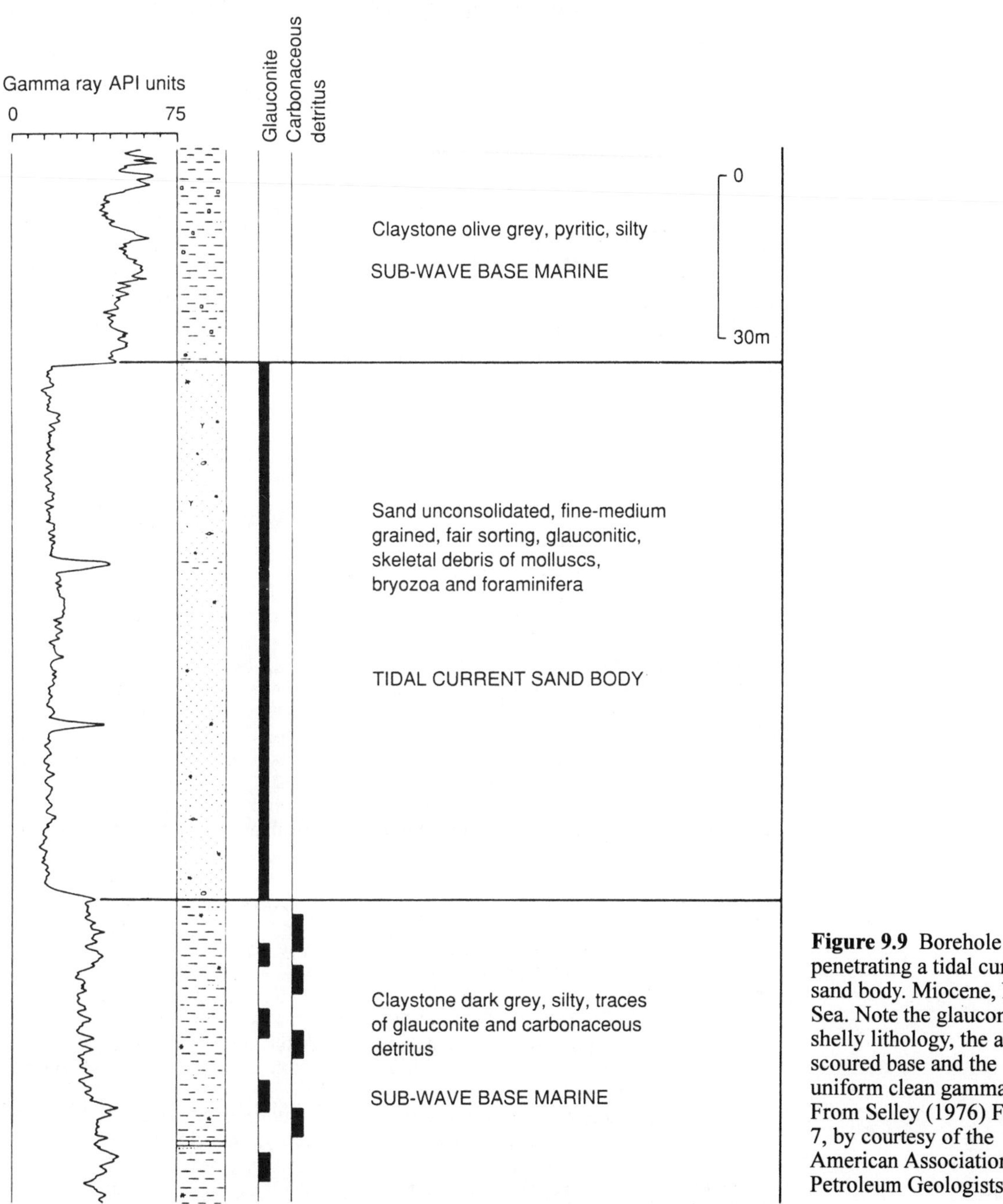

Figure 9.9 Borehole penetrating a tidal current sand body. Miocene, North Sea. Note the glauconitic, shelly lithology, the abrupt scoured base and the uniform clean gamma log. From Selley (1976) Figure 7, by courtesy of the American Association of Petroleum Geologists.

offshore cross-bedding, such as those recorded from the Cambro-Ordovician of Jordan, may be attributable to rip currents (e.g. Yagishita, 1994).

Bimodal bipolar palaeocurrents, attributable to tidal ebb and flow are known, as in the Miocene shoreline of Libya, and described in the previous chapter. Sadly many studies of shallow marine sands, ancient and modern, reveal polymodal cross-bedding. Dipmeter logs commonly show polymodal dips in shallow marine sands. This may be for two reasons. It may, indeed, reflect polymodal cross-bedding. But shallow marine sands are often so clean, that the microlog may be unable to detect resistivity variations in the formations from which dips can actually be calculated.

Finally palaeontology may be similarly disappointing. A transported fragmented shallow marine fauna may be found in beach, barrier island and continental shelf sands. With little difference in water depth, the *Scolithos* ichnofacies is common to all three environments.

REFERENCES

The account of the Cambro-Ordovician sandstones of the Sahara and Arabia was based on the author's own field work and:

Amireh, B. S., Schneider, W. and A. M. Abed (1994) Evolving fluvial-transitional-marine deposition through the Cambrian sequence of Jordan, *Sed. Geol.*, **89**, 65–90.

Balducci, A. and Pommier, G. (1970) Cambrian oil field of Hassi Messaoud, Algeria. In M. T. Halbouty (ed.), *Geology of Giant Petroleum Fields*, Amer. Assoc. Petrol. Geol., **14**, 477–88.

Biju-Duval, B. (1974) Examples de depots fluvio-glacières dans l'Ordovician supérieur et le Précambrian supérieur du Sahara central, *Soc. Pet. Nat. d'Aquitaine (SNPA) Bull. Cent. Rech. Pau.*, **8**, 209–26.

Doughty, C. (1888) *Travels in Arabia Deserta*, Cambridge University Press.

Husseini, M. l. (1990) The Cambro-Ordovician Arabian and adjoining plates: a glacio-eustatic model, *Journ. Petrol. Geol.*, **13**, 267–88.

—— (1991) Tectonic and Depositional Model of the Arabian and Adjoining Plates during the Silurian–Devonian, *Bull. Amer. Assoc. Petrol. Geol.*, **73**, 1117–31.

Lloyd, J. W. and Pim, R. H. (1990) The hydrogeology and groundwater resources development of the Cambro-Ordovician sandstone aquifer in Saudi Arabia and Jordan, *J. Hydrogeol.*, **121**, 1–20.

McGillivray, J. G. and Husseini, M. l. (1992) The Palaeozoic petroleum geology of central Arabia, *Bull. Amer. Assoc. Petrol. Geol.*, **76**, 1475–90.

Selley, R. C. (1970) Ichnology of Palaeozoic sandstones in the Southern Desert of Jordan: a study of trace fossils in their sedimentologic context. In J. C. Harper and T. P. Crimes (eds), *Trace Fossils*, Lpool. Geol. Soc., 477–88.

—— (1972) Diagnosis of marine and non-marine environments from the Cambro-ordovician sandstones of Jordan, *Jl. Geol. Soc. Lond.*, **128**, 109–17.

Other references cited in this chapter were:

Ager, D. V. (1973) Storm deposits in the Jurassic of the Moroccan High Atlas, *Palaeogeog. Palaeoclimatol. Palaeoecol.*, **15**, 83–93.

Bouma, A. H., Berryhill, H. L., Knebel, H. J. and Brenner, R. L. (1982) Continental shelf. In P. A. Scholle and D. Spearing (eds), *Sandstone Depositional Environments*, Amer. Assoc. Petrol. Geol. Tulsa, 281–328.

Emery, K. O. (1968) Relict sediments on continental shelves of the world, *Bull. Amer. Assoc. Petrol. Geol.*, **52**, 445–62.

Evans, W. E. (1970) Imbricate linear sandstone bodies of Viking Formation in Dodsland-Hoosier area of southwestern Saskatchewan, *Canada Bull. Amer. Assoc. Petrol.*, **54**, 469–86.

Harms, J. C. (1975) Stratification and sequence in a prograding shoreline, *Soc. Econ. Min. Pal. Short Course*, **2**, 81–102.

McCave, I. N. (1985) Recent shelf clastic sedimentats. In P. J. Brenchly and B. P. J. Williams (eds), *Sedimentology: Recent Advances and Applied Aspects*, Blackwell, Oxford, 49–66.

McManus, D. A. (1975) Modern versus relict sediment on the continental shelf, *Bull. Geol. Soc. Amer.*, **86**, 1154–60.

Selley, R. C. (1976) Sub-surface environmental analysis of North Sea sediments, *Bull. Amer. Assoc. Petrol. Geol.*, **60**, 184–95.

Sharma, G. D. (1972) Graded sedimentation on Bering shelf. In *24th Internat. Geol. Cong. Montreal.*, Section 8, 262–71.

Shearman, D. J. (1963) Recent anhydrite, gypsum, dolomite and halite from coastal flats of the Arabian shore of the Persian Gulf, *Proc. Geol. Lond.*, **1067**, 63–5.

—— (1966) Origin of marine evaporites by diagenesis, *Bull. Inst. Min. Met.*, **76**, section B., 82–6.

—— and Fuller, J. G. C. M. (1969) Phenomena associated with calcitization of anhydrite rocks, Winnepegosis Formation, Middle Devonian of Saskatchewan, Canada, *Proc. Geol. Soc. Lond.*, **1658**, 235–9.

Shinn, E. A. (1984) Tidal Flat. In P. A. Scholle, D. G. Bebout and C. H. Moore (EDS), *Carbonate Depositional Environments*, Amer. Assoc. Petrol. Geol., **33**, 171–210.

Stanley, D. J. and Swift, D. H. P. (eds) (1976) *Marine Sediment Transport and Environmental Management*, Wiley Interscience, New York.

Stride, A. H. (1970) Shape and size trends for sandwaves in a depositional zone of the North Sea, *Geol. Mag.*, **107**, 469–78.

—— (1982) *Offshore Tidal Sands*, Chapman & Hall, London.

Swie-Djin, N. (1976) Marine transgressions as a factor in the formation of sandwave complexes, *Geol. en Mijnb.*, **55**, 18–40.

Swift, D. J. P., Duane, D. B. and Orrin, H. P. (1973) *Shelf Sediment Transport: Process and Pattern*, Wiley, Chichester.

Tebbutt, G. E., Conley, C. D. and Boyd, D. W. (1965) Lithogenesis of a distinctive carbonate rock fabric, *Univ. Wyoming Contrib. Geol.*, **4**, 1–13.

Tooms, J. S., Summerhayes, C. P. and McMaster, R. L. (1971) Marine geological studies on the northwest African margin: Rabat-Dakar. In *The Geology of the East Atlantic Continental Margin*, 4, Africa. Inst. Sci. Rept. **70**(16), 11–25.

Wilson, J. L. (1975) *Carbonate Facies in Geologic History*, Springer-Verlag, Berlin.

Yagishita, K. (1994) Planar cross-bedding associated with rip currents of Upper Cretaceous formations, northeast Japan, *Sed. Geol.*, **93**, 155–63.

Youssef, M. I. (1958) Association of phosphates with synclines and its bearing on prospecting for phosphates in Sinai, Egypt, *J. Geol.*, **2**, 75–87.

10 Carbonate shelves

INTRODUCTION

The previous chapter began with an introduction to continental shelves, pointing out the problems of applying studies of modern shelves to ancient ones, and reviewing the processes and products of continental shelf sedimentation. This material will not be repeated here, but must be digested before this chapter may be enjoyed to the full.

Carbonate sediments are very different from terrigenous siliciclastic ones. Carbonate sediment forms within the basin, and indeed often within the environment in which it is deposited, whereas terrigenous sand is transported in to the basin from an extraneous source. Almost all carbonate sediment is of organic origin. All life begins with the photosynthesis of plants, so the rate of carbonate sediment formation is closely related to light, and thus water depth. In open oceans carbonate skeletal material forms from pelagic phytoplankton, largely of algal origin. This material forms the basis for food chains in which the phytoplankton are devoured by the zooplankton, which are in turn gobbled up by fish. Moving shorewards there comes a point where the sea floor rises up in to the photic zone that is bathed in sunlight.

In regions where there is negligible influx of terrigenous sand and clay photosynthesis by benthonic algae on the sea floor provide the basis for abundant life. In this zone carbonate sediment formation is at its peak. Moving further up towards the land, however, the nutrients brought in by the open ocean waters are all used up by the plants and animals. Thus, though the sea floor is shallower and sunnier, the rate of carbonate sediment formation declines. Thus as time passes a gently sloping continental shelf will develop what is termed an 'accretionary ramp' of carbonate sediment. This ramp will gradually prograde in a seaward direction (Figure 10.1).

One of the best known modern carbonate accretionary ramps occurs along the north-east coast of Arabia (Purser, 1973, Wilson and Jordan, 1983). A drop in sea level may cause the erosion of a wave cut bench and sea cliff in the emergent ramp. When sea level rises again the optimum locus of carbonate sediment formation will be on the crest of the drowned sea cliff, for it is here that the photic zone first impinges on the sea bed, and it is here that the maximum nutrients from the ocean are available.

The crest of a drowned sea cliff is a favoured site for reef development in particular. Repeated emergence and submergence may cause the gentle slope of the accretionary ramp to be modified in to a second carbonate

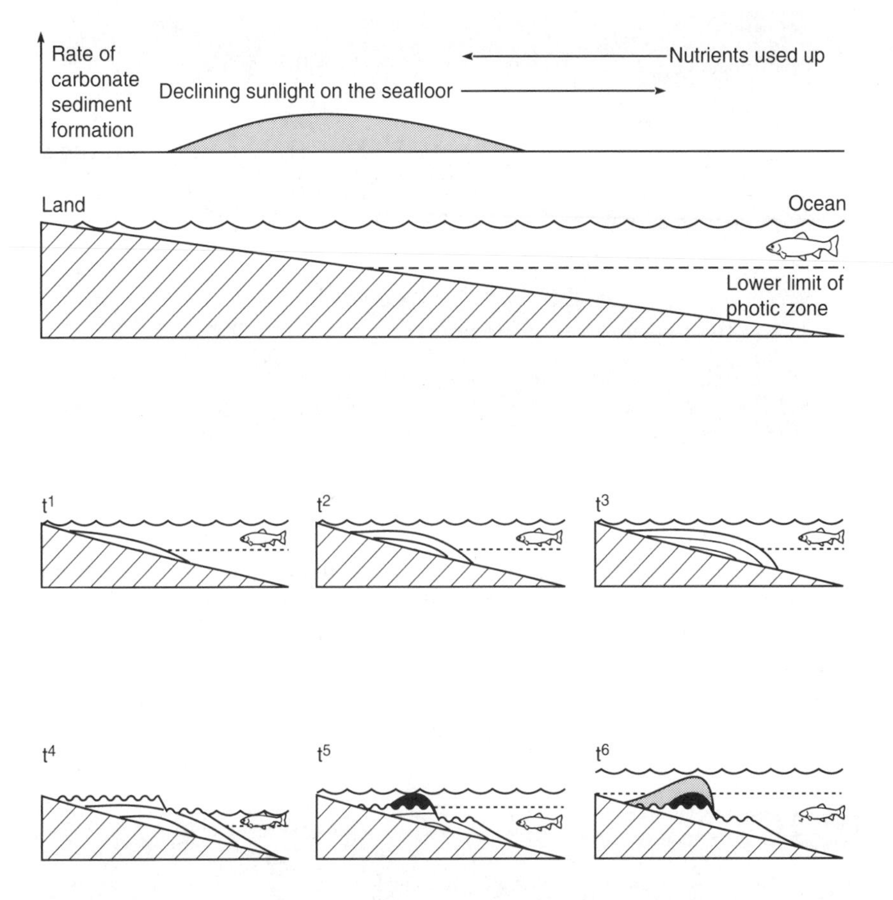

Figure 10.1 Upper: cartoon to show how the optimum zone of carbonate sediment formation on a gently sloping shelf is related to sunlight and the supply of nutrients. Lower: cartoons to show the formation of a carbonate accretionary ramp (t^1), its progradation (t^{2-3}), its emergence and dissection following a drop in sea level (t^4), and the evolution of the accretionary ramp in to a 'drop off' setting (t^{5-6}). Note that the latter can also form on the rim of a tectonic platform. The modern north-east coast of Arabia is an example of an accretionary ramp. The modern Bahama Banks is an example of a tectonic 'drop-off' setting, albeit modified by Pleistocene sea level changes.

model termed the 'drop-off' model. This may also develop at the edge of a tectonic platform. The modern Bahama Banks is an excellent example of a drop-off carbonate terrain. In this case the platform edge, though originally of tectonic origin, has undergone extensive modification during Pleistocene sea level fluctuations (Halley et al., 1983).

Thus one of the main differences between terrigenous and carbonate sedimentation is that the carbonate sediments form within the basin, as discussed above. Furthermore there are many different types of carbonate grain. Space does not permit a detailed discussion of carbonate petrography. The following concise account must suffice.

Carbonate sediment can be deposited in deep water where the sea floor is below the photic zone. The carbonate sediment cannot form on the sea floor, but is introduced from two sources. The skeletons of plankton may drift down from the shallow photic zone to settle out as a fine mud on the sea floor. This is how some oceanic oozes develop, and other carbonate muds on basin floors and shelves alike (Figure 10.2).

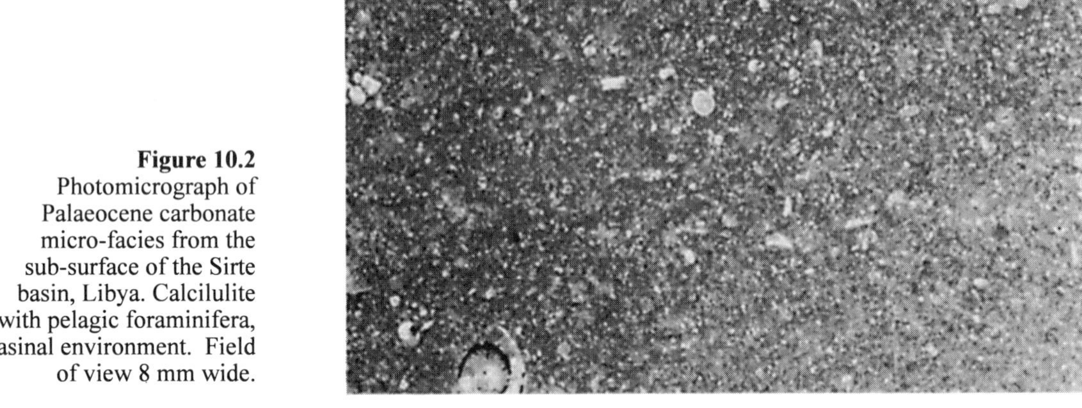

Figure 10.2
Photomicrograph of Palaeocene carbonate micro-facies from the sub-surface of the Sirte basin, Libya. Calcilulite with pelagic foraminifera, basinal environment. Field of view 8 mm wide.

Coarser carbonate detritus can be transported from shallow water by turbidity flows to be deposited in deep water. In this case turbidites containing shallow water biota and other grain types are interbedded with muds containing an open marine fauna, something that may confuse an innocent palaeontologist (Figure 10.3). Such sediments are termed 'allodapic' limestones (Meischner, 1965).

Moving up the shelf, towards the land, carbonate sediment formation rate increases, for reasons already explained. When the sea floor is within the

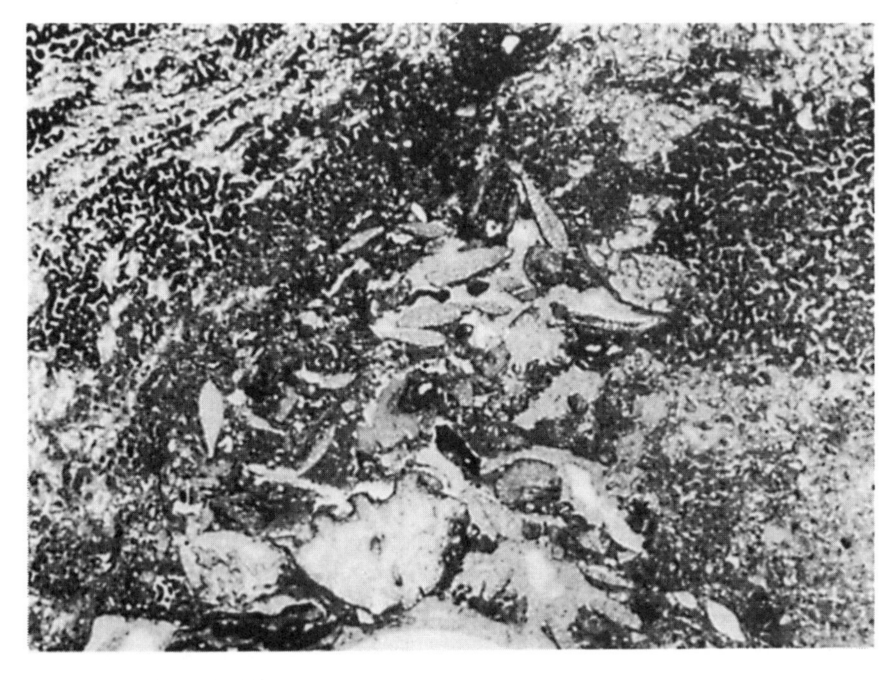

Figure 10.3
Photomicrograph of Montpelier Formation (Miocene), Jamaica. This is an allodapic wackestone deposited in a deep water trough. lt contains a derived shallow water fauna that includes large benthonic foraminifera and Rudist fragments.

photic zone benthonic as well as pelagic biota develop. In this shallow high-energy environment, however, wave action will smash up shells and other skeletal remains, to produce shell sands, often referred to as bioclastic or coquina sands. These will be best developed in the median part of the ramp, dying out to seaward and landward. Another important grain type of the ramp edge is the ooid, or oolith. These are spherical carbonate grains, with concentric growth rings around a nucleus of terrigenous or carbonate sand (Figure 10.4).

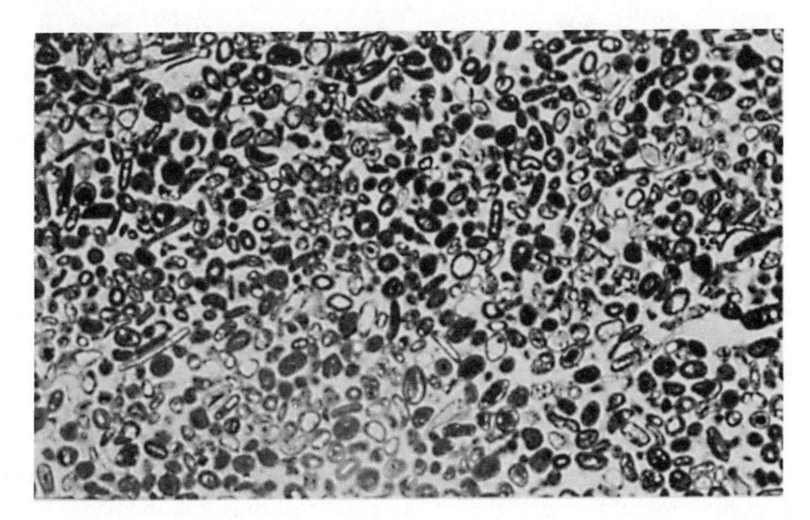

Figure 10.4
Photomicrograph of Palaeocene carbonate micro-facies from the sub-surface of the Sirte basin, Libya. Grainstone of ooids and bioclasts, high-energy shelf environment. Field of view 6 mm wide.

Ooids tend to form where water of different salinity or temperature mix. Thus they are found at the edges of banks, as in the Bahamas, and in tidal channels linking lagoons with the open sea, as in the coast of the United Arab Emirates. Skeletal and oolitic sands may be transported and deposited in migrating sand waves, similar in scale and geometry to those of terrigenous shelves described in the previous chapter (Illing, 1954; Newell *et al.*, 1960; Purdy, 1963; Ball, 1967; and Enos, 1983). In this shallow sunny high energy zone of corals, algae and other sedentary colonial lime-secreting organisms may thrive sufficiently to form reefs. Reefs are such an important carbonate environment that they will be dealt with separately in the following chapter.

Muddy sediments are deposited in the shallow sheltered inner shelf, behind the high energy zone of reefs, and carbonate sands, These may contain skeletons of lagoonal beasts, but they are characterized by an abundance of structureless lime mud peloids. Many of these exhibit a bullet shape, and can be recognized as the excreta of the lagoonal biota. Faecal pellet muds are thus diagnostic of inner shelf lagoonal deposits (Figure 10.5).

On the tidal flats to the landward side of the lagoon repeated emergence aids rapid lithifaction of the carbonate sediment, and rapid erosion of newly

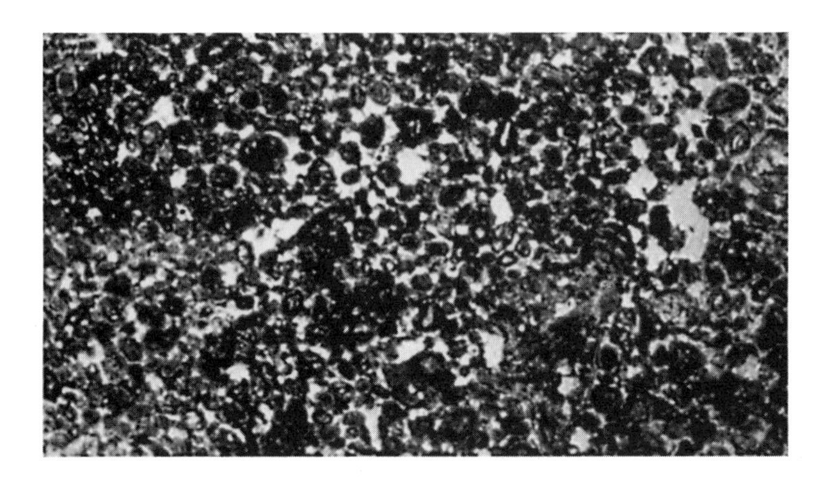

Figure 10.5 Photomicrograph of Palaeocene carbonate micro-facies from the subsurface of the Sirte basin, Libya. Fenestral limestone of lime mud pellets with sparse lime mud matrix. Uncompacted state of pellets suggests early diagenesis possibly due to penecontemporaneous subaerial (intertidal) exposure. Low-energy inner shelf. Field of view 7 mm wide.

cemented limestone. These intraformational conglomerates (intraclasts for short), thus characterize intertidal deposits. Lime is also precipitated in the intertidal zone by algae, giving rise to laminated algal limestones, and rounded algal stromatolites. A particularly characteristic feature of intertidal carbonate sediments is the presence of irregular pore systems that parallel bedding. These are variously described as 'fenestral porosity' (Tebbutt *et al.*, 1965), 'birds-eye' structure, or loferite (Fischer, 1964). This feature has variously been attributed to the buckling of sediment laminae during subaerial exposure, to deformation due to the escape of biogenic methane, and to the leaching of organic algal mucilage.

If salinities in the lagoon are high, dolomites and evaporites may form. There has been considerable speculation whether these are precipitated directly on the floor of lagoons or whether they are due to penecontemporaneous replacement of carbonate mud within tidal flats. In Recent arid shorelines, such as those of the Baja California and the Trucial Coast of the Arabian Gulf, wide salt flats (sabkhas) are developed. Evaporite minerals form in these environments today (Shearman, 1963 and 1966; Holser, 1966; Evans *et al.*, 1969; Kinsman, 1969, Shinn, 1983). As the sun beats down, capillarity draws lagoonal brines into the pore spaces of the sabkha carbonates. Here, as the fluids evaporate and concentrate, they replace the host sediment to form dolomite, gypsum, anhydrite, halite and other evaporite minerals.

The foregoing is a very brief account of the correlation between carbonate grain types and environment. It highlights another difference between carbonates and terrigenous sands. This is that it may be possible to diagnose the depositional environment of a limestone by means of petrography alone, since the combination of grain type and texture are so closely linked to environment and process.This is something that cannot be achieved with terrigenous sands, whose petrography is closely related to the nature of the source from which it was derived.

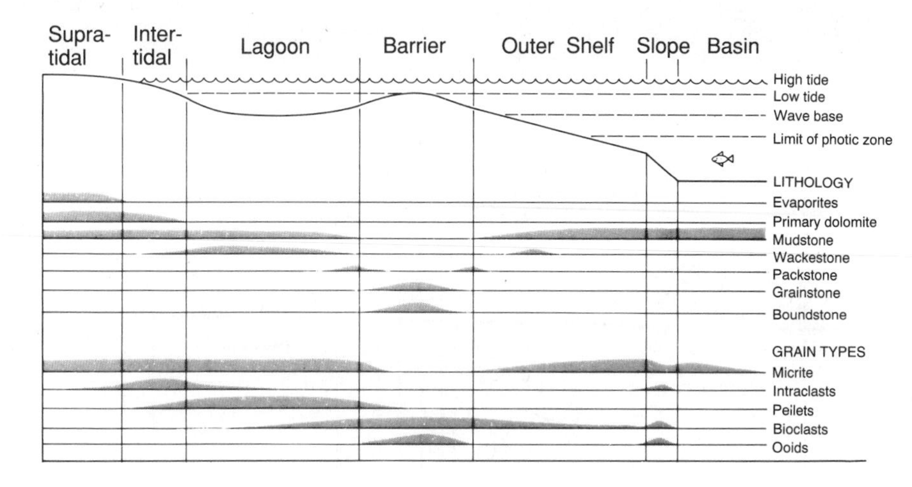

Figure 10.6 Cartoon through an idealised carbonate shelf showing the lateral variations in grain type and texture.

Thus Figure 10.6 summarizes the correlation between grain type and texture for a carbonate shelf, and Figure 10.7 shows the vertical sequence of carbonate rock types produced by the progradation of a single ramp genetic increment. Further more detailed accounts of carbonate depositional systems will be found in Wilson (1975), Reeckman and Friedman (1982) and Read (1985). As mentioned earlier reefs are such important environments that they deserve a chapter to themselves. The case history to be used to illustrate carbonate shelves is a deeper water one.

CASE HISTORY OF CARBONATE SHELF SEDIMENTATION: THE CHALK (UPPER CRETACEOUS) OF NORTH-WEST EUROPE

Description

A good example of ancient outer shelf deposits occur in the Cretaceous rocks of North-west Europe, Middle East, North Africa, and the USA (Figure 10.8). Much of North-west Europe is covered by a fine grained limestone called the Chalk (Figure 10.9). Chalk is composed largely of the skeletons of a group of blue-green algae termed the nannoplankton. These secrete a skeleton called a coccosphere (Figure 10.10). As their name suggests, these have a pelagic lifestyle. When they die their skeletons sink to the sea bed, sometimes disaggregating into their constituent plates, termed coccoliths (Figure 10.11). Chalk limestones also contain calcispheres (microscopic balls of unknown origin), small pelagic globigerinid foraminifera, and a sparse macrofauna. This includes echinoids, brachiopods, lamellibranchs, sponges,

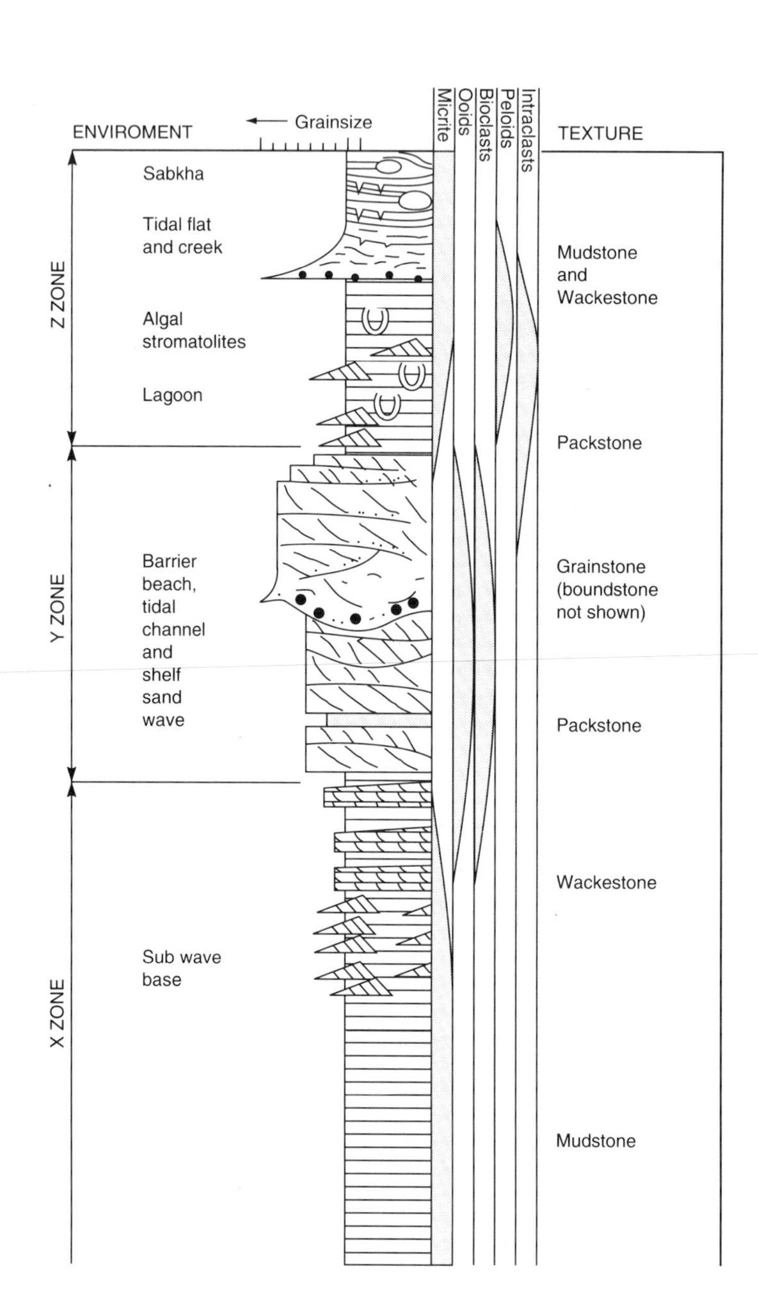

Figure 10.7 Cartoon to show the vertical variation in grain type and texture within a genetic increment of a prograding accretionary ramp.

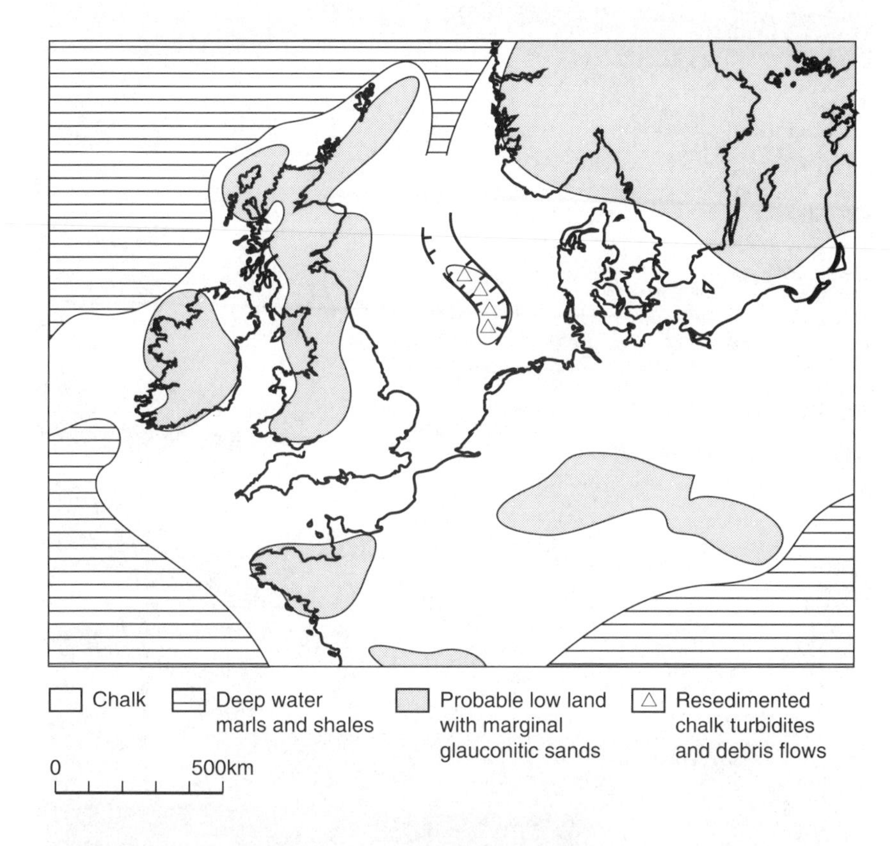

Figure 10.8 Map to show the original distribution of the Chalk of northwest Europe (simplified from Ziegler, 1982). This was deposited as a fine mud over a deeply drowned continental shelf. Little land remained above the sea, and contributed small amounts of sand that was deposited in glauconitic shoals around the islands. Chalk dissolved in deeper water below the continental shelf edge, where only clays were deposited.

Figure 10.9 Photograph of the Upper Cretaceous chalk cliffs of Beer, Devon, showing the regular bedding of hardground horizons.

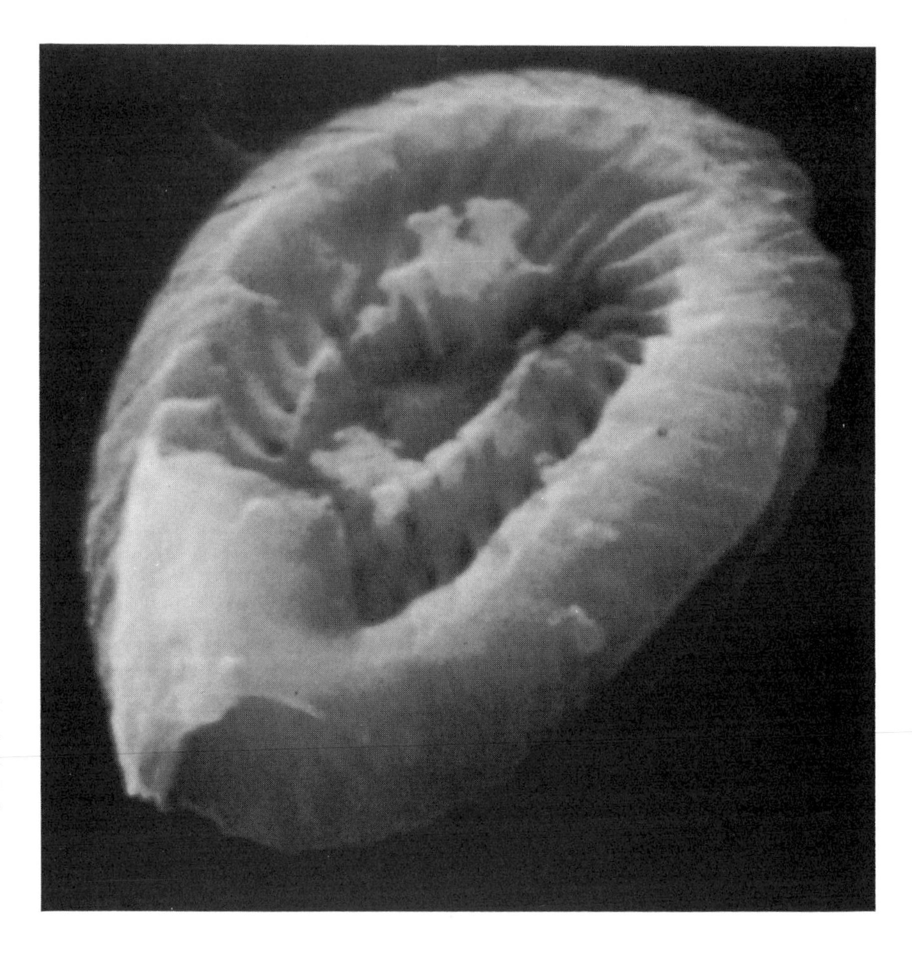

Figure 10.10 Scanning electron micrograph of a single coccosphere from the chalk showing constituent coccoliths. Lower Chalk, Folkestone × 11 000. By courtesy of J. Young.

bryozoa and solitary corals. Chalk is often extensively bioturbated, with the ichnogenera *Thalassinoides* and *Zoophycus* (Bromley, 1967). Chert nodules and beds are common, known by the natives of southern England as 'flint'. Chert often preferentially replaces burrows (Figure 10.9).

The uniform depositional nature of the Chalk sea is demonstrated by the existence of laterally persistent 'hardground' horizons. These are intraformational erosion surfaces that are commonly bored and phosphatized. They are encrusted by the shells of beasts that grew on a rocky substrate. Each erosion surface is overlain by a thin layer of argillaceous chalk with an intraformational basal conglomerate of chalk class. Individual 'hardground' horizons can be traced for hundreds of kilometres across North-west Europe (Jefferies, 1963). Locally, however, rifts developed across the shelves, as in the Central Graben of the North Sea. These were infilled with chalk reworked from the shelves and deposited from turbidity currents and debris flows (Kennedy, 1986).

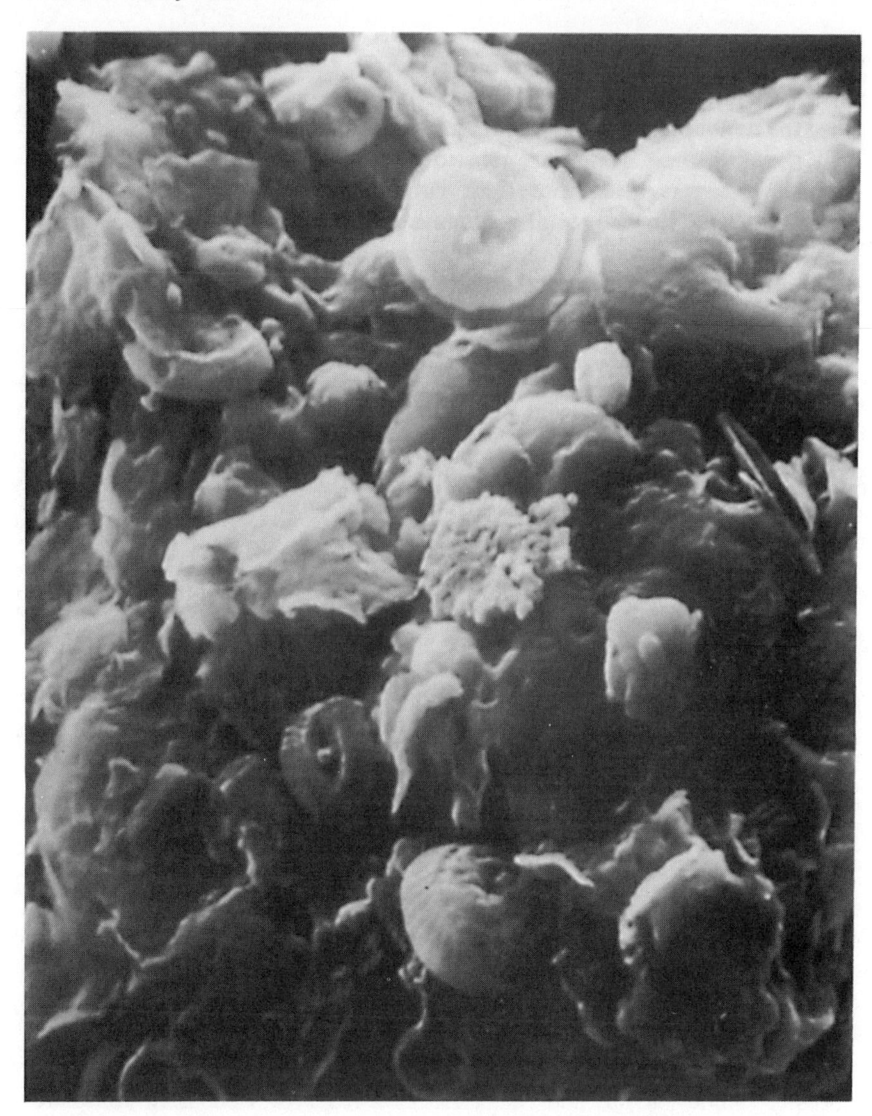

Figure 10.11 Scanning electron micrograph of chalk showing the coccolithic nature of the deposit. Lower Chalk, Folkestone × 2500. By courtesy of J. Young.

Interpretation

The fine grainsize of the Cretaceous Chalk indicates deposition in a low-energy environment. Cross-lamination and cross-lamination indicative of traction current activity are notably absent. The fauna is obviously marine. The absence of benthonic algae implies that the sea floor was below the limit of the photic zone. Detailed palaeontological studies suggest water depths of between 200–600 m (Hakansson *et al.*, 1974).

The widespread distribution of the Chalk, and the remarkable continuity of intimate details of its stratigraphy, indicate that these sediments were deposited during a phase of tectonic quiescence, when there was a global

Figure 10.12 Photograph of chalk showing preferential chertification of burrows. Annis's Knob, Beer, Devon.

marine transgression across the continental shelves (Hancock, 1989). This has been interpreted to indicate a uniform warm climate. Cretaceous chalks thus provide a good example of an outer shelf environment.

ECONOMIC ASPECTS

Carbonate shelf deposits are of great economic significance. Many of the world's major petroleum accumulations occur in shelf carbonates of Iraq, Iran, Libya and the Williston basin (see Dunnington, 1958; Falcon, 1958; Colley, 1964). These accounts show that petroleum is largely reservoired in the shoal carbonates and associated reefs. These often retain their primary porosity (some Libyan and Saudi Arabian reservoirs occur in completely unconsolidated carbonate sand). Alternatively, they may have been cemented, but secondary porosity can form by dolomitization and leaching.

The nearshore low-energy facies generally has less favourable reservoir characteristics. Though primary porosity may have been high, permeability is normally low, because of the fine grainsize of these muddy sediments. Subsequent cementation may destroy all porosity. Fracturing, dolomitization and meteoric leaching can be important, however, in upgrading the reservoir potential of this facies. The caprocks of shelf carbonate reservoirs are generally sabkha evaporites, though micritic chalks can also fill this role.

Micrites are seldom prospective. They may have porosity, but because of their fine grainsize, they generally lack permeability unless they are extensively fractured. The Cretaceous Austin Chalk of Texas produces petroleum from fracture systems adjacent to faults. Ekofisk and associated fields of the North Sea produce from Cretaceous chalk reservoirs. This is where chalk has been fractured due to uplift over Zechstein (Permian) salt domes (Byrd, 1975; Scholle, 1977; D'Heur, 1984).

The regionally persistent stratigraphy of shelf carbonates is of great significance since it allows the formation of laterally extensive reservoirs. For example, shoal limestones of the Upper Jurassic Arab-D in Saudi Arabia are host to the Ghawar oil field which is over 30 km long. The gentle but regionally persistent dips of carbonate shelf deposits allow the local accumulation of hydrocarbon reservoirs from vast catchment areas. It can be seen, therefore, that carbonate shelf deposits are of some interest in the search for oil and gas. Wilson (1975) and Reeckmann and Friedman (1982) give further details of the application of facies analysis to the exploration for petroleum in carbonate reservoirs.

Carbonate shelf deposits are economically interesting for other reasons since they can contain evaporites and phosphates. The origin of phosphates is a complex and controversial problem and the environmental parameters which control their formation are not well understood (e.g. Bromley, 1967). Valuable phosphate deposits occur in Upper Cretaceous shelf carbonates of the Middle East. These are found in a belt stretching from Syria (the Soukhne Formation) through the Wadi Sirhan of Saudi Arabia to Jordan, Palestine, Egypt and Morocco (Tooms *et al.*, 1971).

Though it is clear from their facies relationships that these phosphates are marine deposits their precise environment of formation seems to be variable. In Egypt, Said attributes them to regressive shoreline conditions (1962, pp. 251, 268) while Youssef attributes them to deposition in syn-sedimentary basins on the deeper parts of the sea bed (Youssef, 1958). In Jordan on the other hand the phosphatic facies is restricted to palaeohighs (Bender, 1968, p. 189). There is an interesting problem here (Notholt, 1980; Baturin, 1982).

DIAGNOSIS OF CARBONATE SHELF DEPOSITS

The techniques for diagnosing shelf deposits vary greatly according to whether they are carbonate or terrigenous. As pointed out in Chapter 1, terrigenous environments are largely diagnosed from sequences of sedimentary structures and grainsize, while carbonate environments are largely diagnosed by petrography. Thus the diagnosis of carbonate rocks is largely made from the microscopic examination of etched surfaces and thin sections from out crops, and, in the subsurface, from well cuttings, and cores, when they are available. The environment of a particular lithology may be determined by its texture and grain-type as shown in Figures 10.6 and 10.7.

Moving from basin centre to basin margin the texture of the carbonate will change from micrite to wackestone to packstone to grainstone (or boundstone in reef shorelines) and back to packstone and micrite in the inner shelf lagoon. Correspondingly the grain type varies from lime mud (with pelagic fossils) to skeletal grains of benthonic fossils and/or ooliths, followed by peloidal grains and intraclasts in the lagoons and tidal flats.

These are of course gross over-simplifications, and environmental diagnosis of carbonate rocks should not be attempted without practical studies of modern carbonate sediments and microscopic petrography of ancient ones.

Seismic sections can delineate progradational reflecting horizons in accretionary ramps, and the more dramatic geometries of carbonate shelf edge 'drop-offs' (Figure 10.13).

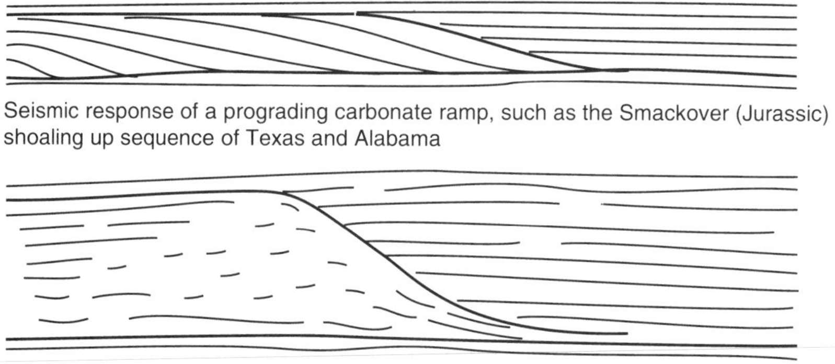

Figure 10.13 Cartoons to show the seismic expression of carbonate accretionary ramp (upper), and a 'drop-off' shelf edge (lower). For a real example of the former see Figure 11.5.

Cross-bedding can be measured in outcrops of carbonate shelf sediments, but they present similar problems of interpetation to those encountered in terrigenous shelf sands, as discussed in the previous chapter. Unimodal, bimodal and polymodal azimuth plots have been recorded similar to those in terrigenous shelf sands. In boreholes the dipmeter and imaging tools are of less use than in sandstones. This is because carbonates commonly undergo more diagenesis than do the terrigenous sands. Thus porosity is largely secondary solution related, so resistivity variations detected by the dipmeter and imaging tools can seldom delineate the direction of dip of the original sedimentary features. There are some exceptions.

REFERENCES

The Cretaceous Chalk case history was based on the author's own observations, together with:

Byrd, W. D. (1975) Geology of the Ekofisk field, offshore Norway. In A. W. Woodland (ed.), *Petroleum and the Continental Shelf of Northwest Europe*, Applied Science Publishers, London, 439–46.

Hakansson, E., Bromley, R. and Perch-Nielsen, K. (1974) Maastrichtian chalk of northwest Europe, a pelagic shelf sediment. In K. J. Hsu and H. C. Jenkyns (eds), *Pelagic Sediments: On Land and Sea*, Blackwell Scientific Publications, Oxford, 221–33.

Hancock, J. M. (1989) Sea level changes in the British region during the Late Cretaceous, *Proc. Geol. Ass. Lond.*, **100**, 565–94.

—— (1990) Cretaceous. In K.W. Glennie (ed.), *Introduction to the Petroleum Geology of the North Sea*, Blackwell Scientific Pubs, Oxford, 255–72.

D'Heur, M. (1984) Porosity and hydrocarbon distribution in the North Sea chalk reservoirs, *Mar. and Pet. Geol.*, **1**, 211–39.

Jefferies, R. (1963) The stratigraphy of the *Actinocamax plenus* subzone (Turonian) in the Anglo-Paris Basin, *Proc. Geol. Ass.*, **74**, 1–34.

Kennedy, W. J. (1986) Sedimentology of Late Cretaceous Palaeocene Chalk reservoirs, North Sea Central Graben. In J. Brooks and K. W. Glennie (eds), *Petroleum Geology of North West Europe*, Graham & Trotman, London, 469–81.

Scholle, P. A. (1977) Chalk diagenesis and its relation to petroleum exploration: oil from chalks, a modern miracle, *Amer. Assoc. Petrol. Geol. Bull.*, **61**, 982–1009.

Tyson, R. V. and Funnell, B. M. (1987) European Cretaceous shorelines, stage by stage, *Palaeogeog. Palaeoclimatol. Palaeoec.*, **59**, 69–91.

Ziegler, P. A. (1982) *Geological Atlas of Western and Central Europe*, Shell Internationale Petroleum Maatschappij B.V, The Hague.

Other references cited in this chapter were:

Ball, M. M. (1967) Carbonate sand bodies of Florida and the Bahamas, *Jl. Sediment. Petrol.*, **37**, 556–91.

Baturin, G. N. (1982) *Phosphorites on the Sea Floor: Origin, Composition and Distribution*, Elsevier, Amsterdam.

Bebout, D. G. and Pendexter, C. (1975) Secondary carbonate porosity as related to Early Tertiary depositional facies, Zelten field, Libya, *Bull. Amer. Assoc. Petrol. Geol.*, **59**, 665–93.

Bender, F. (1968) Der Geologie von Jordanien, *Beitrager Reg. Geol. Erde. Bd.*, **7**, Borntrager, Berlin.

Bromley, R. G. (1967) Marine phosphorites as depth indicators. In A. Hallam (ed.), Depth Indicators in Marine Sedimentary Environments, Marine Geol. Sp. Issue, **5**, 503–10.

Colley, B. B. (1964) Libya: Petroleum Geology and development, *Sixth World Petrol. Cong. Proc.*, Section **1**, 1–10.

—— (1971) Stratigraphy and lithofacies of Lower Paleocene rocks, Sirte Basin, Libya. In C. Gray (ed.), *Symposium on the Geology of Libya*, University of Libya, Tripoli, 127–40.

Dunnington, H. V. (1958) Generation, migration, accumulation, and dissipation of oil in Northern Iraq. In L. G. Weeks (ed.), *Habitat of Oil*, Amer. Assoc. Petrol. Geol., 1194–1252.

Enos, P. (1983) Shelf. In P. A. Scholle, D. B. Bebout and C. H. Moore (eds), *Carbonate Depositional Environments*, Amer. Assoc. Petrol. Geol., **33**, 267–96.

Evans, G., Schmidt, V., Bush, P. and Nelson, H. (1969) Stratigraphy and geologic history of the sabkha, Abu Dhabi, Persian Gulf, *Sedimentology*, **12**, 145–59.

Falcon, N. L. (1958) Position of oil fields of southwest Iran with respect to relevant sedimentary basins. In L. G. Weeks (ed.), *Habitat of Oil*, Amer. Assoc. Petrol. Geol., 1279–93.

Fischer, A. G. (1964) The Lofer cyclothem of the Alpine Triassic, *Kansas Geol. Surv. Bull.*, **169**, 107–49.

Halley, R. B., Harris, P. M. and Hine, A. C. (1983) Bank margin In P. A. Scholle, D. G. Bebout and C. H. Moore (eds), *Carbonate Depositional Environments*, Amer. Assoc. Petrol. Geol., **33**, 463–506

Holser, W. T. (1966) Diagenetic polyhalite in recent salt from Baja, California, *Am. Mineralogist*, **51**, 99–109.

Illing, L. V. (1954) Bahaman calcareous sands, *Bull. Amer. Assoc. Petrol. Geol.*, **38**, 1–95.

Kinsman, D. J. (1969) Modes of formation, sedimentary association and diagnostic features of shallow-water and supratidal evaporites, *Bull. Amer. Assoc. Petrol. Geol.*, **53**, 830–40.

Kirkland, D. W. and Evans, R. (1973) Marine evaporites: origins, diagenesis and geochemistry, *Benchmark Papers in Geology*, Dowden, Hutchinson & Ross Inc., Stroudsburg, Penn.

Lees, A. (1973) Platform carbonate deposits, *Bull. Centre Rech. Pau.*, **7**, 177–92.

Meischner, K. D. (1965) Allodapische Kalke, turbidite in Riff-Nahen Sedimentations Becken. In A. Bouma and A. Brouwer (eds), *Turbidites*, Elsevier, Amsterdam, 156–91.

Newell, N. D., Purdy, E. G. and Imbrie, J. (1960) Bahaman oolitic sand, *Jour. Geol.*, **68**, 481–97.

Notholt, A. J. G. (1980) Economic phosphatic sediments: mode of occurrence and stratigraphic distribution, *Jl. Geol. Soc. Lond.*, **137**, 793–805.

Purdy, E. G. (1963) Recent calcium carbonate facies of the Great Bahama Bank, *Jour. Geol.*, **71**, 334–55 and 477–97.

Purser, B. H. (ed.) (1973) *The Persian Gulf (Holocene Carbonate Sedimentation and Diagenesis in a Shallow Epicontinental Sea)*, Springer-Verlag, Berlin.

Read, J. F. (1985) Carbonate depositional models, *Bull. Amer. Assoc. Petrol. Geol.*, **69**, 1–29.

Reeckmann, A. and Friedman, G. M. (1982) *Exploration for Carbonate Petroleum Reservoirs*, J. Wiley & Sons, Chichester.

Said, R. (1962) *The Geology of Egypt*, Elsevier, Amsterdam.

Scholle, P. A. (1977) Chalk diagenesis and its relation to petroleum exploration: oil from chalks, a modern miracle, *Amer. Assoc. Petrol. Geol. Bull.*, **61**, 982–1009.

Shearman, D. J. (1963) Recent anhydrite, gypsum, dolomite and halite from coastal flats of the Arabian shore of the Persian Gulf, *Proc. Geol. Lond.*, **1067**, 63–5.

—— (1966) Origin of marine evaporites by diagenesis, *Bull. Inst. Min. Met.*, **76**, section B, 82-6.

Shinn, E. A. (1983) Tidal flat. In P. A. Scholle, D. G. Bebout and C. H. Moore (eds), *Carbonate Depositional Environments*, Amer. Assoc. Petrol. Geol., **33**, 171–210.

Tebbutt, G. E., Conley, C. D. and Boyd, D. W. (1965) Lithogenesis of a distinctive carbonate rock fabric, *Univ. Wyoming Contrib. Geol.*, **4**, 1–13.

Tooms, J. S., Summerhayes, C. P. and McMaster, R. L. (1971) Marine geological studies on the northwest African margin: Rabat-Dakar. In *The Geology of the East Atlantic Continental Margin*, 4, Africa. Inst. Sci. Rept. **70**(16), 11–25.

Wilson, J. L. (1975) *Carbonate Facies in Geologic History*, Springer-Verlag, Berlin.

—— and Jordan, C. (1983) Middle shelf. In P. A. Scholle, D. G. Bebout and C. H. Moore (eds), *Carbonate Depositional Environments*, Amer. Assoc. Petrol. Geol., **33**, 297–344.

Youssef, M. I. (1958) Association of phosphates with synclines and its bearing on prospecting for phosphates in Sinai, Egypt, *J. Geol.*, **2**, 75–87.

11 Reefs

INTRODUCTION

As mentioned in the previous chapter, reefs are a particularly important feature of carbonate shelves, and they deserve their own chapter. The term 'reef' was originally applied to rocky prominences on the sea floor on which ships could be wrecked. Coral reefs are a particular form of this navigational hazard found today in tropical waters. Geologists have applied the term reef to lenses made of the calcareous skeletons of sedentary organisms. In many cases, however, it is not possible to demonstrate either that an ancient 'reef' was a topographic high on the sea floor or, if it was, that it was wave resistant. Cummings (1932) classified calcareous skeletal deposits into:

Bioherm 'A reef, bank, or mound; or reeflike, moundlike, or lens-like or otherwise circumscribed structures of strictly organic origin, embedded in rocks of different lithology' (*ibid.,* p. 333).

Biostrome 'Purely bedded structures such as shell beds, crinoid beds, coral beds, et cetera, consisting of and built mainly by sedentary organisms, and not swelling into moundlike or lens-like forms . . . which means a layer or bed' (*ibid.*, p. 334).

Subsequently the term 'reef' has been applied very loosely to lenses of carbonate, often even on the basis of seismic data prior to drilling. Dunham (1970), Braithwaite (1973), and Heckel (1974) have reviewed the problems of reef identification and classification. It is extremely difficult to place a lens of carbonate rock into one of the rigidly defined classes that some nomenclatural schemes propose. This is not only because there is a limited amount of rock available for study, in the form of cuttings and cores, but also because diagenesis may have destroyed the critical evidence of fauna and texture on which a definition of 'reef' may hinge. For the purposes of this book the following definitions will be used:

Carbonate build-up A lens of carbonate rock (a descriptive term). Carbonate build-ups may be divided into two genetic types.

Reef A carbonate build-up of skeletal organisms which at the time of formation was a wave resistant topographic feature which rose above the general level of the sea floor.

Bank A carbonate build-up which was a syn-depositional topographic high of non-wave resistant material, e.g. an oolite shoal, a coquina bank, or a mound of crinoid debris.

This chapter describes ancient carbonate build-ups which have been interpreted as reefs. First, though, the features of Recent reefs will be summarized. Coral reefs are one of the best documented of all present-day environments. The following brief synthesis is based on the references listed at the end of this chapter.

Summary of Recent reefs

The majority of reefs growing at the present time occur in shallow tropical seas. The factors which restrict reef growth are variable but in general they form best in less than 25 fathoms of water where the salinity is between 27 and 40 per thousand and the temperature seldom drops below about 20°C (Shepard, 1963, p. 351). The inevitable exception to the rule is where coral reefs form in cold water 35 fathoms deep off the Norwegian coast (Teichert, 1958). Reefs of calcareous algae occur in non-marine waters (Fouch and Dean, 1982). The biota of Recent reefs is exceedingly diverse (Jones and Endean, 1973).

The resistant reef framework is composed of corals, calcareous algae, hydrocorallines, and bryozoa. Corals are generally actually subordinate to the other groups (Yonge, 1973). Other organisms associated with reefs include calcareous sponges, foraminifera, echinoids, lamellibranchs, gastropods, and sabellariid (carbonate-secreting) worms. In cross-section a reef complex can be broadly sub-divided into four geomorphological units (Figure 11.1).

A reef generally shelters a shallow water lagoon from the open sea. The floor of the lagoon is covered by carbonate mud in its deeper parts and sands in shallower turbulent regions. Lagoonal sediments are composed of faecal pellets, foraminiferal sands, coralgal sands of comminuted corals and

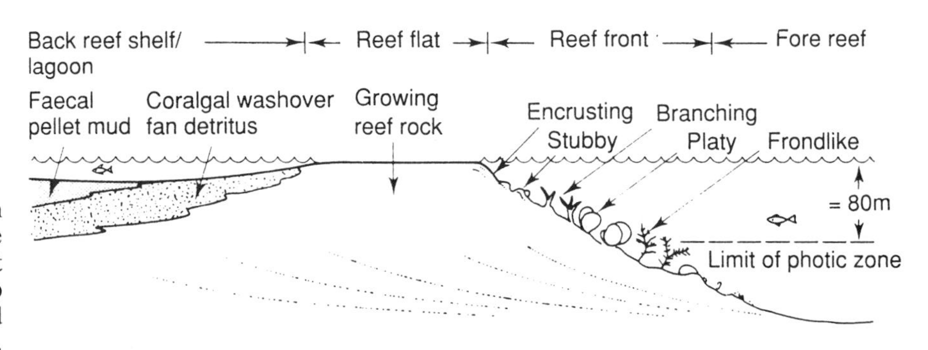

Figure 11.1 Cartoon through a reef to show the distribution of sediment types, and the relationship between ecology and environment.

calcareous algae, together with other skeletal sands and finely divided carbonate mud. Scattered through the lagoon may be irregular patch reefs. Grainsize increases across the lagoon towards the reef and locally becomes conglomeratic due to organic debris broken off the reef and carried into the lagoon by storms. The reef itself is, by definition, composed of a resistant framework of calcareous organic skeletons. The top of the reef is flat since the creatures cannot survive prolonged sub-aerial exposure. In addition the upper surface is constantly scoured and planed off by wave action and dissected by seaward trending surge channels. The tops of these sometimes become overgrown to form submarine tunnels.

The reef framework itself is often highly porous, figures of up to 80° porosity have been cited (Emery, 1956). The seaward edge of the reef, the reef front, is a submarine cliff with a talus slope at its base. This slope is made up of organic debris broken off from the reef front; grainsize decreases down slope into deeper water. Immediately at the foot of the reef front boulders of reef rock may be present, grading into sand and then mud in deeper water. The talus slope supports a fauna similar to that of the reef but the quieter conditions allow more delicate branching corals and calcareous algae to form. The reef talus has poorly developed seaward dipping bedding. Slumps and turbidites have been described from modern and ancient reef flanks. Exhumed ancient reefs sometimes display the progradational bedding of the fore-reef talus slope (Figure 11.2). This can be detected on seismic data and on well logs, as discussed later in the chapter.

It is important to put reefs in perspective as far as their size is concerned. Reefs may perhaps be best thought of as machines for making carbonate sediment. The biogenic carbonate material which grows in the reef environment is constantly being attacked by the sea. Thus relatively small areas of actual reef seem to be preserved within vast formations of coralgal sands and muds derived from them. This point is illustrated by the Great Barrier Reef of Australia. Though this modern reef tract extends for some 2000 km along the coast, few individual reefs are more than 20 km in length (Maxwell, 1968; Hopley, 1982).

Recent reefs are classified into three main kinds by their geometry. Inevitably there are transitions between the three types. Fringing reefs are linear in plan and stretch parallel to coasts with no intervening lagoons (Figure 11.3a).

Fringing reefs can form where low rainfall means that little freshwater and mud is brought into the sea to inhibit the growth of reef colonial organisms. Good examples occur along the desert shores of the Gulf of Akaba on the Red Sea (Friedman, 1968). Barrier reefs are linear too, but a lagoon separates them from the land (Figure 11.3b).

This may be narrow or, in the case of the Great Barrier Reef of Australia, an open sea many hundreds of kilometres wide (Maxwell, 1968). Atolls are sub-circular reefs enclosing a lagoon from the open sea (Figure 11.3c).

(a)

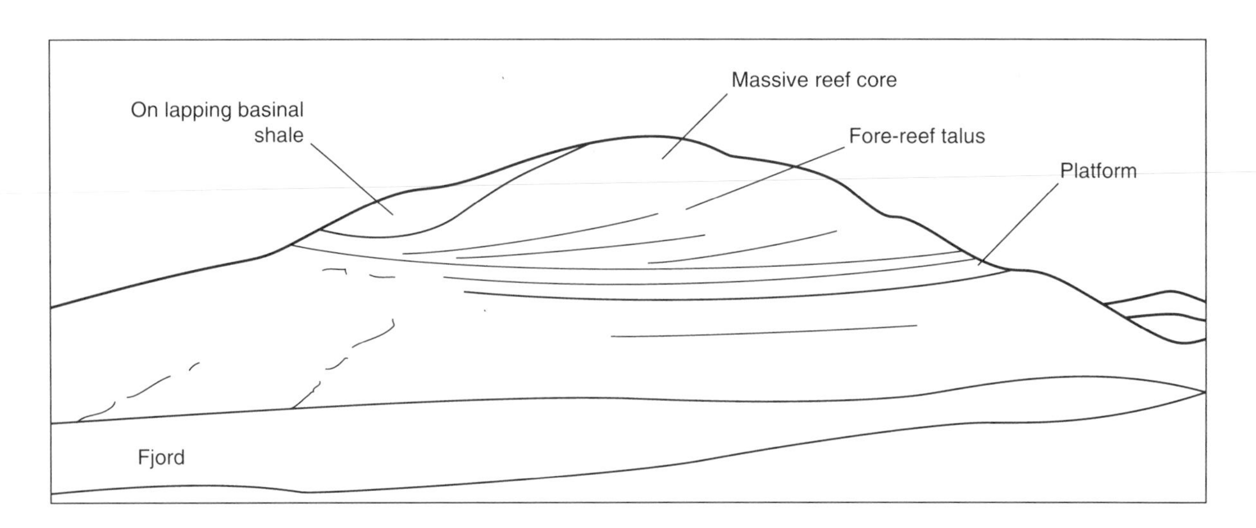

Figure 11.2 Photographs and sketches of exhumed ancient reefs to show features of their geometry. (a) Gotlandian (Silurian) reef, North Greenland. Fore-reef talus dips indicate progradation from right to left. Polar bear on ice flow provides scale. (b) The El Capitan reef (Permian), Guadalupe Mountains, west Texas (photo by courtesy of J. C. Harms, P. N. McDaniel and I. C. Pray).

(b)

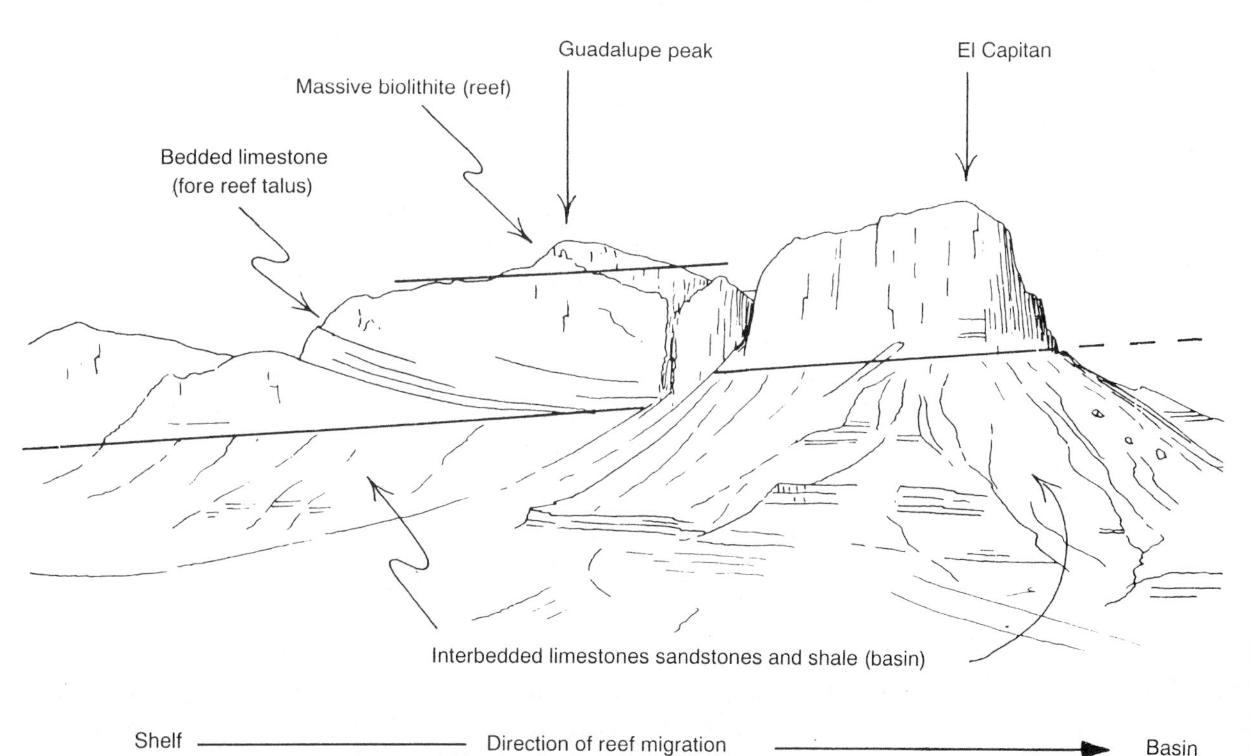
Guadalupe peak
El Capitan
Massive biolithite (reef)
Bedded limestone
(fore reef talus)
Interbedded limestones sandstones and shale (basin)
Shelf
Direction of reef migration
Basin

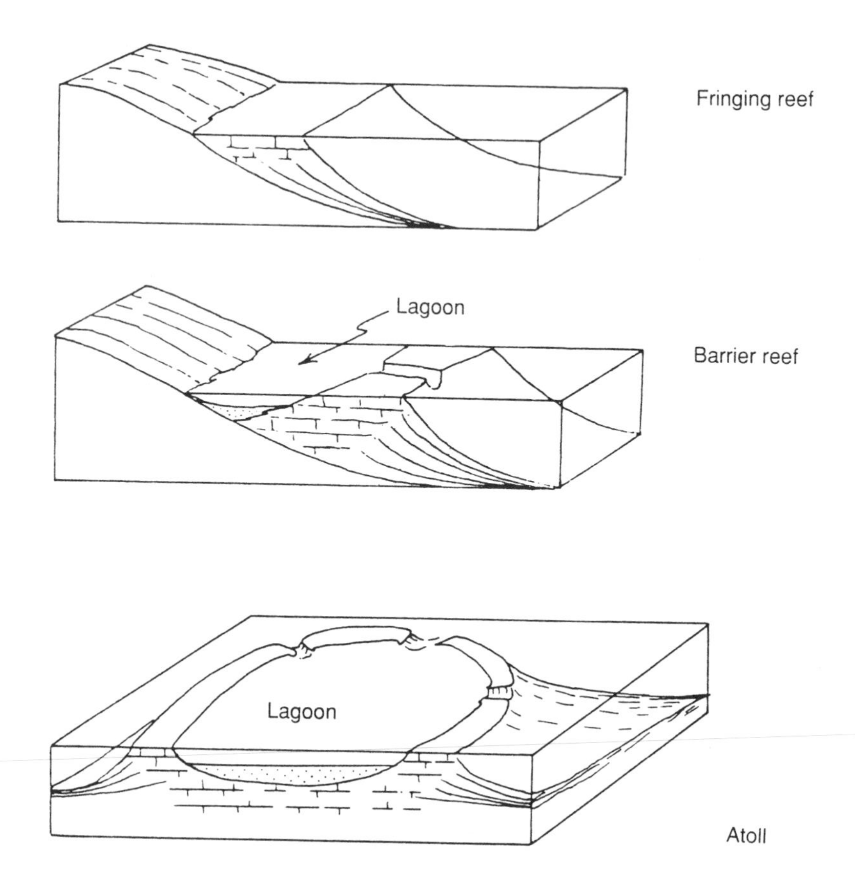

Figure 11.3 Illustrative of the three main types of present-day reefs.

This kind of reef is abundant in the Pacific Ocean. Bikini Atoll is a typical example (Emery, Tracey and Ladd, 1954). The classic theory for atoll formation was put forward by Darwin in 1842. This proposed that barrier reefs originally formed around volcanic islands. As the island sank under its own weight the barrier would build upwards to keep pace with the relative rise in sea level. Ultimately after all trace of the volcanic island had vanished beneath the sea a circular coral reef, or atoll, would remain. Studies of Recent reefs and the concepts derived therefrom can be usefully applied to the study of ancient reefs.

Before describing case histories of fossil reefs it is necessary however to make three cautionary remarks. First, the reef-building organisms of the past were not always of the same groups as those of today. The roles of different groups vary in time and place. For example, calcareous algae have in some situations merely acted as binding agents holding the skeletons of the reef framework organisms together. In other times and situations algae formed the dominant framework of reefs.

The second point to note is that Recent reefs have suffered subaerial exposure during the Pleistocene Period. Thus their present-day topography is in part a drowned one. In particular it should be noted that most modern reef fronts are old sea cliffs.

The third point concerns the relationship between Recent and ancient reef geometries. True fringing reefs appear to be rare in the geological record. Linear reef complexes are common but either are of the barrier type, separating open marine deposits from lagoonal facies, or lie on structural highs between basinal troughs.

Fossil atolls are rare. Sub-circular reef complexes have been described as atolls but are not based on volcanic piles. Examples include the Horseshoe Atoll and Scurry-Snyder reefs (Pennsylvanian) of west Texas (Waite, 1993) and the El Abra reef of Mexico (Wilson and Ward, 1993). The second major type of ancient reef, in addition to barriers, is essentially a circular reef core rimmed by reef slope talus. The core may provide shelter for a lagoon on its leeward side. This type of structure is termed a patch reef, or if markedly conical in vertical profile, a pinnacle reef. A case history of an ancient barrier reef will now be described.

THE BU HASA OIL FIELD, ABU DHABI

Description

The Bu Hasa oil field was discovered in onshore Abu Dhabi. It produces from the Shuaiba Formation (Aptian) at the top of the Thamama Group. The Bu Hasa field occupies a local culmination on a Cretaceous carbonate shelf margin that developed between the Arabian shield to the south and the Gulf trough to the north (Figure 11.4).

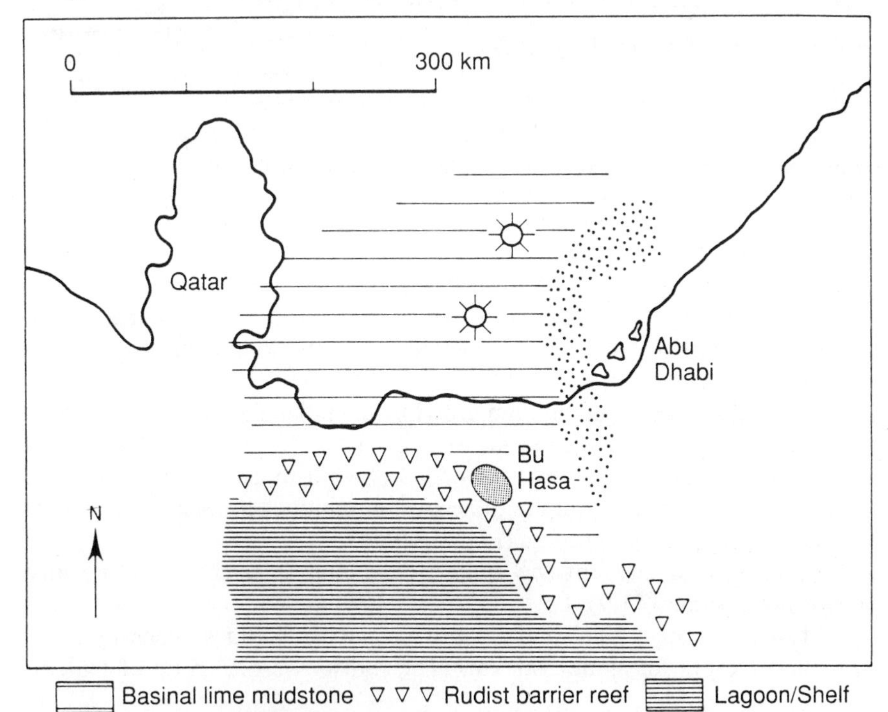

Figure 11.4 Map to shown the regional setting of the Cretaceous Bu Hasa reef oil field, Abu Dhabi. Based on Harris *et al.* (1968), Twombley and Scott (1975) and Hassan *et al.* (1975).

Seismic data show that the Shuaiba Formation prograded out across the shelf over the flat-lying Kharaib Formation. There is clear evidence of emergence of the post-Shuaiba surface, with the Bab Formation onlapping its seaward-dipping slope.(Figure 11.5).

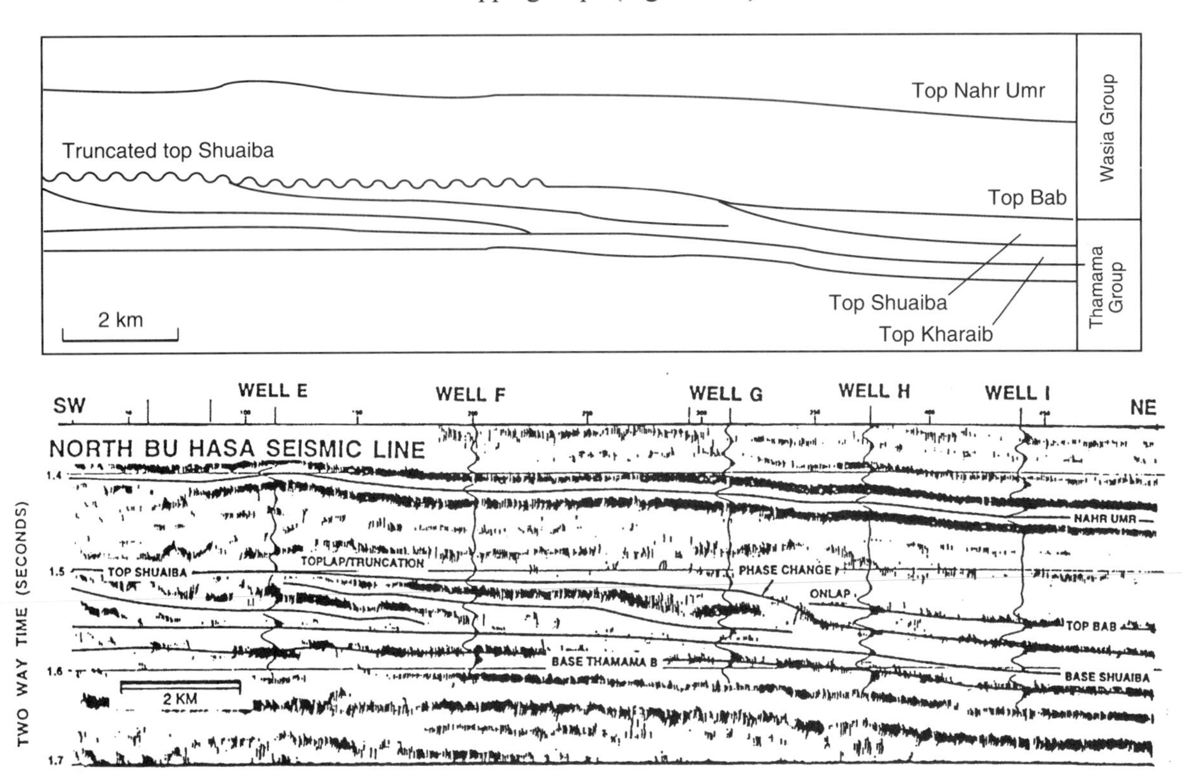

Figure 11.5 Seismic line and interpretation through the Bu Hasa field, Abu Dhabi, UAE. Note evidence for basinward progradation of the Shuaiba Formation reef. Evidence for post-Shuaiba emergence is provided by the truncation of the progrades, and by the onlap of the Bab Formation. The development of solution porosity in the upper part of the Shuaiba reservoir is related to this emergent phase, which resulted in meteoric flushing, and the formation of secondary solution porosity. Developed from Calavan *et al.* (1992).

Detailed petrographic studies defined a large number of carbonate microfacies. A simplified cross-section showing their distribution across the field is shown in Figure 11.6.

The Bu Hasa carbonate buildup growth was initiated on a laterally extensive biostrome formed by the encrusting calcareous alga *Bacinella*. This unit is overlain on the shelf by Lithocodium boundstone (Figure 11.7).

This passes basinward into argillaceous, bituminous lime mudstones with occasional pelagic foraminifera (Figure 11.8).

The basinward edge of the *Lithocodium* unit is overlain by a rudist boundstone (Figure 11.9).

This is about 60 m thick and 6 km wide and can be traced laterally along the shelf margin for some 200 km. (The rudists were a group of lamelli-

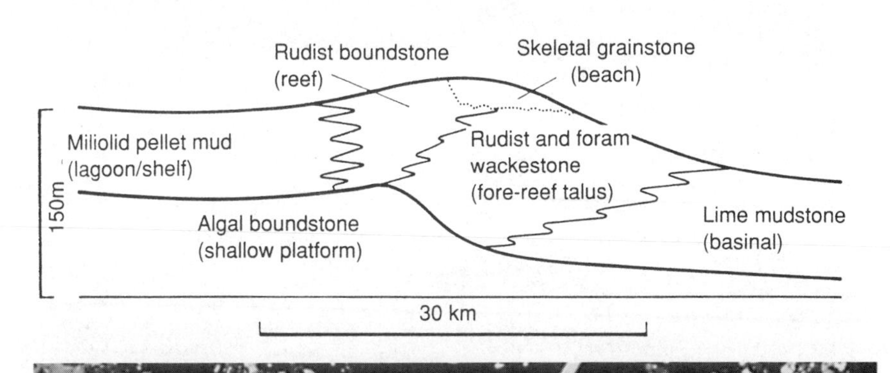

Figure 11.6 Cross-section to show the facies and environments (in brackets) of the Bu Hasa reef oil field, Abu Dhabi. The skeletal grainstone on the reef rim provides evidence of emergence. It probably developed as a beach deposited on a wave-cut bench during the post-Shuaiba drop in sea level. Based on Harris *et al.* (1975), Twombley and Scott (1975) and Hassan *et al.* (1975).

Figure 11.7 Photomicrograph of the algal boundstone on which the Bu Hasa reef developed.

Figure 11.8 Photomicrograph of the lime mudstone deposited in the basin to the north of the Bu Hasa reef.

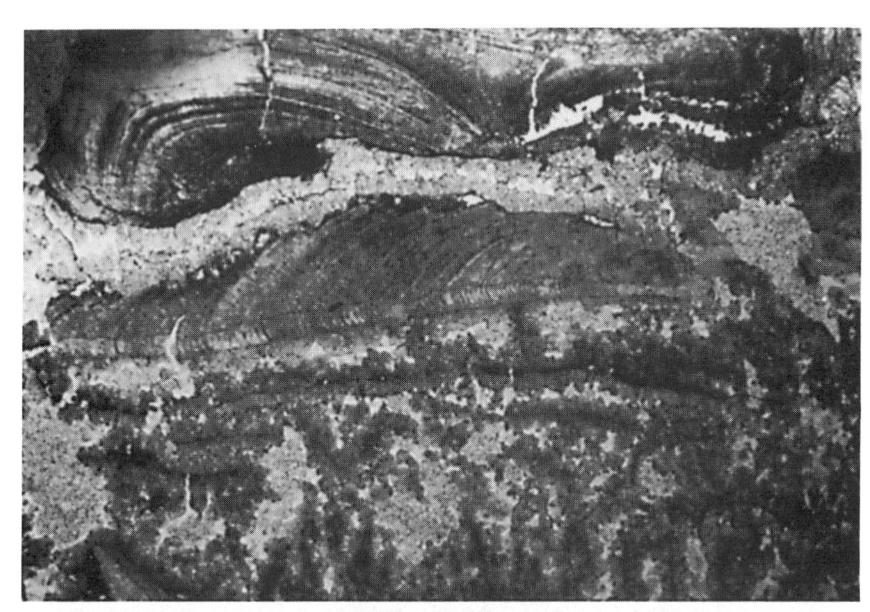

Figure 11.9
Photomicrograph of Rudist reef rock from the Bu Hasa field.

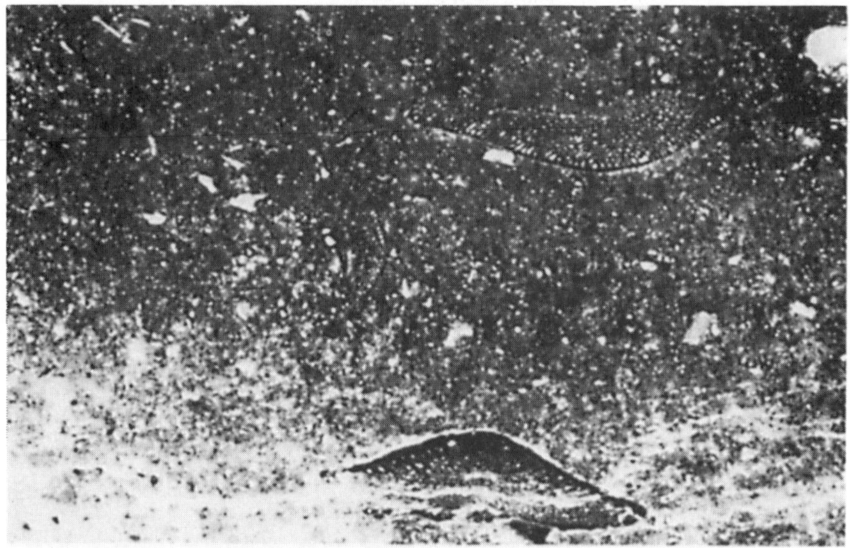

Figure 11.10
Photomicrograph of fore-reef skeletal wackestone with Orbitoline forams and Rudist debris.

branchs that assumed a coralline habit, with one valve resembling the calyx and the other a protective lid. They were extremely successful reef builders in the Cretaceous Period, and provide petroleum reservoirs in Libya, Mexico and USA, as well as in the Gulf.) The Bu Hasa rudist boundstones pass basinward into skeletal wackestones, with fragments of rudists and large benthonic Orbitoline forams (Figure 11.10). These wackestones pass basinward into lime mudstones.

The rudist boundstone passes south towards the Arabian shield into faecal pellet muds, with miliolid foraminifera. (Figure 11.11).

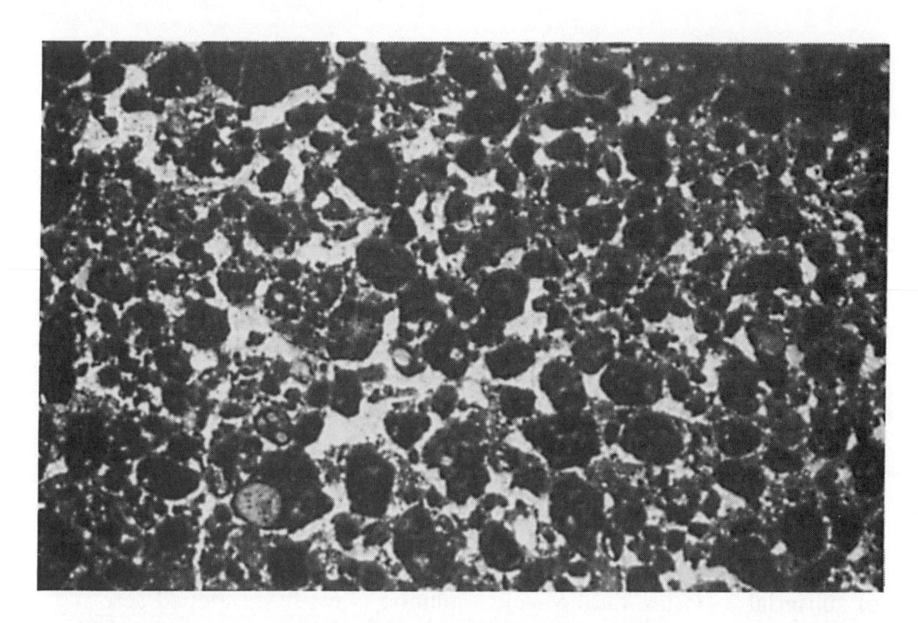

Figure 11.11
Photomicrograph of faecal pellet packstone from the shelf behind the Bu Hasa barrier reef.

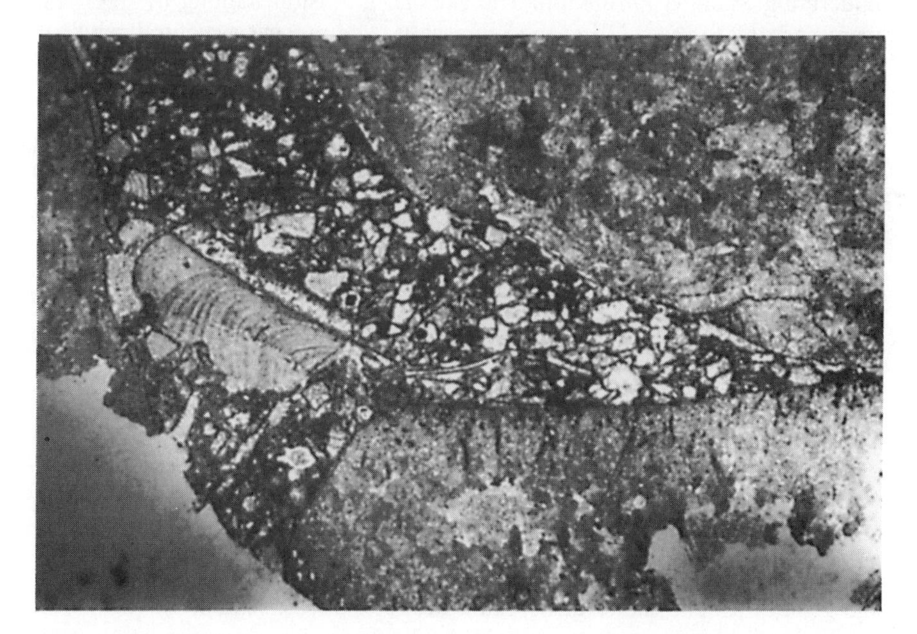

Figure 11.12
Photomicrograph of Rudist grainstone from the seaward rim of the Bu Hasa reef.

Locally the basinward crest of the rudist boundstone is replaced by a detrital rudist grainstone (Figure 11.12).

The top of the Shuaiba Formation is marked by a regional unconformity. It is overlain by the organic-rich shales of the Nahr Umr Formation (Albian).

Interpretation

The *Bacinella* and *Lithocodium* boundstones were clearly deposited in shallow water, within the photic zone required for algal development. The lateral transition of the *Lithocodium* boundstone into the lime mudstones suggest the initiation of topographic relief on the sea bed. The lime mudstones, with their pelagic forams, must have been deposited in quieter deeper water below the limit photic zone. As with so many reefs, the seaward limit of the shelf favours the establishment and ebullient reproduction of carbonate-secreting organisms. Thus the seaward rim of the *Lithocodium* boundstone appears to have provided ideal conditions for the development of a rudist barrier reef. The *Orbitolina* wackestones on the seaward side of the reef presumably represent the fore-reef talus that graded out into the basinal lime mudstones. The Miliolid-bearing faecal pellet muds behind the reef are analogous to the lagoonal deposits of recent carbonate shelf environments.

The skeletal grainstones along the seaward rim of the reef front are perhaps beach deposits. The pre-Albian unconformity marks an extensive event of subaerial exposure during which solution porosity developed across the underlying Shuaiba Formation. The skeletal grainstones along the reef rim were perhaps formed during the regression. The present structural relief of the Bu Hasa dome measures some 35 by 20 km. It is not clear to what extent this relief is structural, depositional or erosional.

The distribution of porosity and permeability in the Bu Hasa field locally cross-cuts the facies boundaries, and is in part due to solution associated with the post-Shuaiba unconformity. The reef core provides excellent reservoir, though much of the porosity is biomoldic, as well as primary intergranular. The lagoonal pellet muds also act as reservoir where matrix-moldic porosity has been leached out adjacent to the unconformity.

This case history thus demonstrates how petrography aids the environmental interpretation of carbonate environments. It also shows, alas, that facies analysis does not necessarily aid the prediction of porosity and permeability in carbonate petroleum reservoirs.

GENERAL DISCUSSION OF REEFS

In this section the following three topics are discussed: factors controlling the geometries and facies distribution of reefs; the association of reefs with evaporites and euxinic shales; and lastly the diagenesis of reef rocks.

Factors controlling reef geometries and facies

The shape of a reef and the distribution of associated facies are controlled by the interplay of sea-level changes, tectonic setting, biota, and oceanography. It is widely held that ancient reefs grew in shallow water (absolute depth unspecified). This conclusion is based on analogy with Recent reefs. Furthermore, the sedimentary facies associated with ancient reefs generally suggest shallow shelf environments.

If it is true that ancient reefs grew in shallow water their growth must have been closely controlled by fluctuations of sea level. Reef organisms cannot build up above the sea since they are killed by prolonged sub-aerial exposure. Similarly they cannot grow at great depths because algae, being plants, only thrive within the photic zone. Many living coelenterates are similarly restricted because they have symbiotic relationships with zooxanthellae (a type of algae) within their living tissues. When the sea level remains static the reef will prograde seaward over its own talus slope, as in the case of the Permian reefs of west Texas. If sea level rises slowly the reef will build essentially upwards with no lateral facies migration, or it will transgress landward over the back reef lagoonal facies. Examples of this occur in the Devonian reefs of the Canning basin, Australia (Playford and Lowry, 1966; Southgate *et al.*, 1993).

A rapid rise in sea level will kill a reef due to great depth. A slow drop in sea level will cause the reef to migrate seaward and downward. However, such sequences are relatively rare since, as the shoreline retreats, old reefs tend to be destroyed by sub-aerial erosion. A rapid drop in sea level will kill a reef instantaneously due to prolonged exposure. Thus fluctuating sea level exerts an important effect on the geometry of a reef and its associated facies. Modern concepts of sequence stratigraphy have been enthusiastically applied to carbonate shelves in general, and reefs in particular (Loucks and Sarg, 1993).

The second main controlling factor of reef geometry is tectonic setting, since this is also important in bringing the sea bed in crucial juxtaposition to sea level. Basically reefs are typical of tectonic shelves where sedimentation is shallow, marine, and free from land-derived clastics. Within this broad realm four main sub-types can be recognized. First, reefs very commonly form at the edge of a shelf where it passes into a deeper basin. Along such trends barrier reefs may form (e.g. The Permian reefs of west Texas (Horvoka *et al.*, 1993; Mutti and Simo, 1993) or discontinuous lines of patch reefs such as the Leduc (Devonian) complex of Canada (Walls, 1983).

Sometimes the edge of the shelf may be a fault with reefs built along the crest of the fault scarp. The Cracoe and Craven reefs of north England may be a case in point (Bond, 1950). Sometimes a shelf is too deep for reef-formation but local syn-sedimentary movement along anticlinal crests bring the sea bed within a depth sufficiently shallow for reefs to develop. The Palaeocene Intisar reefs of Libya are one example of this (Terry and Williams, 1969, p. 45), the Clitheroe reefs in the Lower Carboniferous Bowland trough of England are another (Parkinson, 1944). Similarly volcanic eruptions on the sea floor can build piles of lava up into shallower depths where their crests may be colonized by reef organisms. The Devonian Hamar Lagdad reef of the Tafilalet basin, Morocco, is of this type (Massa, 1965).

As already mentioned, ancient atoll-capped volcanoes are rare. This is possibly because such features are essentially oceanic and unlikely to be preserved in areas at present accessible to the land-based geologist. The fourth arrangement of reefs on shelves is where patch reefs are distributed at random over a wide area. Such archipelagoes, or buckshot patterns as they are sometimes colourfully called, occur in the Niagaran (Silurian) beds on the edge of the Canadian Shield (Lowenstam, 1950).

The third controlling factor of reef geometry and facies to consider is the biota. The essential control of water depth has been noted already. One of the curious features of reefs is that though they range in age from PreCambrian to the present day, and show similar geometries and subfacies, their fauna varies through time. PreCambrian algal biostromes have been classified into several types according to their geometry and internal anatomy. Opinions are divided as to whether these differences are of stratigraphic or ecological control (Logan, Rezak, and Ginsburg, 1964).

Lower Cambrian rocks in many parts of the world contain reefs built by curious beasts of uncertain affinities termed Archaeocyathines. Reefs are widespread in Palaeozoic rocks throughout the world. The framework of these was built principally by stromatoporoids, rugose corals, hydrocorallines, and bryozoa. Rugose and tabulate corals largely became extinct at the end of the Palaeozoic Era, while the alcyonaria and hexacorals suddenly developed at that point in time and occupied the same ecological niche.

In Early Cretaceous times a particular group of lamellibranchs, the Rudistids, became important reef builders. In these forms one valve became shaped like the calyx of a simple coral, while the other formed the lid. Rudistid reefs of Early Cretaceous age occur in the Edwards Limestone of Texas and the Golden Lane of Mexico (Wilson and Ward, 1993), also in the Alps, Libya and Iran (Henson, 1950). The Richtofenid group of brachiopods followed a similar pseudo-coralline reefal habit in the Late Palaeozoic.

Through the vast span of geological time there seem to have been few aquatic invertebrate groups which have not at some time or another developed colonial reef-forming species. At almost all times, however, the calcareous algae have been important reef organisms (Harlan Johnson, 1961). Generally their role has been to bind together reef frameworks composed of various other organisms. Sometimes, though, they form the framework of reefs largely unaided by other groups. Algal reefs generally tend to be thinner and laterally more continuous than those of other creatures. Thus, since a reef is composed almost entirely of fossils, palaeontology and palaeoecology are vital to an understanding of their depositional environment. There is commonly a close correlation between palaeontology and reef sub-environment (Figure 11.13).

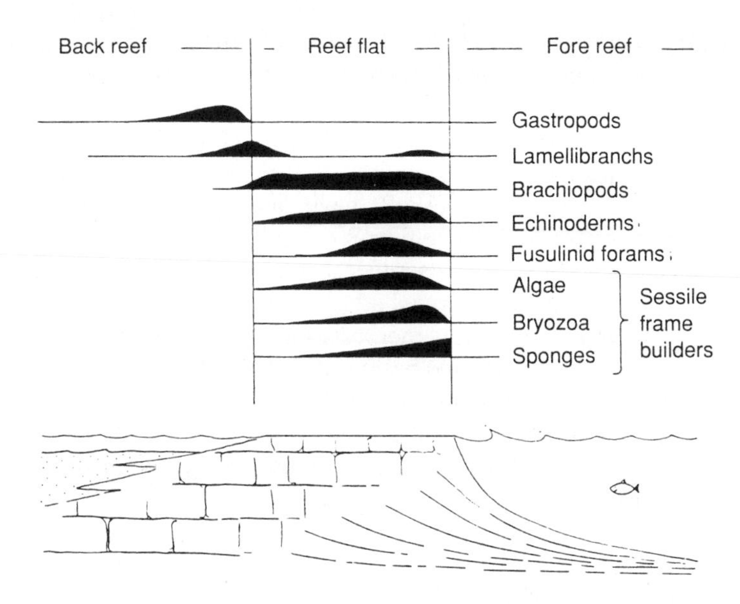

Figure 11.13 Inferred life distribution of west Texas Permian reef organisms. From Newell, N. D. *et al.* (1953).

Furthermore, since the biota of a reef controls its diagenesis and hence porosity development, an understanding of reef biota is of some economic significance.

The reef:evaporite:euxinic shale association

It has been noted for many years that reefs are commonly associated with evaporites and black euxinic shales. Weeks (1961) in drawing attention to this phenomenon cited 19 examples. This association is of considerable economic importance, because it is associated with petroleum and certain types of mineral deposits (Melvin, 1991).

There has been considerable speculation on the origin and development of such basins. The conventional explanation is that the reefs grew around the margin of a restricted sea where upper waters of normal salinity encouraged reef growth. Dense brines in deeper water inhibited benthonic activity allowing foul-smelling muds to form and ultimately precipitated evaporite minerals on the deeper parts of the basin floor. As the basin filled up the evaporite facies transgressed shorewards, killing the reefs and trapping within them hydrocarbons which had migrated from the basinal muds (Hunt, 1967, p. 237).

Critical to this concept is the demonstration that marginal reef growth was synchronous with basinal evaporite precipitation (for examples, see Henson, 1950, Figure 11 and Heybroek, 1965, pp. 28–9). This is not always possible to prove, however, and an alternative mechanism is sometimes possible. It has already been described how evaporites can form at the present time due to diagenesis of inter-tidal and supratidal sabkhas (p. 214). These deposits are not only petrographically similar to those of 'basinal' evaporites, but many of the textures are identical too, such as organic laminations, contortions and anhydrite nodules.

Thus Shearman and Fuller (1969) have discussed the possibility that inter-reef evaporites of the Devonian Winnepegosis Formation of Canada may be diagenetic sabkha deposits rather than basin floor precipitates. If this is correct then the evaporites formed after the reefs were killed by a drop in sea level. Similarly, Permian evaporites of the Delaware basin advanced through time from the shelf to the basin (Hills, 1968, Figs 4–7). This is what one would expect from a prograding sabkha. One would expect the reverse effect if the evaporites formed on the basin floor.

This prompts the question, have tectonic basins (gently subsiding areas infilled with sediment) been confused with sedimentary basins (basins which at the time of sedimentation were topographic sub-marine depressions)? Clearly the origin of basinal evaporites and their time relationships with reefs deserves careful evaluation (see also Kirkland and Evans, 1973).

Diagenesis of reefs

Strictly speaking the diagenesis of reefs is post-depositional and therefore falls outside the study of depositional environments which is the theme of this book. However, reef rock diagenesis is critical to an understanding of their economic significance. It deserves some attention, therefore. The diagenesis of carbonates in general, and of reefs in particular, is a large and complex topic which has been discussed at length in the literature (see for example Bathurst, 1975; Langres, Robertson and Chilingar, 1972; Reeckmann and Friedman, 1982). The following account is therefore necessarily brief.

Reefs show three interesting and rather unusual properties which are of great significance to their post-depositional history. First, they are formed with a very high primary porosity. Second, they are lithified at their time of formation, compaction is slight and initially primary porosity is preserved. Third, reefs are formed of chemically unstable minerals (dominantly aragonite and calcite). These can undergo diverse chemical changes due to reactions with circulating pore fluids. Basically two types of diagenetic changes can be recognized in reefs: mineralogical and textural. These are obviously inter-related.

Initially, aragonitic skeletal material alters to its stable polymorph calcite. Theoretically this should result in an increase in volume and a concomitant decrease in porosity and pemeability (Hoskin, 1966). This process takes place at different rates in different areas of the reef complex due to the different stabilities of the various carbonate particles (Friedman, 1964). Simultaneously, or later than the aragonite: calcite change, the carbonates may be enriched with magnesium from the sea water and converted to dolomite. Theoretically this should in contrast result in an overall contraction of the total rock volume of as much as 13% causing intercrystalline porosity between dolomite crystals (Chilingar and Terry, 1964).

These chemical changes are associated with textural modification of the reef fabric. First, primary porosity may be reduced shortly after deposition due to the infiltration of finely divided carbonate mud produced by the fragmentation of calcareous algae. A reduction of porosity may also be achieved through infilling of the framework by sparry calcite accompanied by recrystallization of the carbonate skeletons and the development of calcite overgrowths. These changes all result in a decrease in porosity and, together with dolomitization, often destroy all signs of the original organic fabric. Subsequently secondary porosity may occur due to solution along fractures accompanied by the formation of moldic and vuggy porosity.

Many of these processes are reversible. Thus the diagenesis of a reef is complex, being a function of its original faunal and lithological composition and of the chemistry of the fluids which subsequently circulated within it. The main points to be noted, however, are that recrystallization may completely destroy the primary organic fabric of a reef. Secondly, though ancient reefs are often highly porous, the type and distribution of the porosity may bear no relation to that which existed at the time of their formation. This is demonstrated by the Bu Hasa oil field case history. These points are critical to an understanding of the economic geology of reefs.

ECONOMIC GEOLOGY OF ANCIENT REEFS

Ancient reef deposits are of great economic significance: because of their porosity they often make good hydrocarbon reservoirs when suitably sealed, and because of their porosity and chemical instability they are liable to replacement by metallic ore minerals.

The association of reefs rimming shale basins and capped by evaporites has been already noted. Considerable attention has been given to the occurrence of oil in such situations. It is generally agreed that the evaporites play a critical role both because their parent brines inhibited oxidation of organic matter, allowing petroleum to form, and because the evaporites themselves make good caprocks (Week, 1961; Sloss, 1959).

The significance of carbonates as oil source rocks in their own right has been frequently discussed (e.g. Hunt, 1967). However, it is generally considered that petroliferous reefs are unlikely to be their own source rocks due to intensive oxidation of organic matter during reef formation. It is interesting to note, however, that in Tertiary reefs of Iran, oil trapped within fossils predates cementation and is different in composition from that which fills the bulk of the reservoir (Henson, 1950). Though oil and gas accumulations in reefs are hard to find and exploit the effort can be rewarding. For additional data on reefal oil reservoirs see Reeckmann and Friedman (1982).

Apart from their ability to trap large quantities of oil and gas, reefs are also host to a particular variety of metalliferous deposit termed Mississippi Valley type after one of the prime examples (see Evans, Campbell and Krouse, 1968; Brown, 1968; Hutchison, 1983, pp. 47–62, Guilbert and Park, 1986). These are telethermal replacement sulphide ore deposits of sphalerite and galena, with subordinate fluorspar, barite, dolomite, and, of course, calcite. Typically they occur replacing reefs and other carbonate build-ups in tectonically stable areas. They are often far from faults and igneous rocks which could have been sources for the metals.

There is evidence that evaporites provide brines to dissolve and transport the metals into the reefs (Amstutz and Bubinicek, 1967, p. 431; Amstutz *et al.*, 1964). The mud around a reef compacts, residual metal-rich solutions are expelled and may escape into adjacent porous reefs where they replace the host rock. As with petroleum, the distribution of minerals within a reef complex varies from place to place. Lower Carboniferous reefs of Ireland have mineralized cores (Derry, Clark, and Gillatt, 1965). The Devonian reefs of the Canning basin, Australia, are mineralized in the fore-reef (Johnstone *et al.*, 1967), while in Alpine Triassic reefs mineralization sometimes occurs in the lagoonal back-reef.

Thus ancient reefs are of considerable economic significance both as hydrocarbon reservoirs and as hosts for mineralization. Environmental interpretation analysis has a role to play in the location and exploitation of these deposits.

DIAGNOSIS OF REEFS

The first problem of identifying an ancient reef lies with the definition. If used in the restricted sense, as a limestone lens composed of skeletal remains in growth position, then it may be very hard to find an ancient reef. As noted earlier, many such structures have undergone such extensive diagenesis, that their original fabric may be invisible.

It is hard to locate reefs since they may be scattered at random over shelves buried beneath horizontal strata often with no surface structural expression. It is hard to identify an oil pool as a reef since diagenesis has often obliterated the original organic fabric. These two problems will now be considered.

Since normal methods of surface geological field work are often inapplicable, reef hunting is largely based on geophysical techniques. This is often largely a matter of luck. Searching for pinnacle reefs only a few kilometres in diameter in several hundred square kilometres of shelf can be like looking for a needle in a haystack. A case in point was the discovery of the Intisar reefs in the Sirte basin of Libya. The second seismic line to be shot in a 1880 sq. km concession just happened to transect the crest of a reef. The discovery well was tested for an initial production of 43 000 bd. The seismic programme was initiated to search for structures. The reef was a shock (Terry and Williams, 1969; Brady *et al.*, 1980). Reefs, being composed of limestone, or occasionally dolomite, are generally denser than the shales and evaporites with which they are commonly overlain. Reefs may thus be identified as a positive Bouguer anomaly on gravity surveys. This method has now largely been superseded by the seismic method.

Reefs can generally be identified from seismic data. Where there is sufficient velocity contrast between the reef and its cover then the top may show up as a reflecting horizon. This is generally the case when a reef is capped by evaporites. With compacted shales however the velocity contrast may be too small to generate a signal. Sometimes a reflector is present over the flank of a reef but dies out over the crest. This implies that the crestal limestone is acoustically slower, and therefore more porous, than that on the flanks. Reefs may be seismically invisible if there is no velocity contrast with the overlying sediments. Domed reflectors above the level of a reef may indicate where it lies.

Reefs themselves tend to have a uniform velocity, and thus generally do not produce any internal reflectors. An exception to this rule may be a 'bright spot' indicative of a fluid contact (generally gas on liquid). Examples have been noted in some of the Upper Permian Zechstein reefs in the North Sea (Jenyon and Taylor, 1983). Normally however reflectors in the flank of a reef die out towards the core. The reflecting horizon beneath a reef may also be interesting. A tight reef may cause a velocity 'pullup' suggesting to the unwary that the reef overlies an anticline. By contrast the reverse situation of an apparent syncline beneath a reef is good news. It signifies a velocity 'pull-down' suggesting a low-velocity interval and hence porosity.

Figure 11.14 illustrates some of these examples. For further details see

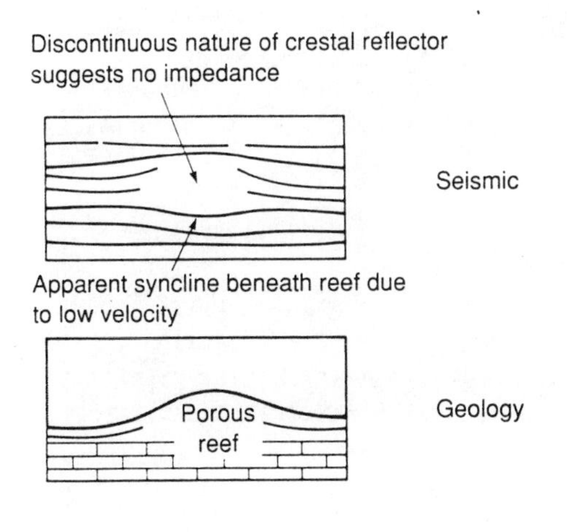

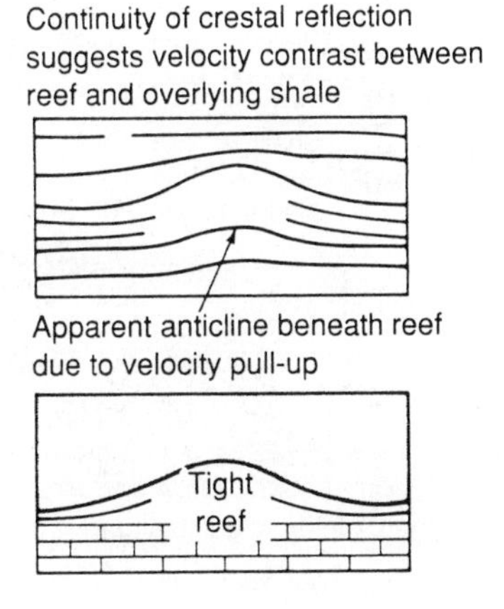

Figure 11.14 Cartoons to show some of the tricks involved in the seismic interpretation of reefs. For explanation see text.

Bubb and Hatledid (1977). With good quality seismic data, and where the velocity contrast is sufficient, the geometry of reefs may be mappable.

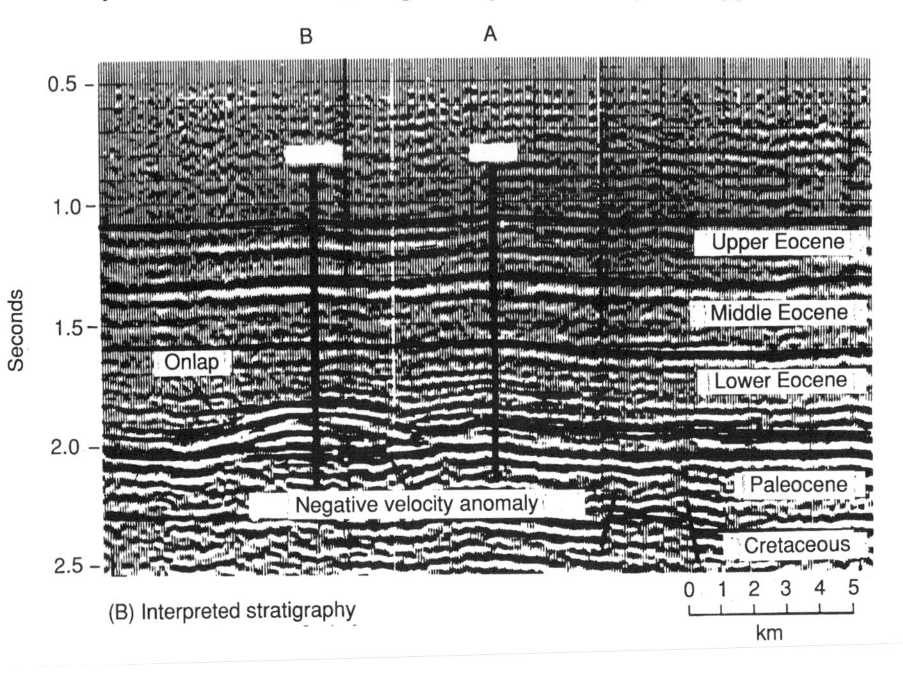

Figure 11.15 Seismic line through one of the Idris/Intisar pinnacle reef oil fields of the Sirte basin, Libya. Note negative velocity anomaly beneath the reef. From Bubb and Hatledid (1977), Figure 5, by courtesy of the American Association of Petroleum Geologists.

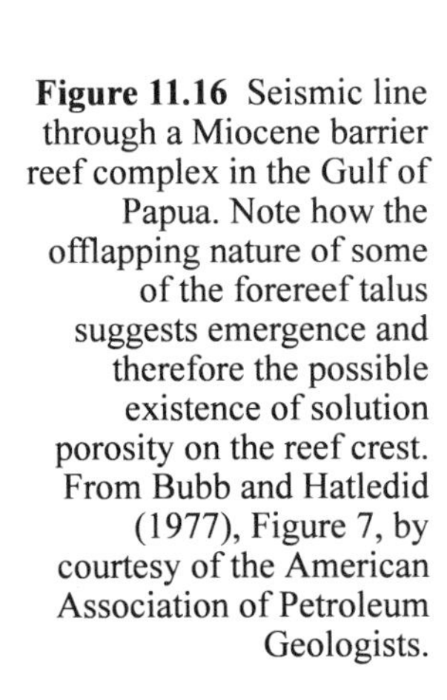

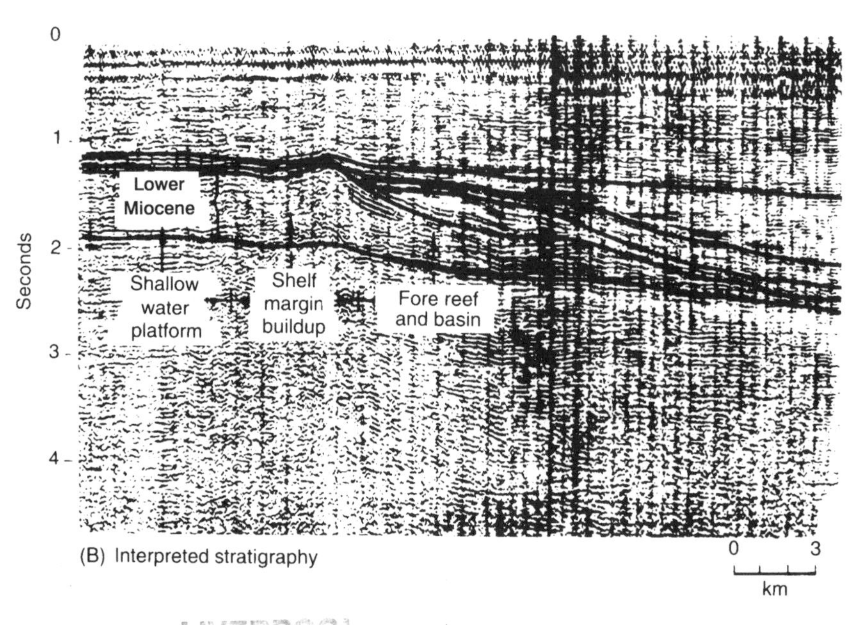

Figure 11.16 Seismic line through a Miocene barrier reef complex in the Gulf of Papua. Note how the offlapping nature of some of the forereef talus suggests emergence and therefore the possible existence of solution porosity on the reef crest. From Bubb and Hatledid (1977), Figure 7, by courtesy of the American Association of Petroleum Geologists.

Figures 11.15 and 11.16 illustrate examples of pinnacle and barrier reef respectively. If geophysical data is absent or of poor quality reefs may be located from well data. Criteria to look for are anomalous thickening of limestones between closely spaced wells, anomalous dips due to reef talus bedding, and faunal changes suggesting the proximity of an organic build-up (e.g. Andrichuk, 1958, p. 90). The reef core, if drilled through, may be unrecognizable due to recrystallization. Where two wells are drilled, one of which passes through an open marine facies and the equivalent interval of the other is lagoonal, it may be profitable to search for a barrier reef in the intervening ground.

The case history of the Bu Hasa oil field demonstrates how petrography holds the key to the subsurface facies analysis of reefs.This example also demonstrates how diagenesis can alter the original distribution of porosity and permeability. Thus, even when a reef has been identified it may be hard to predict the distribution of optimum reservoir properties within it. In some cases production comes from the reef core itself (e.g. the Leduc and Intisar reefs of Canada and Libya). In other situations the reef core may be barren and production restricted to the fore-reef facies as in Tertiary reefs of Iran (Henson, 1950). Back-reef lagoonal facies are generally poor reservoirs. In pelletal limestones porosity may be high but permeability low (Stout, 1964, p. 334). This may be increased by fracturing. Oil production comes from lagoonal sands of the Permian Texas reef complex where they interfinger up dip with evaporites and dolomites (Mutti and Simo, 1993).

As mentioned earlier, gamma and SP log motifs are of little use in the diagnosis of carbonate environments. An upward-decreasing API curve may be present on the gamma curve where a fore-reef talus progrades transitionally over basinal shales (refer back to the outcrop examples illustrated in Figure 11.2). The overlying reef will show the typical pure carbonate log profile of a steady low API curve (Figure 11.17).

The dipmeter is particularly useful, however, in establishing the geometry of a reef once it has been found. The reef core often shows a random 'bag o' nails' motif due to the fractured and recrystallized fabric. Exhumed ancient reefs show the progradational bedding of the fore-reef talus. Depositional dip is often clearly seen in the fore-reef talus slope, with a characteristic upward increasing blue pattern. The reef core lies in the opposite direction to the progradational dip (Figure 11.17). Where two adjacent wells have both inadvertently penetrated the flank of a pinnacle reef, its crest may be found by drawing two lines reciprocal to the fore-reef dip direction (Jageler and Matuszak, 1972, p. 128).

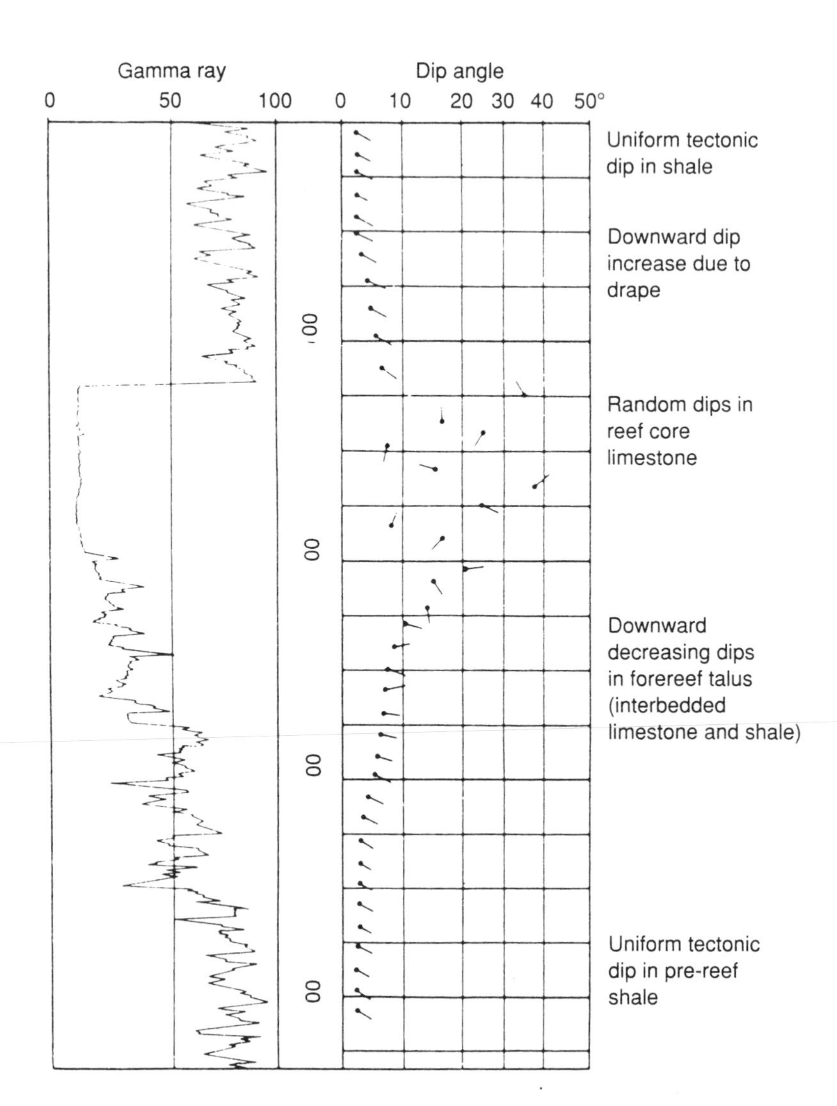

Figure 11.17 Gamma ray and dipmeter log for a well drilled on the south-east side of a reef.

REFERENCES

This is an exceedingly brief bibliography of works on recent reefs:

Emery, K. O., Tracey, J. I. and Ladd, H. S. (1954) Geology of Bikini and nearby atolls, Pt. I, *Geology*, US Geol. Surv. Prof. Pap., 260-A.

Fairbridge, R. W. (1950) Recent and Pleistocene coral reefs of Australia, *J. Geol.*, **58**, 336–401.

Friedman, G. M. (1968) Geology and geochemistry of reefs, carbonate sediments and waters, Gulf of Akaba (Elat) Red Sea, *J. Sediment. Petrol.*, **38**, 895–919.

Ginsburg, R. N., Lloyd, R. M., Stockman, K. W. and McCallum, J. S. (1963) Shallow water carbonate sediments. In M. N. Hill (ed.), *The Sea*, vol. III, Interscience, NY, 554–82.
Guilcher, A. (1965) Coral reefs and lagoons of Mayotte Island, Comoro Archipelago, Indian Ocean, and of New Caledonia, Pacific Ocean. In W. F. Whittard and Bradshaw, R. (eds), *Submarine Geology and Geophysics*, Butterworths, London, 21–45.
Hopley, D. (1982) *The Geomorphology of the Great Barrier Reef*, J. Wiley, Chichester.
James, N. P. (1983) Reef. In P. A. Scholle, D. G. Bebout and C. H. Moore (eds), *Carbonate Depositional Environments*, Amer. Assoc. Petrol. Geol., **33**, 345–462.
Jones, O. A. and Endean, R. (1973) *Biology and Geology of Coral Reefs*, 3 vols, Academic Press, London.
McKee, E. D. (1958) Geology of Kapingamarangi Atoll, Caroline Islands, *Bull. Geol. Soc. Amer.*, **69**, 241–77.
—— , Chronic, J. and Leopold, E. B. (1959) Sedimentary belts in lagoon of Kapingamarangi Atoll, *Bull. Amer. Assoc. Petrol. Geol.*, 43, 501–62.
Maxwell, W. G. E. (1968) *Atlas of the Great Barrier Reef*, Elsevier, Amsterdam.
Shepard, F. P. (1963) Submarine geology. In *Coral and Other Organic Reefs*, Harper, London, chapter 12, 349–70.
Wiens, H. J. (1962) *Atoll Environment and Ecology*, Yale Univ. Press.
Yonge, C. M. (1973) The nature of reef-building (hermatypic) corals, *Bull. Mar. Sci.*, **23**, 1–15.

The account of the Bu Hasa reef case history was based on:

Alsharhan, A. S. (1985) Depositional environments, reservoir units evolution, and hydrocarbon habitat of Shuaiba Formation, Lower Cretaceous, Abu Dhabi, United Arab Emirates, *Bull. Amer. Assoc. Petrol. Geol.*, **69**, 899–912.
—— (1987) Geology and reservoir characteristics in giant Bu Hasa Field, Abu Dhabi, U.A.E., *Bull. Amer. Assoc. Petrol. Geol.*, **71**, 1304–18.
—— and Nairn, A. E. M. (1993) Carbonate platform models of Arabian Cretaceous reservoirs. In J. A. T. Simo, R. W. Scott and J-P. Masse (eds), *Cretaceous Carbonate Platforms*, Amer. Assoc. Petrol. Geol. Mem., **56**, 173–84.
Calavan, C. W., Hagerty, R. M., Mitchell, J. C. and S. R. Schutter (1992) *Integrated Reservoir Study and Modelling of the Shuaiba Reservoir, Bu Hasa Field, Abu Dhabi, United Arab Emirates*, Soc. Pet. Eng. Paper No. 24509.
Harris, T. J., Hay, J. T. C. and Twombley, B. N. (1968) Contrasting limestone reservoirs in the Murban Field, Abu Dhabi. In *Second Regional Technical Symposium*, Soc. Pet. Eng. Dahran, 149–82.
Hassan, T. H., Mudd, G. C. and Twombley, B. N. (1975) The stratigraphy and sedimentation of the Thamama Group (Lower Cretaceous) of Abu Dhabi, *Ninth Arab Petroleum Congress*, 107, B-3.
Twombley, B. N. and Scott, J. (1975) Application of geological studies in the development of the Bu Hasa field, Abu Dhabi, *Ninth Arab Petroleum Congress*, 133, B-l.

Other references mentioned in this chapter were:

Amstutz, G. C. and Bubinicek, L. (1967) Diagenesis in sedimentary mineral deposits. In S. Larsen and G. V. Chilingar (eds), *Diagenesis in Sediments*, Elsevier, Amsterdam, 417–75.
——, Randohr, P., El. Baz, F. and Park, W. C. (1964) Diagenetic behaviour of sulphides. In G. C. Amstutz (ed.), *Sedimentology and Ore Genesis*, Elsevier, Amsterdam.

Andrichuk, J. M. (1958) Stratigraphy and facies analysis of Upper Devonian reefs in Leduc, Settler and Redwater areas, *Bull. Amer. Assoc. Petrol. Geol.*, **42**, 1–93.

Bathurst, R. G. C. (1975) *Carbonate Sediments and Their Diagenesis*, 2nd edn, Elsevier, Amsterdam.

Bissell, H. J. and Wolf, K. H. (1967) Diagenesis of carbonate rocks. In G. Larsen and G. V. Chilingar (eds), *Diagenesis in Sediments*, Elsevier, Amsterdam, 197–322.

—— , Mannon, R. W. and Rieke, H. (1972) *Oil and Gas Production from Carbonate Rocks*, Elsevier, Amsterdam.

Bond, G. (1950) The Lower Carboniferous reef limestones of northern England, *J. Geol.*, **58**, 430–87.

Brady, T. J., Campbell, N. D. H. and Maher, C. E. (1980) Intisar 'D' Oil Field, Libya. In M. T. Halbouty (ed.), *Giant Oil Fields of the Decade 1968–78*, Amer. Assoc. Petrol. Geol., **30**, 543–64.

Braithwaite, C. J. R. (1973) Reefs: just a problem of semantics?, *Bull. Amer. Assoc. Petrol. Geol.*, **57**, 1100–16.

Brown, J. S. (ed.) (1968) Genesis of stratiform lead-zinc-barite-puorite deposits (Mississippi Valley type deposits), *Economic Geology*, Monograph 3, Econ. Geol. Pub. Co. Blacksburg, Va.

Bubb, J. N. and Hatledid, W. G. (1977) Seismic recognition of carbonate buildups. In C. E. Payton (ed.), *Seismic Stratigraphy – Applications to Hydrocarbon Exploration*, Amer. Assoc. Petrol. Geol., **26**, 185–204.

Chilingar, G. V., Bissell, H. J. and Fairbridge, R. W. (eds) (1967) *Carbonate Rocks*, Part A, Elsevier, Amsterdam.

Chilingar, G. V. and Terry, R. D. (1964) Relationship between porosity and chemical composition of carbonate rocks, *Petrol. Eng.*, B-S4, 341–2.

Cummings, E. R. (1932) Reefs or bioherms?, *Bull. Ceol. Soc. Amer.*, **43**, 331–52.

Darwin, C. (1842) *The Structure and Distribution of Coral Reefs*, John Murray, London.

Derry, D. R., Clark, G. R. and Gillatt, N. (1965) The Northgate base-metal deposit at Tynagh, Co. Galway, Ireland, *Econ. Geol.*, **60**, 1218–37.

Dunham, R. J. (1970) Stratigraphic reef versus ecological reef, *Bull. Amer. Assoc. Petrol. Geol.*, **54**, 1931–2.

Emery, K. O. (1956) Sediments and water of Persian Gulf, *Bull. Amer. Assoc. Petrol. Geol.*, **40**, 2354–83.

—— , Tracey, J. L. and Ladd, H. S. (1954) Geology of Bikini and nearby atolls, Pt. I, *Geology*, US Geol. Surv. Prof. Pap. 260-A.

Evans, T. L., Campbell, F. A. and Krouse, H. R. (1968) A reconnaissance study of some Western Canada Pb-Zn deposits, *Econ. Geol.*, **63**.

Fouch, T. D. and Dean, W. E. (1982) Lacustrine Environments. In P. A. Scholle and O. Spearing (eds), *Sandstone Depositional Environments*, Amer. Assoc. Petrol. Geol. Tulsa, 87–114.

Friedman, G. M. (1964) Early diagenesis and lithifaction in carbonate sediments, *J. Sediment. Petrol.*, **34**, 777–812.

—— (1968) Geology and geochemistry of reefs, carbonate sediments and waters, Gulf of Akaba, *J. Sediment. Petrol.*, **38**, 895–919.

Guilbert, J. M. and Park, C. F. (1986) *The Geology of Ore Deposits*, W H Freeman, New York.

Harlan Johnson, J. (1961) *Limestone-building Algae and Algal Limestones*. Colorado School of Mines.

Heckel, P. H. (1974) Carbonate build-ups in the geologic record: a review. In L. F. Laporte (ed.), *Reefs in Time and Space*, Soc. Econ. Pal. Min., Spec. Pub. 18, 90–154.

Henson, F. R. S. (1950) Cretaceous and Tertiary reef formations and associated sediments in the Middle East, *Bull. Amer. Assoc. Petrol. Geol.*, **34**, 215–38.
Heybroek, F. (1965) The Red Sea Miocene Evaporite Basin. In *Salt Basins Around Africa*, Inst. Pet., London, 17–40.
Hills, J. M. (1968) Permian Basin field area, west Texas and Southeastern New Mexico. In R. M. Mattox (ed.), *Saline Deposits*, Geol. Soc. Amer., Sp. Pap. **88**, 17–28.
Horvoka, S. D, Nance, H. S. and Kerans, C. (1993) Parasequence geometry as a control on permeability evolution: examples from the San Andres and Grayburg Formations in the Guadalupe Mountains, New Mexico. In R. G. Loucks and J. F. Sarg (eds), *Carbonate Sequence Stratigraphy*, Amer. Assoc. Petrol. Geol., **57**, 493–514.
Hoskin, C. M. (1966) Coral pinnacle sedimentation, Alacran Reef Lagoon, Mexico, *J. Sediment. Petrol.*, **36**, 1058–74.
Hunt, J. M. (1967) The origin of petroleum in carbonate rocks. In G. V. Chilingar, H. J. Bissell and R. W. Fairbridge (eds), *Carbonate Rocks*, Part B, Elsevier, Amsterdam, 225–51.
Hutchison, C. S. (1983) *Economic Deposits and their Tectonic Setting*, Macmillan, London.
Jageler, A. H. and Matuszak, D. R. (1972) Use of well logs and dipmeters in stratigraphic trap exploration. In R. E. King (ed.), *Stratigraphic Oil and Gas Fields*, Amer. Assoc. Petrol. Geol., Spec. Pub., **10**, 107–35.
Jenyon, M. K. and Taylor, J. C. M. (1983) Hydrocarbon indications associated with North Sea Zechstein shelf features, *Oil and Gas Jl.*, 5 Dec., 155–60.
Johnstone, M. H., Jones, P. J., Koop, W. J., Roberts, J., Gilbert-Tomlinson, J., Veevers, J. J. and Wells, A. T. (1967) Devonian of Western and Central Australia. In D. H. Oswald (ed.), *Internat. Symp. Devn. System*, Calgary, **1**, 599–612.
Kinsman, D. J. (1969) Modes of formation, sedimentary associations and diagnostic features of shallow-water and supratidal evaporites, *Bull. Amer. Assoc. Petrol.*, **53**, 830–40.
Kirkland, D. W. and Evans, R. (1973) *Marine Evaporites*, Benchmark Papers in Geology, Dowdon, Hutchinson & Ross, Stroudsburg, Penn.
Langres, G. L., Robertson, J. O. and Chilingar, G. V. (1972) *Secondary Recovery and Carbonate Reservoirs*, Elsevier, Amsterdam.
Laporte, L. F. (ed.) (1974) *Reefs in Time and Space*, Soc. Econ. Pal. Min., Sp. Pub., **18**, Tulsa.
Logan, B. R., Rezak, R. and Ginsburg, R. N. (1964) Classification and environmental significance of algal stromatolites, *J. Geol.*, **72**, 68–83.
Loucks, R. G. and Sarg, J.F. (eds) (1993) Carbonate Sequence Stratigraphy, Amer. Assoc. Petrol. Geol., **57**.
Lowenstam, H. A. (1950) Niagaran Reefs of the Great Lakes Area, *J. Geol.*, **58**, 430–87.
Massa, D. (1965) Observations sur les series Devonniennes des confins Algero-Marocains du Sud, CFP Mem. No. 8, Paris.
Melvin, J. L. (ed.). (1991) *Evaporites, Petroleum and Mineral Resources*, Elsevier, Amsterdam.
Mutti, M. and Simo, J. A. T. (1993) Stratigraphic patterns and cycle-related diagenesis of Upper Yates Formation, Permian, Guadalupe Mountains. In R. G. Loucks and J. F. Sarg (eds), *Carbonate Sequence Stratigraphy*, Amer. Assoc. Petrol. Geol., **57**, 515–45.
Parkinson, D. (1944) The origin and structure of the Lower Visean reef knolls of the Clitheroe District, Lancashire, *Quart. Jl. Geol. Soc. London*, **99**, 155–68.

Playford, P. E. and Lowry, D. C. (1966) Devonian reef complexes of the Canning Basin, Western Australia, *Geol. Surv. West. Aust. Bull.*, **118**.

Reeckmann, A. and Friedman, G. M. (1982) Exploration for Carbonate Petroleum Reservoirs, J. Wiley & Sons, Chichester.

Shearman, D. J. and Fuller, J. G. C. M. (1969) Phenomena associated with calcitization of anhydrite rocks, Winnepegosis Formation, Middle Devonian of Saskatchewan, Canada, *Proc. Geol. Soc. Lond.*, 235–9.

Sloss, L. L. (1959) Relationship of primary evaporites to oil accumulation, *Proc. Fifth Wld. Petrol. Congr.*, Section 1, 123–35.

Southgate, P. N., Kennard, J. M., Jackson, M. J., O'Brian, P. E. and Sexton, M. J. (1993) Reciprocal Lowstand and Highstand Carbonate Sedimentation, Subsurface Devonian Reef Complex, Canning Basin, Western Australia. In R. G. Loucks and J. F. Sarg (eds), *Carbonate Sequence Stratigraphy*, Amer. Assoc. Petrol. Geol., **57**, 157–80.

Stout, J. L. (1964) Pore geometry as related to carbonate stratigraphic traps, *Bull. Amer. Assoc. Petrol. Geol.*, **48**, 329–37.

Teichert, C. (1958) Cold and deep-water coral banks, *Bull. Amer. Assoc. Petrol. Geol.*, **42**, 1064–82.

Terry, C. E. and Williams, J. J. (1969) The Idris 'A' Bioherm and Oilfield, Sirte Basin, Libya – its commercial development, regional Palaeocene geological setting stratigraphy. In P. Hepple (ed.), *The Exploration for Petroleum in Europe and North Africa*, Inst. Petrol., London, 31–48.

Toomey, D. F. (ed.) (1981) *European Fossil Reef Models*, Soc. Econ. Pal. Min. Sp. Pub., **30**.

Waite, L. E. (1993) Upper Pennsylvanian seismic sequences and facies of the Eastern and Southern Horseshoe Atoll, Midland Basin, West Texas. In R. G. Loucks and J. F. Sarg (eds), *Carbonate Sequence Stratigraphy*, Amer. Assoc. Petrol. Geol., **57**, 213–40.

Walls, R. A. (1983) Golden Spike Reef Complex, Alberta. In P. A. Scholle, D. G. Bebout and C. H. Moore (eds), *Carbonate Depositional Environments.*, Amer. Assoc. Petrol. Geol., Memoir No. 33, 445–53.

Weeks, L. G. (1961) Origin, migration and occurrence of petroleum. In G. R. Moody (ed.), *Petroleum Exploration Handbook*, McGraw Hill, New York.

Wilson, J. L. (1974) Characteristics of carbonate-platform margins, *Amer. Assoc. Petrol. Geol. Bull.*, 810–24.

—— and Ward, W. C. (1993) Early Cretaceous Carbonate Platforms of Northeastern and East-Central Mexico. In J. A. T. Simo, R. W. Scott and J-P. Masse (eds), *Cretaceous Carbonate Platforms*, Amer. Assoc. Petrol. Geol., **56**, 35–50.

12 Deep water sands

DEEP WATER SANDS DEFINED

Recent deep water sands may be defined as those deposited below the mud line, that is to say below the level where wave, storm and tidally generated currents operate. Deep water sands occur in both fresh water and in marine settings of varied water depth. For marine environments deep water normally means below level of the continental shelf (>200 m).

To define what is meant by an ancient deep water sands is far more difficult. It is extremely hard, and generally of only academic importance, to determine palaeobathymetry accurately. There are several criteria that may be used to conclude that a sand was deposited in deep water (as defined above). Some of these criteria are negative, some positive. There should not be any features indicating subaerial exposure, or shallow water deposition. Thus a deep water sand must obviously lack such features as desiccation cracks, rain prints, vertebrate tracks, rootlet horizons and palaeosols.

Shallow water features should also be lacking, though these are more equivocal than subaerial ones. Shallow water criteria include *in situ* algal colonies (because algae only thrive in the photic zone), shallow water trace fossils (principally vertical burrows), and extensive cross-bedding (on the assumption that traction currents in deep water are of insufficient velocities to generate megaripples). The positive criteria that indicate a deep water environment for a sand are very dubious. Examination of deep water sediments shows that their fauna is largely composed of transported shallow-water beasts and of pelagic free swimming ones.

Body fossils of deep marine animals are rare, though a trace fossil assemblage of invertebrate surface tracks and trails is quite common. The regional and stratigraphic position of a formation may be used to infer a deep water origin. One should suspect a deep water origin for sandstones in basin centres, especially if they are overlain by, or pass shelfward into, slope deposits (as indicated by abundant slump bedding) succeeded by shallow marine deposits. A classic example of this is provided by the Lower Carboniferous sequence at Mam Tor, Derbyshire (Figure 12.1).

Here basinal black shales (the Edale Shales) pass up via graded sandstones (the Mam Tor Series) through multistory channel sands (the Shale Grit), into shallow marine bioturbated sands and shales (the Grindslow Shales), which are in turn overlain by the fluvial Millstone Grit. Walther's Law suggests a deep water origin for the Mam Tor sandstones (Figure 12.2).

Though the process by which they were deposited may be equivocal,

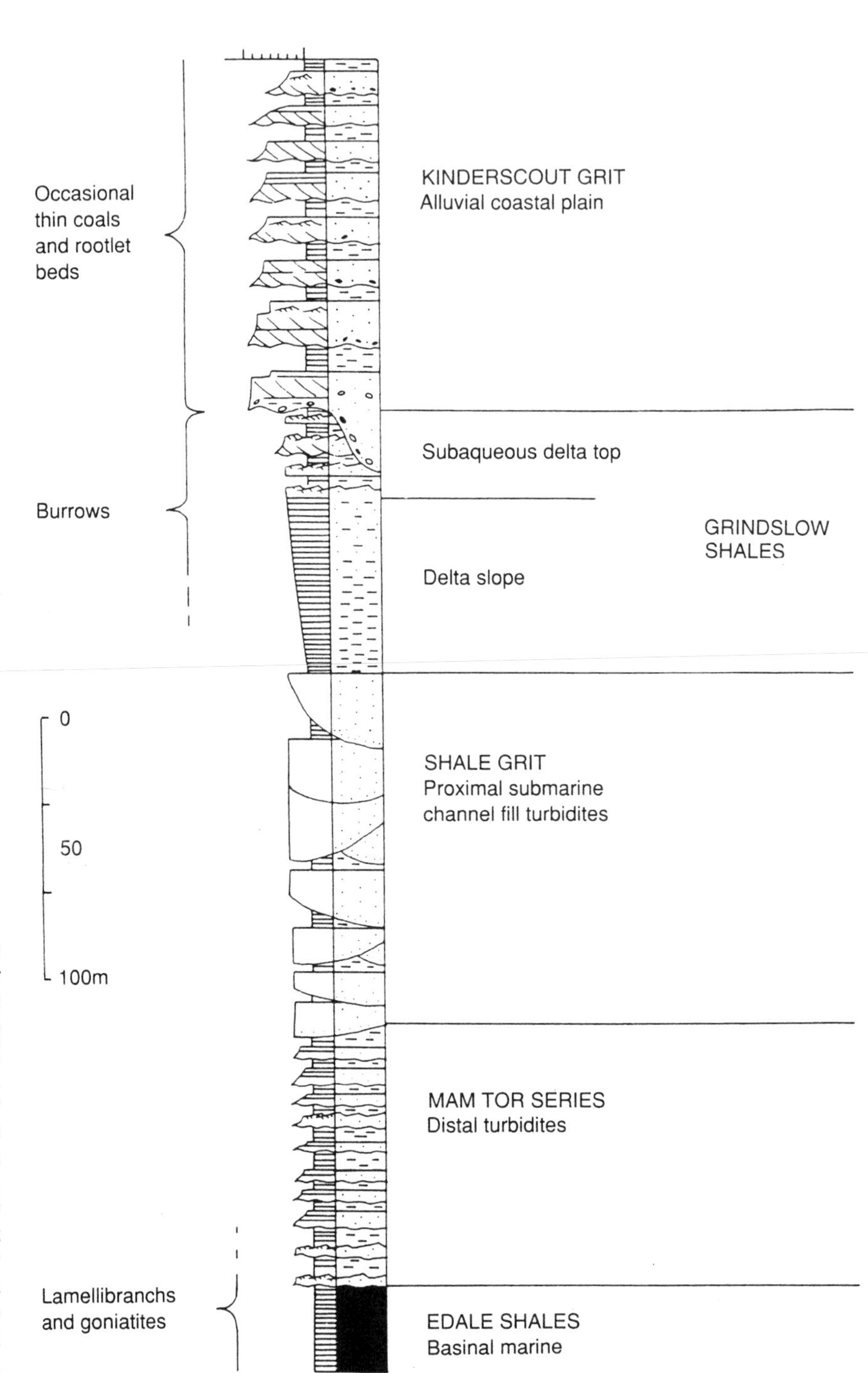

Figure 12.1 Stratigraphic section through the Lower Carboniferous sequence from the basinal marine Edale Shales to the fluvial Millstone Grit, Derbyshire. Though the sedimentary processes that deposited the graded beds of the Mam Tor series may be debated, there is no doubt that they were deposited at the foot of a prograding delta. They are therefore deep sea sands. Based on data due to Collinson, Reading and Walker.

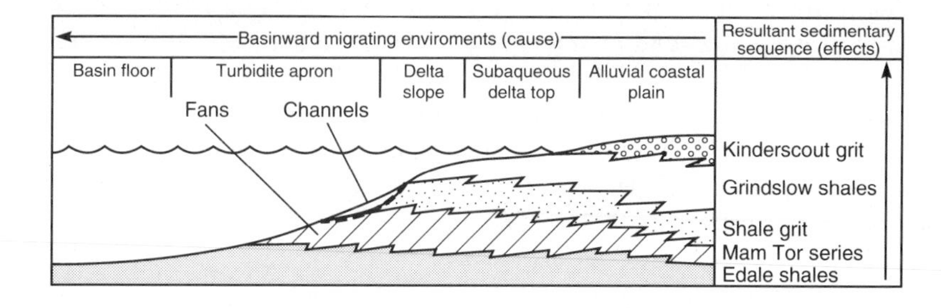

Figure 12.2 Cross-section diagram to show the relationship between facies and environments in the Lower Carboniferous section of Derbyshire, a classic example of Walther's Law.

there can be no doubt that they were deposited in deep water at the foot of a prograding delta. In the subsurface seismic data may be useful. Sandstones encountered when drilling into the toeset of a major seismically defined prograde must have been deposited in water depths at least as great as the interval from topset to toeset.

Large volumes of ancient deep water sandstones abound in the stratigraphic record. Many of these formations were formerly referred to as 'flysch', a term which fortunately seems to be going out of fashion. Petrographically many flysch sandstones are greywackes. The sedimentary structures of flysch sandstones suggest that they are 'turbidites', having been deposited from turbidity flows, a particular type of density current. These three terms deserve definition and discussion.

Flysch Thick sequences of interbedded sands and shales. The sandstones generally have erosional bases and are internally graded. The shales contain a marine fauna. First defined in the Alps, this word has been applied to similar rocks in geosynclines (now termed fore-arc basins) of all ages and in all parts of the world (for further data see Dzulinsky and Walton, 1965; Hsu, 1970).

Turbidity current: 'A current flowing on consequence of the load of sediment it is carrying and which gives it excess density' (Kuenen, 1965, p. 217).

Turbidite A sediment deposited by a turbidity current.

Greywacke A poorly sorted sandstone with abundant matrix, feldspar, and/or rock fragments.

Flysch sandstones are largely turbidites. Many flysch sandstones, but by no means all, are greywackes. Because of this, the terms 'flysch', 'turbidite', and 'greywacke' have been used as synonyms. This is misleading. 'Flysch' describes a facies, 'greywacke' is a petrographic term for a particular kind of rock, and 'turbidite' is a genetic term describing the process which is thought to have deposited a sediment. Not only are these three words quite different but they are all extremely loosely defined in the literature.

This chapter begins with a brief review of modern deep water sands,an account of turbidites, and an attempt to summarize their characteristics and origin. A case history is then described, in which an outcrop study in Scotland develops a sedimentary model that can be applied to petroleum exporation in the North Sea. The chapter continues with a discussion of the economic importance of deep water sands, and it concludes with a review of their diagnostic criteria in the subsurface.

DEEP WATER SEDIMENTARY PROCESSES AND PRODUCTS

Studies of modern continental margins show that their physiography is analogous to that of desert wadis and alluvial fans (Figure 12.3). The continental shelf edge is dissected by submarine channels. The source of these channels lies adjacent to either major river mouths or even coastal beaches. The channels continue down through the continental slope and debouch their sediment load onto a submarine fan, analogous to a subaerial alluvial fan. At the fan apex the submarine channel splits into a radiating complex of small channels, which are often flanked by levees. These channels become more numerous, shallower and wider down the fan. There are also changes down the fan in the sediment facies. The coarsest sand, and sometimes gravel, is deposited in the proximal (near source) part of the fan. Grainsize declines distally down fan as the submarine fan sands grade out into pelagic muds of the oceanic basin floor (Figure 12.3). For a more detailed account see Howell and Normark (1982), Bouma *et al.* (1985), Shanmugam and Moiola (1988).

Sand is transported in to deep water by a variety of gravity-related processes. Once in deep water traction currents seldom attain sufficient velocity to transport sand in megarippled bed forms, but only to redistribute it by lower velocity ripple trains. The initial gravitational processes range from slumps and debris flows of boulders, with a poorly sorted matrix, via grain flows to turbidity flows. These will now be briefly described.

When transported out in to a basin sediment may be deposited on a slope at the angle of repose. Some slopes are accretionary muddy slopes, as on the edge of a delta, others are fault scarps, both may be dissected by basinward trending channels. Vibration, often of seismic origin, will trigger movement on the slope, soft sediment will slide and slump down the slope. Boulders, with assorted poorly sorted detritus, will likewise tumble down in to the depths and come to rest as inchoate debris flows. These processes may operate across a whole delta front or fault scarp, or be confined to channels that dissect them. In some instances clean continental shelf sands may be swept into channel heads by longshore drift. These clean sands may then flow down the channels like an underwater avalanche.

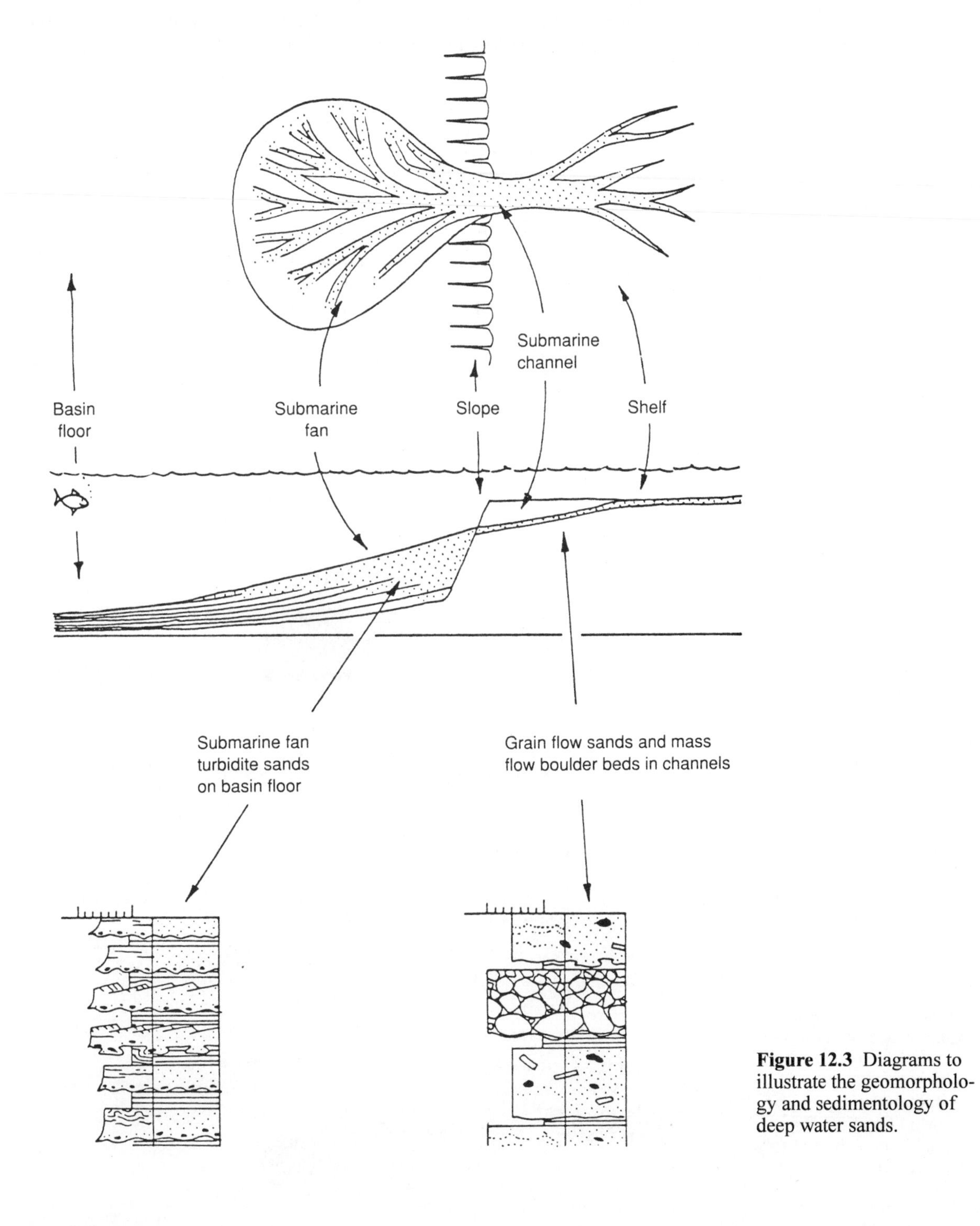

Figure 12.3 Diagrams to illustrate the geomorphology and sedimentology of deep water sands.

The confining walls of the channel keeps the pore pressure high, so the sand flows by fluidization and grain flow (Bagnold, 1966, Stauffer, 1967). Grain flows are normally massive sands with sharp tops and bases, which may be loaded, but seldom erosional. Internally grain flows are typically massive, with faint lamination that is often disrupted into 'dish structure' caused by the vertical escape of excess pore water. Grain flows sometimes contain scattered clasts. Pebbles and boulders do not concentrate at the bottom, as in traction current infilled channels, or as in turbidites.

As grain flows move out of the submarine channels and out on to the basin plain they become muddier and grade in to turbidity flows that lose energy and settle out as turbidites. Turbidites are identified by no single feature, but by the sum of many criteria. The most noticeable feature of turbidites is that they are vertically graded. Each bed generally has a sharp base and an upward gradation into shale. Detailed examination of turbidites has shown that they tend to contain a regular vertical sequence of structures, termed the eponymous Bouma sequence (Figure 12.4).

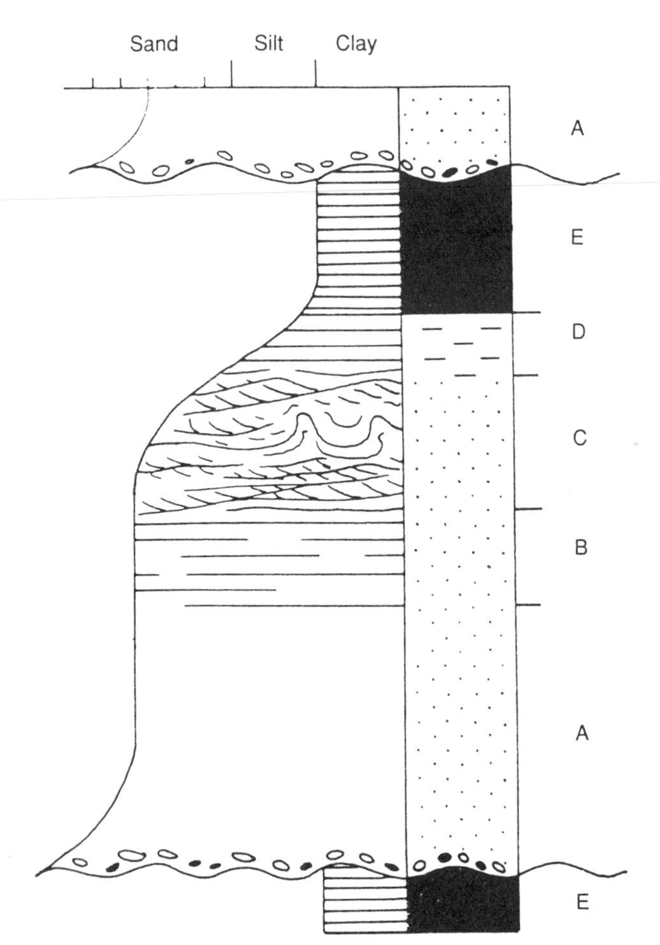

Figure 12.4 Generalized sequence through a turbidite unit. Alphabetic letters code a vertical sequence of five different units (see text). Modified from Bouma, 1962, Figure 8, by courtesy of Elsevier Publishing Co.

This suite of structures has been interpreted in the following way (Walker, 1965; Harms and Fahnestock, 1965; Hubert, 1967). First a current scoured a variety of structures on a mud surface. Sedimentation of the turbidite then took place under waning current conditions. First the massive A unit was deposited, perhaps as antidunes, under an upper flow regime; shooting flow deposited the next laminated B unit, and a lower flow regime deposited the cross-laminated C unit. Various explanations have been offered for the upper laminated silt (Walker, 1965). Perhaps it heralds a return to shooting flow, but since the grainsize is much finer, the actual current velocity was probably less than that which caused the lower laminated unit. The upper pelitic E unit indicates resumption of the low-energy environment which prevailed before the turbidite was laid down.

It is important to note that the sequence of structures in Figure 12.4 is seldom completely developed in any one turbidite. The previous interpretation is based on the study of many turbidite sequences. Variations on this motif are discussed by Bouma (1962), Walker (1965), and Van der Lingen (1969). Systematic vertical variations in the frequency of bed forms within sequences of turbidites have been documented by Walker (1967, 1970). There is a tendency for only the upper units to be developed at the base of a turbidite sequence. Moving up the section, progressively lower units are present within each turbidite until complete A–E sequences appear towards the top of the series. The lower incomplete turbidites, termed 'distal', are believed to have been deposited far from the source. The upper complete units are termed 'proximal' and are believed to have been deposited near the source. A thick sequence of turbidites may thus record a gradual advance of the sediment source into the depositional area.

There has been considerable discussion about the texture of turbidites. Modern and ancient turbidites are sometimes poorly sorted, and sometimes well sorted. The greywacke nature of many ancient turbidites may be due not so much to depositional matrix, as to the development of a 'pseudomatrix' formed by the breakdown of unstable mineral grains during burial (Cummings, 1962). Carbonate turbidites have been described from ancient and Recent sediments. Turbidites have been described from a very wide range of depositional environments, not only deep water, but also from ephemeral lakes, as described in Chapter 5 (p. 122). They are not even restricted to sedimentary rocks, but have also been described from layered gabbros (Irvine, 1965).

Once sand has been carried out onto the basin plain by the various processes described above it may be preserved as a submarine fan of turbidites. Sometimes, however, no sooner is a sand deposited in deep water, than it is subjected to reworking by ocean bottom currents. Studies of modern oceans reveals the existence of major current gyres that rotate around ocean basins in response to the Coriolis force. These currents tend to flow parallel to the continental rise. They are thus referred to as geostrophic or contour currents. Contour currents are seldom sufficiently powerful that they general megaripples, nor can they transport coarse sand and gravel. Their velocity is, however, sufficient to transport trains of ripples, and thus to rework turbidites and to deposit cross-laminated fine sand. Contour currents are episodic, due to disturbances in the oceanic circulation termed 'benthic storms'. These give rise to interbedded sequences of laminated pelagic mud, and cross-laminated sand. The latter are termed contourites, naturally, and they have been recognized in deep water sediments ancient and modern (Stow and Faugeres, 1993).

UPPER JURASSIC DEEP SEA SANDS OF SUTHERLAND, SCOTLAND

Description

Mesozoic rocks of the Moray Firth basin crop out at several points along the coast of Sutherland. The most complete section is exposed in a two or three kilometres wide strip that extends for some 30 km from Golspie in the south to Helmsdale in the north (Figure 12.5).

Figure 12.5 View looking north from Brora, Sutherland. The Helmsdale fault separates the low lying coastal strip of Jurassic sediments from the Palaeozoic basement hills to the west. The scale is provided by the Clynelish whisky distillery in the foreground.

The Mesozoic strata are downfaulted against Caledonide metamorphics and intrusives together with Devonian Old Red Sandstone (Figure 12.6).

The Helmsdale fault is a branch of the Great Glen fault and marks the present boundary of the Moray Firth basin. The Helmsdale fault controlled the deposition of the Jurassic deep sea sediments to be described.

The Sutherland section begins with red fanglomerates and alluvium of Permotriassic age. These deposits are overlain by an almost complete Jurassic section ranging from Liassic to Portlandian (Figure 12.7).

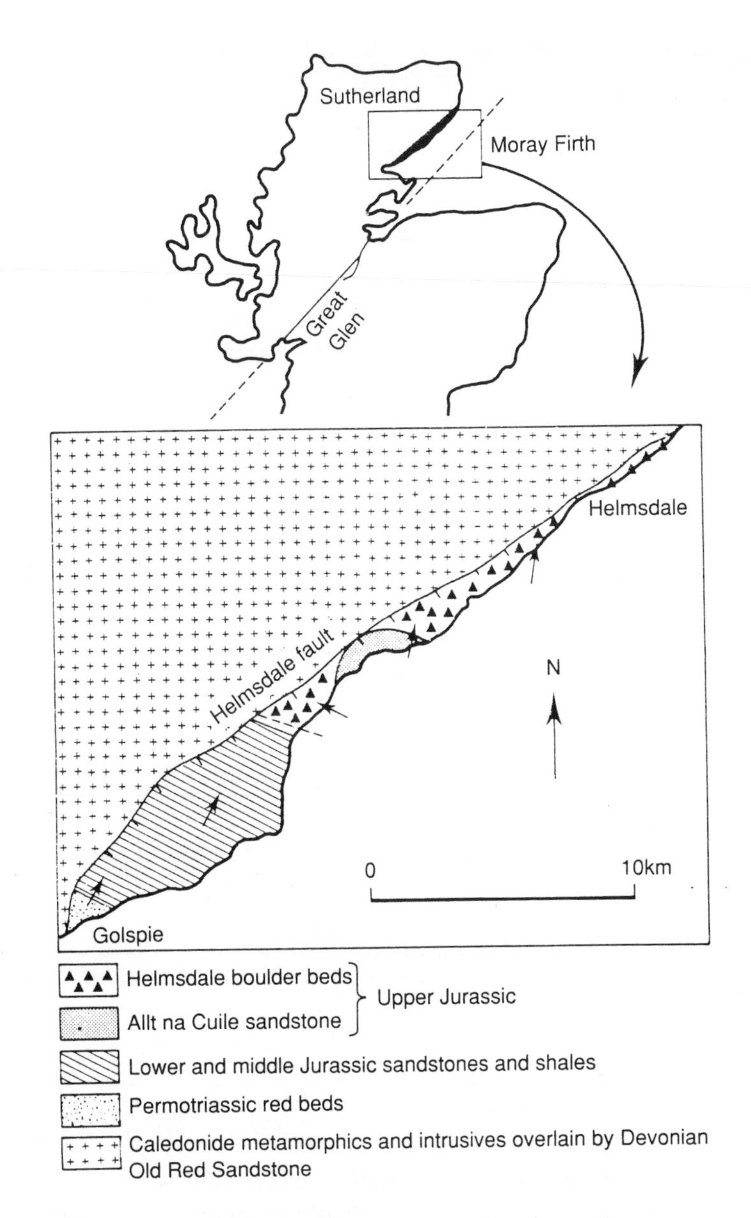

Figure 12.6 Geological sketch map of the Helmsdale outlier, Sutherland.

The Lower and Middle Jurassic sediments are of a shallow water origin, including fluvio-deltaic sands, with coals, rootlet horizons, and shallow water burrows. These are interbedded with offshore shales. The first sign of deep water sedimentation appears in the Allt na Cuile Sandstone (Early Kimmeridgian). This consists of massive multistorey channel sands with occasional scattered granules and clasts. The top of each channel is often flat-bedded (Figure 12.8).

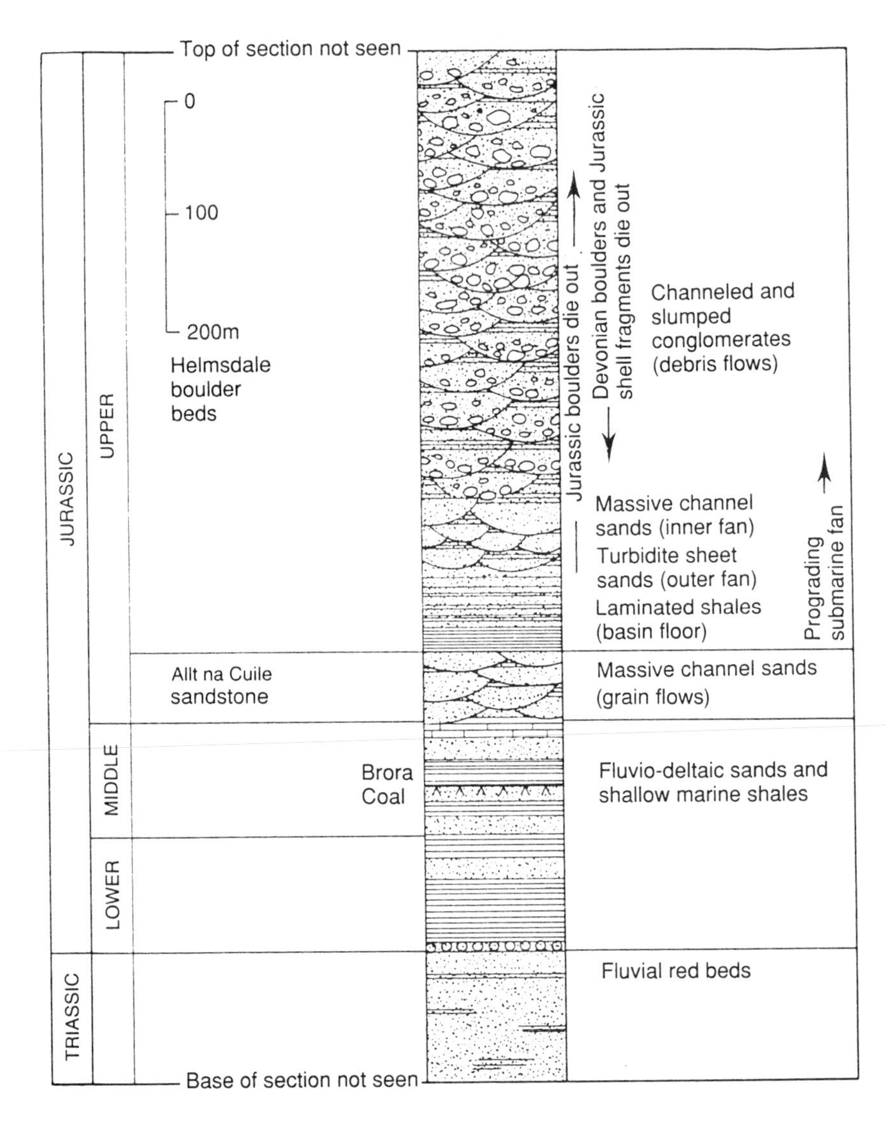

Figure 12.7 Stratigraphic section, facies and deduced environments of the Mesozoic sediments of the Sutherland coast. Accurate formation thicknesses are only approximate due to extensive faulting. Pre-Upper Jurassic stratigraphy diagrammatic.

The Allt na Cuile Sandstone is overlain by laminated black carbonaceous shales with a sparse marine fauna and abundant plant detritus (Figure 12.9).

These beds pass gradually up into thinly interbedded shales and sandstones. The sandstones have sharp, sometimes erosional, bases and gradational tops. Internally the sandstones are massive, laminated or occasionally cross-laminated. They look like turbidites (Figure 12.10).

The laterally extensive turbidite sheet sands are locally overlain by thicker coarser massive sandstones. These are generally channeled, with intraformational shale pebble conglomerates (Figure 12.11).

Figure 12.8 Photograph of Allt na Cuile Sandstone outcrop, showing multistorey massive channel sands, interpreted as grain flows. Lothbeg Point, Sutherland.

Figure 12.9 Laminated black shales of basinal marine origin. Lothbeg, Sutherland.

Figure 12.10 Graded turbidite sands and shales interpreted as outer fan deposits. Lothbeg, Sutherland.

Figure 12.11 Massive channel sands cut into turbidite sands and shales. Interpreted as inner fan deposits. Lothbeg, Sutherland.

The topmost 500 m or so of the exposed section consists of the Helmsdale boulder beds (Figure 12.12). These are spectacularly large conglomerates with clasts ranging in size up to that of a highland croft. The conglomerates are interbedded with black shales and thin turbidite sands.

There is an interesting vertical change in the nature of the boulder beds. Many of the clasts in the lower part are clearly reworked shallow water Jurassic sandstone. These sands are cross-bedded and occasionally contorted, suggesting that they were emplaced in a semi-consolidated state.

Moving up the sequence however these clasts die out to be replaced by pebbles and boulders of Devonian Old Red Sandstone. The matrix of the boulder beds also changes upwards, becoming more calcareous, often consisting of a skeletal packstone. This is made of a diverse fauna which includes ammonites, brachiopods, oysters, corals, echinoids and crinoid debris. Throughout the boulder beds there is abundant slumping and occasional intrusion by sandstone dykes (Figs 12.13a,b and 12.14).

Figure 12.12 Upper Jurassic Helmsdale boulder beds (fine grain variety), Lothberg, Sutherland. Interpreted as debris flows shed from the adjacent Helmsdale fault.

Figure 12.13a and b. Examples of slumping in interlaminated sands and shales within the Helmsdale boulder beds, Kintradwell, Sutherland. These features indicate deposition on a slope.

Figure 12.14 Sandstone dyke penetrating Upper Jurassic sediments, Kintradwell, Sutherland.

The top of the Helmsdale boulder beds is not exposed, though shallow marine Lower Cretaceous sandstones are known to crop out on the sea floor of the Moray Firth.

Interpretation

The Permo-Triassic continental beds, and the paralic Lower to Middle Jurassic beds will not be considered now. It is the Upper Jurassic section that can be used to illustrate the characteristics of deep sea sands. The Lower and Middle Jurassic sediments show the characteristics of shallow water deposition. Anomalous features are shown by the overlying massive channelled Allt na Cuile Sandstone. The absence of cross-bedding suggests that the sands may not have been deposited by traction currents. Their massive nature, together with the presence of occasional scattered clasts, suggests that they may have been deposited by channelized grainflows.

This conclusion demonstrates the initiation of a slope from shallow to deeper water. There is no doubt that this slope was caused by the uplift of the Helmsdale fault. The laminated black shales which overly the Allt na Cuile sands were clearly deposited in a sub-wave base marine environment, but the abundance of plant debris suggests that this basin lay close to the coast. The grain flow sand:basinal shale contact marks a major subsidence of the sea floor. The basinal shales are overlain by turbidite sheet sands followed by thicker pebbly channel sands. Using Walther's Law this vertical succession of facies appears to have been deposited by the progradation of a submarine fan across the basin floor. The turbidite sheet sands were deposited on the outer fan, and overlain by inner fan channel sands.

The overlying Helmsdale boulder beds show all the hallmarks of submarine debris flows (see Lowe, 1982). These were clearly derived from the adjacent Helmsdale fault scarp to the west. It is apparent that the gentle slope responsible for the Allt na Cuile grain flows became gradually steeper, until a submarine fault scarp broke through to the surface. Initially Jurassic sediment was eroded from the upthrown block to be redistributed as turbidite sands and debris flows on the adjacent basin floor.

Once this had been removed the lithified Devonian Old Red Sandstone contributed clasts to the debris flows. The hard rocky substrate of the upthrown block permitted a diverse benthonic fauna to thrive and contribute its detritus to the basinal deposits. Faulting continued after the deposition of the Upper Jurassic boulder beds. It is a curious fact that though the boulder beds are locally faulted against the Helmsdale granite they contain no clasts of it. This suggests that post-Jurassic movement may have had a transverse component. The dramatic sequence of events outlined above is illustrated in Figure 12.15.

LATE JURASSIC DEEP WATER SANDS OF THE VIKING GRABEN

The Helmsdale fault was only one of a multitude of faults that were active in the Late Jurassic, in the North Sea basin (Figure 12.16).

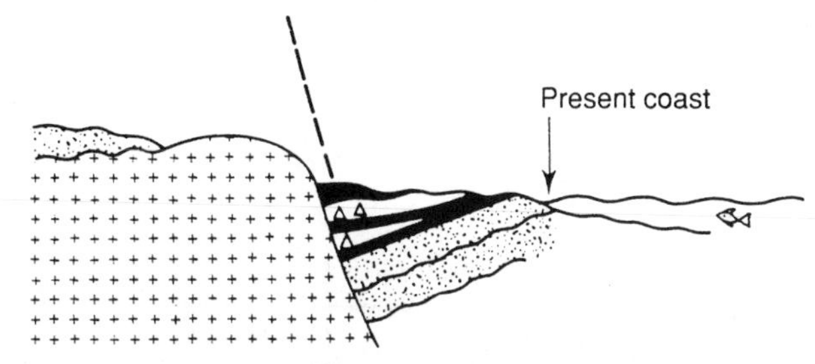

4

Post-Lower Cretaceous uplift. All Jurassic and much Devonian cover denuded from the upthrown block to expose the Helmsdale Granite

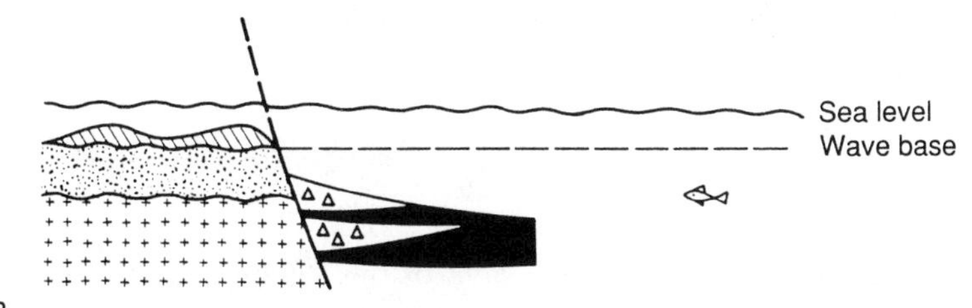

3

Upthrown block moves above wave base. Jurassic skeletal shoal sands move across rocky Devonian substrate to form matrix of beds of Devonian boulders

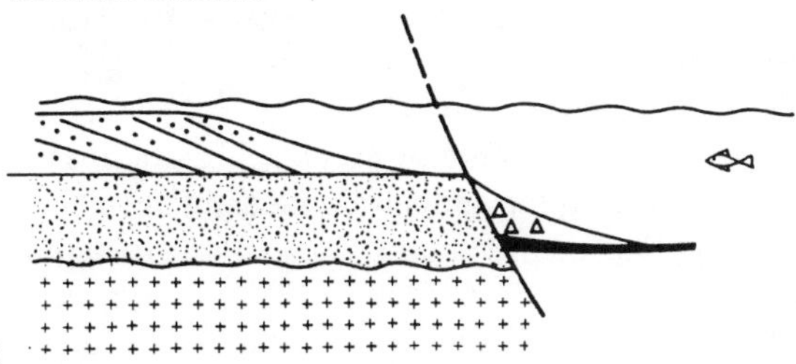

2

Fault scarp breaks through to sea bed. Carbonaceous deltaic sand shed off as turbidites and consolidated blocks mixed with Devonian boulders

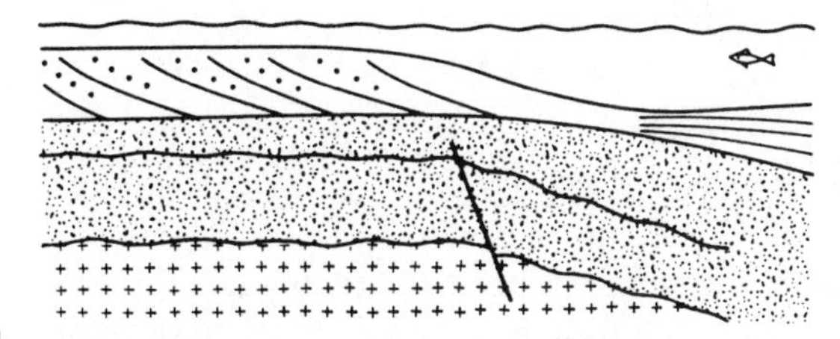

1

Movement begins at depth on line of old Devonian Helmsdale fracture. Tilted sea bed causes grain flow slides.

Figure 12.15
Geophantasmograms to illustrate the relationship between the Helmsdale fault and its associated Jurassic sediments (from Neves and Selley, 1975; by permission of the Norwegian Petroleum Society).

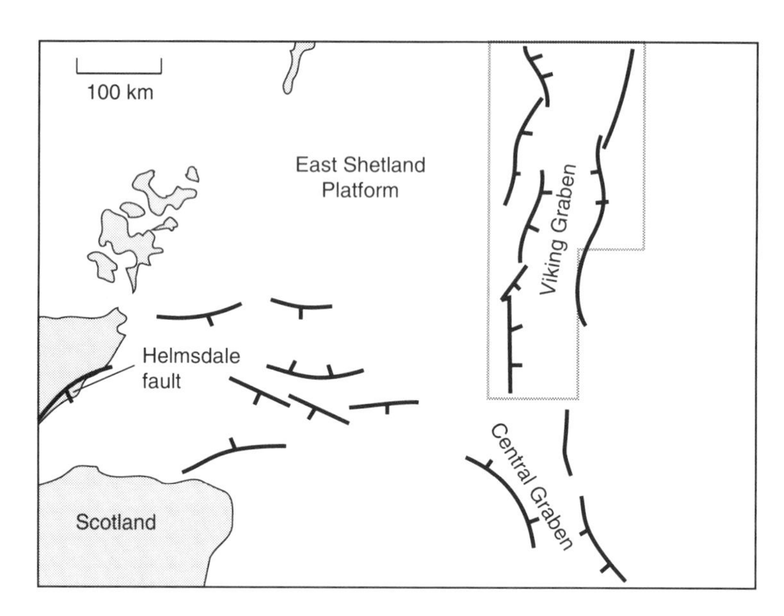

Figure 12.16 Map to show the relationship between the Helmsdale fault and the coeval faults and fault-derived sediments of the Viking Graben of the North Sea. Inset detail of Figure 12.17.

This was when the major rift system of the Viking and Central grabens failed as the Atlantic Ocean began to open.

Several oil fields produce from Upper Jurassic deep sea sands and boulder beds that were deposited adjacent to fault scarps analogous to the one just described from Helmsdale. These include Crawford, Brae, Tiffany, Toni and Thelma. So the sedimentary model developed from the onshore outcrop is applicable to the offshore subsurface situation.

Figure 12.17 shows the location of some of these fields, and Figure 12.18 shows a seismic line across the Viking graben boundary fault and its foot-wall fans. It is worth noting that one of the oil companies operating in the area ignored the Helmsdale sequence, and identified the Brae reservoir, not as a submarine fan, but as a fan delta (Harms *et al.*, 1981). This depositional model implied that the reservoir sands would not extend out in across the basin. They thus relinquished the acreage in which the Miller oil field now produces from distal fan turbidites (Garland, 1993).

ECONOMIC ASPECTS OF DEEP WATER SANDS

Deep water sands are obviously not potential hydrocarbon reservoirs when they are volcaniclastic turbidites deposited within fore-arc basins and involved in orogenesis. Incipient metamorphism destroys porosity, and breaks down hydrocarbons, while structural deformation allows the escape of any pore fluids. Deep sea sands in non-orogenic situations are often highly productive of oil and gas when they occur at the foot of deltas or in fault-bounded troughs with restricted marine circulation.

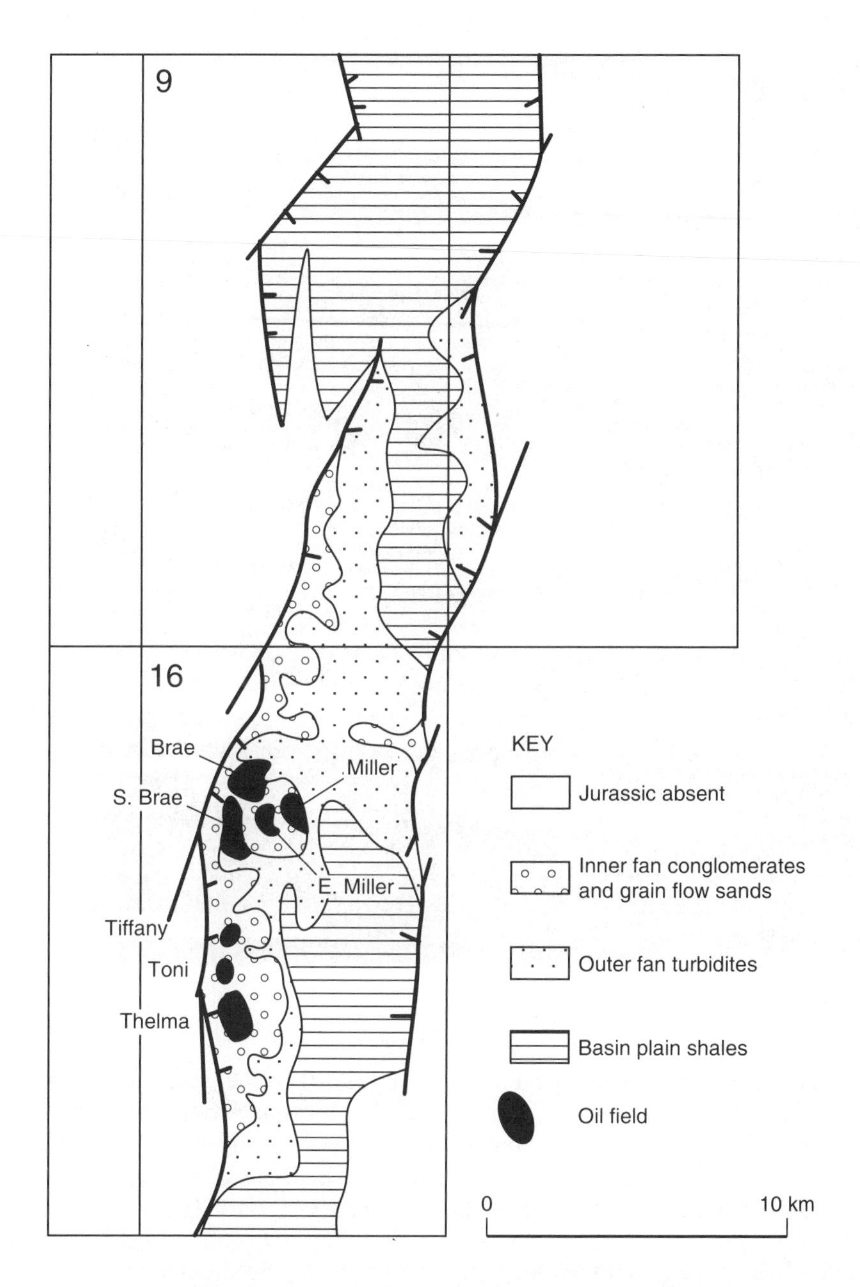

Figure 12.17 Late Jurassic (Kimmeridgian) palaeogeographic map of the southern Viking Graben, showing the distribution of deep sea sediments, and the petroleum accumulations that they contain. Developed from Harris and Fowler (1987).

In these situations the pelagic basin floor muds may act as source beds which generate oil and gas. The hydrocarbons can migrate updip through the turbidite fan sands with which they interfinger. Oil and gas may be trapped both in structural traps, and stratigraphically where submarine channel sands are sealed updip by impermeable slope muds. The turbidite fans

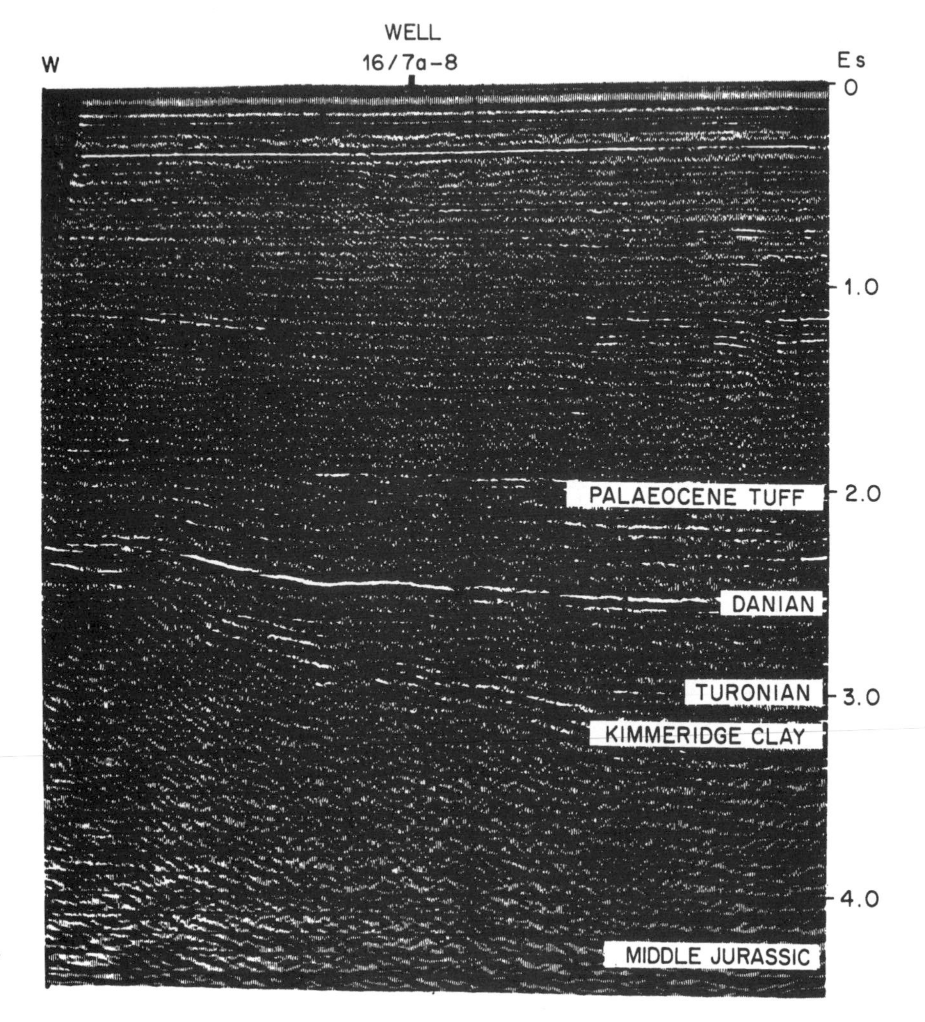

Figure 12.18 Seismic section through Upper Jurassic submarine fan boulder beds of the Brae oil field on the margin of the Viking Graben, North Sea. This reservoir is similar to the Helmsdale boulder beds that crop out on the coast of Sutherland (from Harms *et al.*, 1981, by courtesy of the Institute of Petroleum).

generally have lower porosities and permeabilities than the grainflow channel sands, but are laterally more extensive. The Jurassic submarine fans of the North Sea have already been mentioned. Palaeocene deep sea sand reservoirs produce in the Forties and Montrose fields (Parker, 1975). The Frigg field produces from a Lower Eocene submarine fan (Heritier *et al.*, 1981).

It has been pointed out that deep sea sand petroleum accumulations correlate with global drops in sea level (Shanmugam and Moiola, 1982). It may be argued that when sea level drops shallow marine sands are eroded off the continental shelf and transported by gravity flows down submarine channels and deposited on fans at the foot of the continental slope. High stands of the sea, on the other hand lead to starved sediment supply, and often lead to the deposition of organic rich muds on the drowned shelf, as discussed in Chapter 13 (Figure 13.4).

An example of prolific oil production in turbidites occurs in the Tertiary basins of California. Here over 10 000 m of turbidites were deposited above, and interbedded with deep water organic-rich muds. Subsequent tectonism has involved only gentle folding. Foraminifera in the oldest shales indicate original basin depths of some 1200 m. Despite poor sorting and relatively low porosity and permeability the turbidite sands are good oil reservoirs. This is because individual beds are of unusually great thickness for turbidites, sometimes in excess of 3 m. Furthermore, multi-storey sand sequences are common, shale between turbidites being absent either due to erosion or to a rapid frequency of turbidity flows.

Three main geometries can be recognized in Tertiary Californian turbidite facies and these can be related to deep-sea sediments now forming off the present-day Californian coast (Hand and Emery, 1964). Shoestring turbidites occupy channels sometimes located along syn-sedimentary synclines. Sullwold (1961) cites an example 50 m thick and 600 m wide, whose associated foraminifera suggest deposition in 700 m of water. The Rosedale Channel (Miocene) of the San Joaquin basin is some 2 km wide and contains some 4000 m of turbidites. It can be traced downslope for about 8 km and its fauna suggests depths greater than 400 m (Martin, 1963). Other examples are discussed by Webb (1965).

Shoestring turbidite bundles such as these formed in submarine channels similar to those of the present-day continental margins. The second turbidite geometry is fan-shaped in plan and lenticular in cross section. The Upper Miocene Tarzana fan identified by Sullwold (1961) is about 100 km wide and 1200 m thick. Associated foraminifera indicate a depositional depth of about 1000 m. Other examples are lobate and branching (Webb, 1965). Geometries such as these can be directly compared to present-day fans formed by sediment debouching from the mouths of submarine canyons on to abyssal plains. The third facies geometry found in Californian Tertiary turbidites is a sheet. This was due to deposition on a basin floor. The Repetto Formation (Lower Pliocene) is an example of this type. It has a maximum thickness of more than 750 m and covers some 2000 km^2 (Shelton, 1967).

Figure 12.19 illustrates the geometry of these three types of deep sea facies. Sedimentology obviously has a role to play in the exploitation of petroleum from deep sea sands as, and when, it is found. It is important first to recognize the environment, second to identify the type of geometry, and finally to predict the trend of an individual turbidite bundle.

SUB-SURFACE DIAGNOSIS OF DEEP SEA SANDS

Because of their economic importance great attention has been paid to the

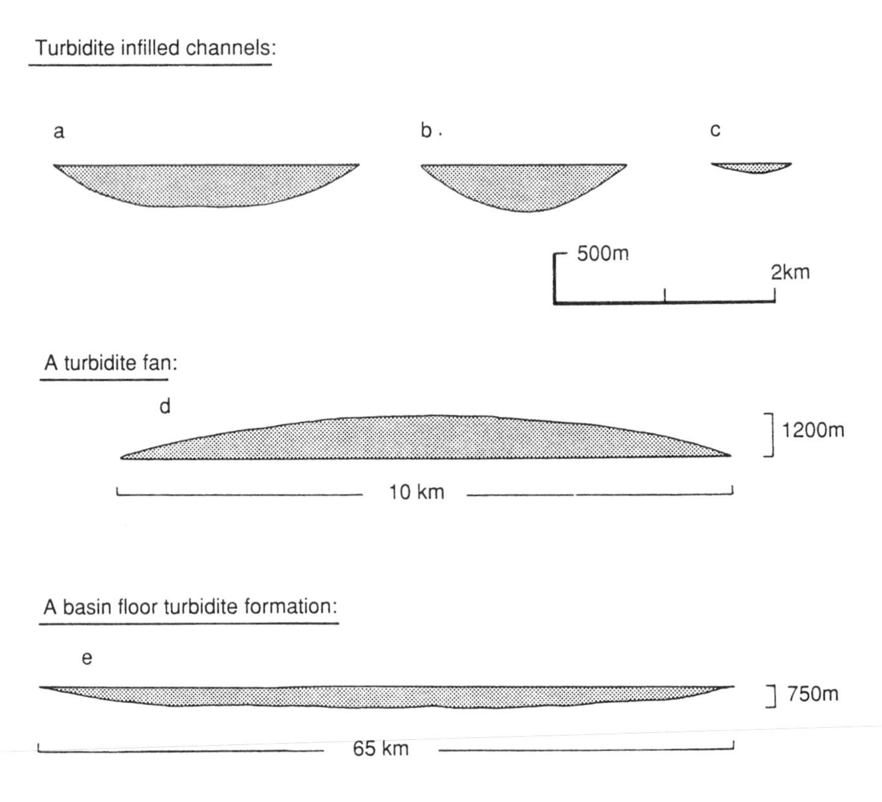

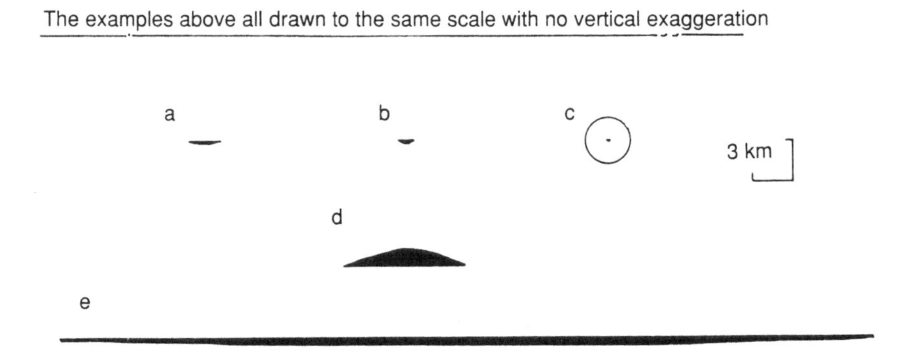

Figure 12.19 Palaeostrike cross-sections of turbidite geometries. Sources of data cited in the text. (a) Contes channel, Annot trough, Alpes Maritimes; (b) Rosedale channel, San Joaquin basin, California; (c) Sansina channel, Los Angeles basin, California; (d) Tarzana fan, Los Angeles basin, California; (e) Repetto turbidite sheet, Los Angeles basin, California.

recognition of deep sea sands both at outcrop (e.g. Stanley and Unrug, 1972), and in the subsurface (Selley, 1976; Parker, 1977; Howell and Normark, 1982; Galloway and Hobday, 1983).

The geometry of submarine channels and fans can be mapped from well logs and seismic data. Seismic lines oriented parallel to the palaeoslope may enable deep sea sands to be identified at the foot of prograding deltas or submarine fault scarps (refer back to Figs 6.16 and 12.18 respectively). Seismic sections aligned parallel to the palaeostrike may show the palaeotopography of the abandoned submarine fan surface (Figure 12.20).

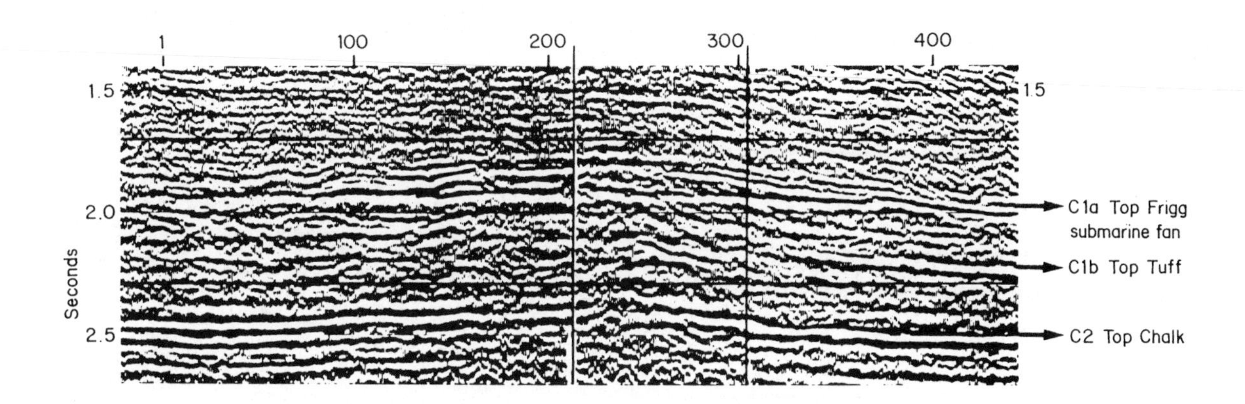

Figure 12.20 Seismic line through the Frigg field showing the palaeotopography of the buried submarine fan reservoir. For location see Figure 12.21. From Heritier *et al.* (1981), by courtesy of the Institute of Petroleum.

Sometimes it may be possible to map the complete geometry of the fan (Figure 12.21). Note though that a fan shape is not a unique feature of submarine fans. It may also occur in alluvial fans and delta lobes.

Lithologically deep water sediments are diverse, ranging from boulder beds to the finest silt. Glauconite, shell debris, mica and carbonaceous detritus are often found mixed together in turbidite sands where the sediment is derived from both deltaic and marine shelf environments. As pointed out earlier, carbonate turbidites occur at the foot of carbonate shelves. An admixture of transported shallow water fauna in the turbidite sands with a pelagic fauna with the '*Nereites*' ichnofacies in the intervening shales is also typical.

When cores are available the typical Bouma A–E sequences may be found in the submarine fan turbidites, and the diagnostic features of grain flows may be found in the submarine channels. There are particularly diagnostic gamma and SP log motifs but they are not unique to deep water sands, and can be confused with deltaic ones. On a gamma log turbidite sequences show a very 'nervous' pattern with the curve swinging to and fro with a large amplitude. This reflects the interbedding of sands and shales. The vertical resolution of both the gamma and SP logs is seldom sufficient actually to show individual graded beds. Both the micro-resistivity log and FMI and FMS imaging tools can pick up the graded beds of turbidites.

Turbidite fan sequences often show an overall upward-sanding motif which reflects the basinward progradation of the fan over basinal shales. Submarine channels which are infilled by grainflow sands show steady

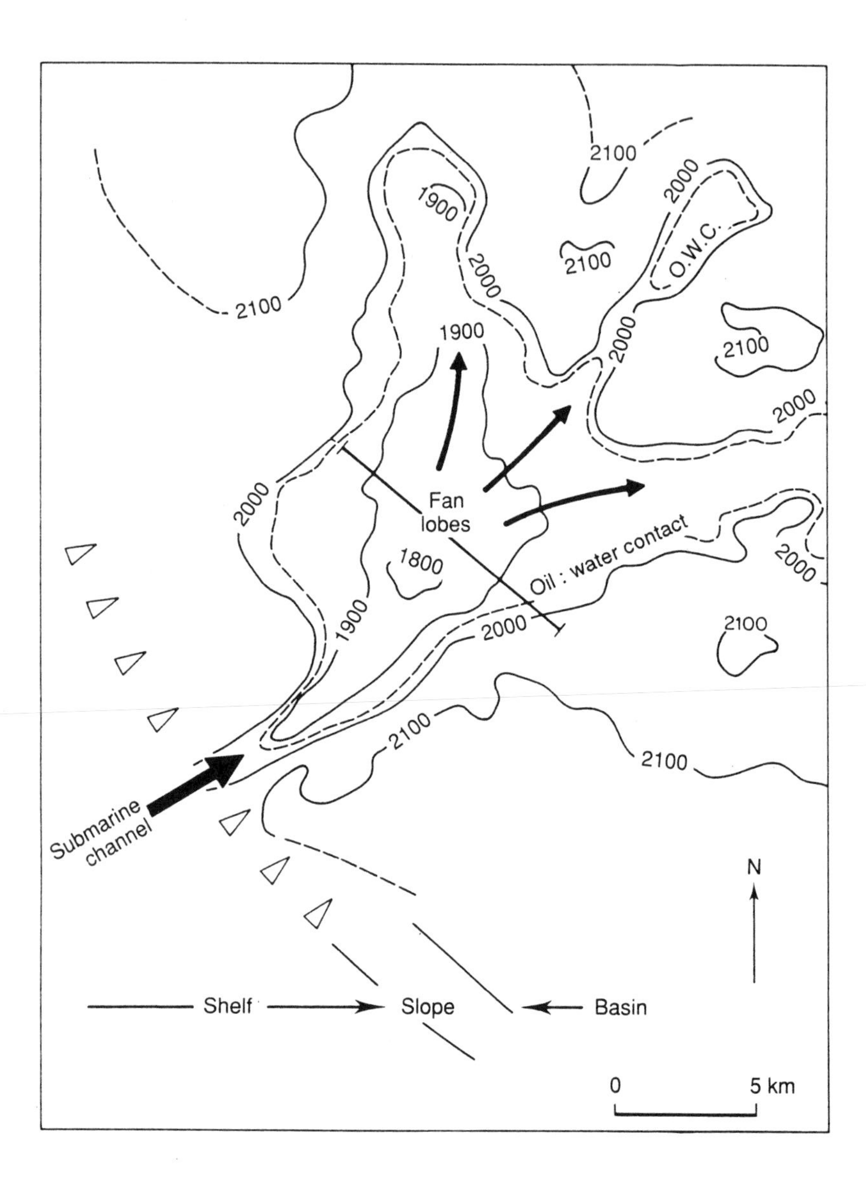

Figure 12.21 Structure contour map (metres) on top of the Frigg oil and gas field of the North Sea. The reservoir is a submarine fan of Lower Eocene age. A-B indicates location of seismic line in Figure 12.20. Based on Heritier *et al.* (1981).

featureless log profiles with abrupt lower surfaces reflecting the erosional base of the channel floor. Submarine channels that are infilled by turbidites, on the other hand, show a spiky upward-fining motif, reflecting the gradual infilling and abandonment of the channel. These motifs are commonly found above progradational fan sequences. Figure 12.22 shows these log motifs and Figure 12.23 illustrates a real example from the North Sea.

(a)

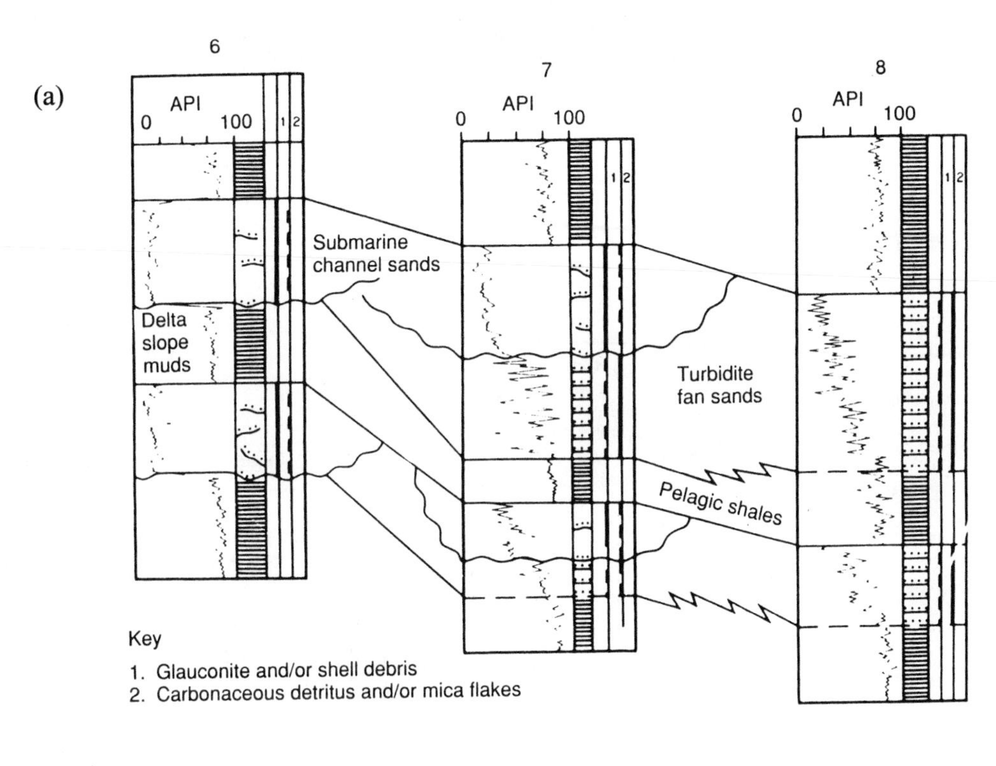

(b)

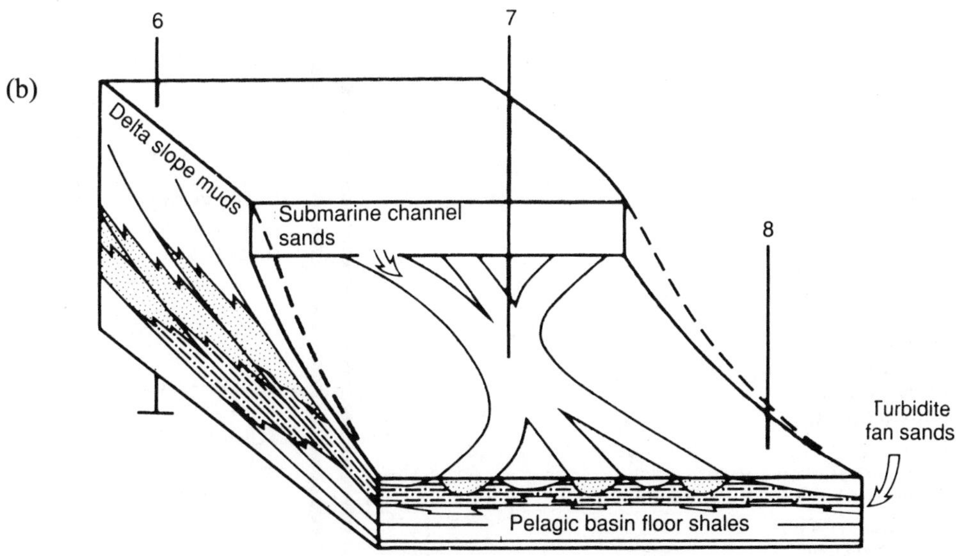

Figure 12.22 Diagrams to illustrate the log motifs (a) and geometry (b) of deep sea sands. Note the similarity to the progradational and distributary channel motifs of deltas (Figure 6.20).

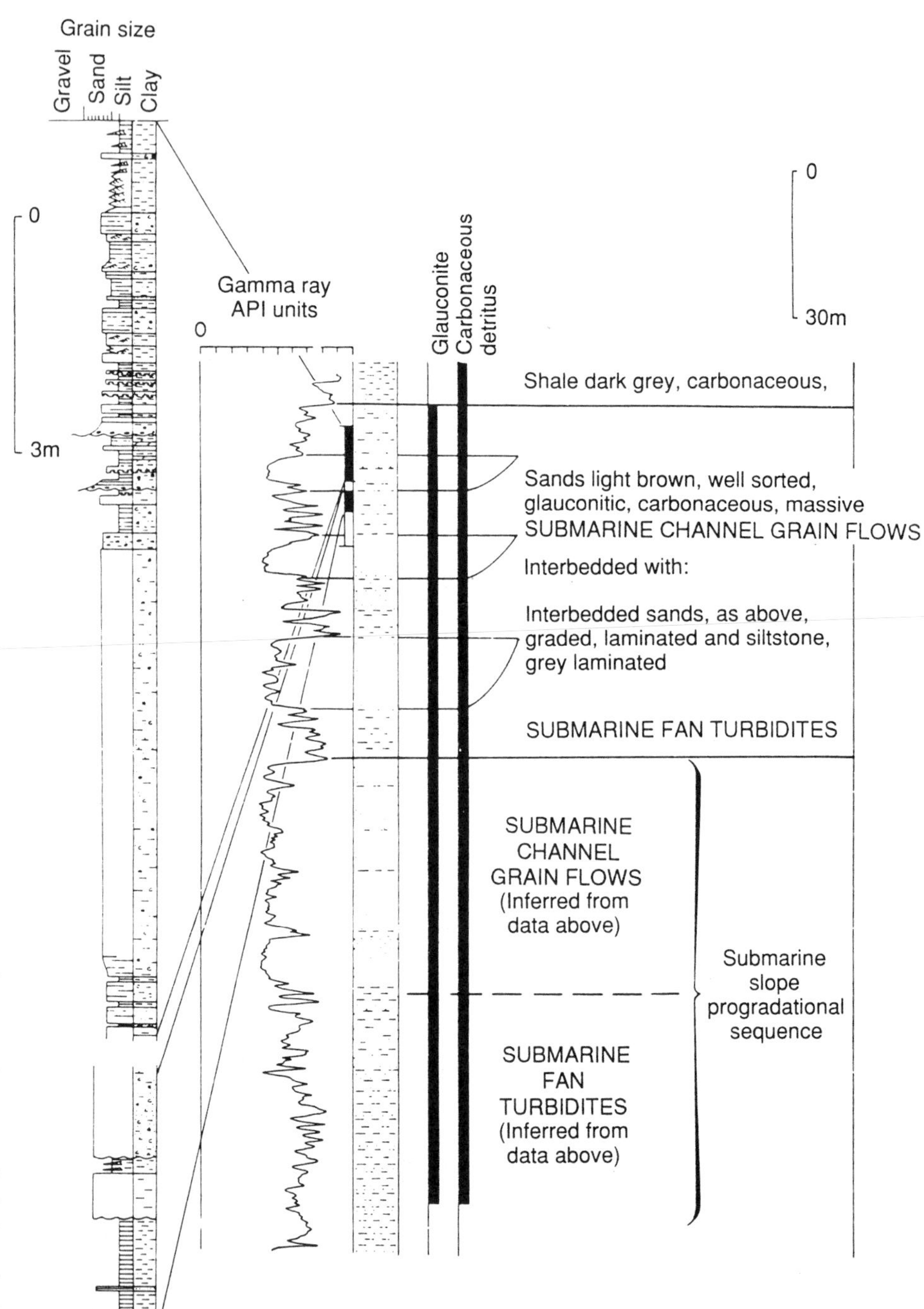

Figure 12.23 Sedimentary core log and gamma log of a deep sea sand from the North Sea. Note how the upper part of the core with turbidite features correlates with an erratic log curve, while the lower part of the core, interpreted as a grainflow, correlates with a steady log curve (from Selley, 1976, Figure 6, by courtesy of the American Association of Petroleum Geologists).

There are also characteristic dipmeter motifs in deep sea sands (Selley, 1976). Within the submarine channels the heterogenous fabric of boulder beds and grainflow sands may yield a random bag o' nails pattern. Turbidite-infilled channels may show upward-declining red patterns reflecting the gradual infilling of the curved channel floor. Dips will point towards the channel axis, as in shallow water channels (see p. 160). As cross-bedding is absent there will not be dips pointing down the channel axis. The progradational fan sequences sometimes show upward-increasing blue motifs analogous to those of delta slopes. The dips of those, as in deltas, also point in the direction of sediment progradation (Figure 12.24).

Vertical rotation of progradational dips may also be identified, similar to that found in delta lobes (p. 159). As in deltas, one well may tell on which side of a fan it has been drilled. Where there are two wells in one fan it may be possible to plot the dips back, and so locate the apex of the fan lobe (Selley, 1989). Though this technique was stumbled on by the author working in deltaic reservoirs, other geologists have had the same geospasm while working with deep sea sands petroleum reservoirs.

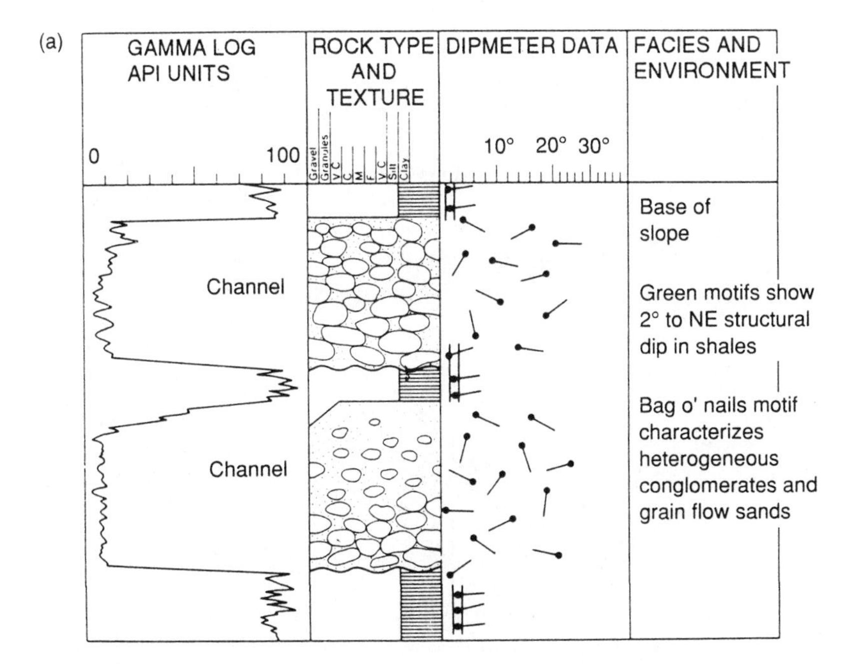

Figure 12.24a Deep sea sand dip motifs for slope grain flow or conglomerate channels. For explanations see text.

(b)

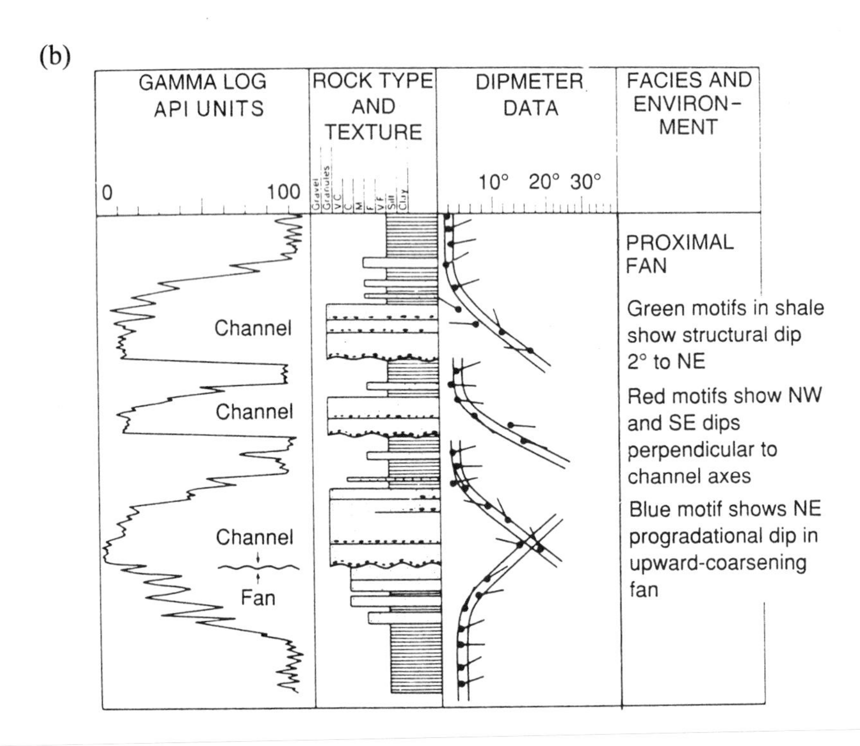

(c)

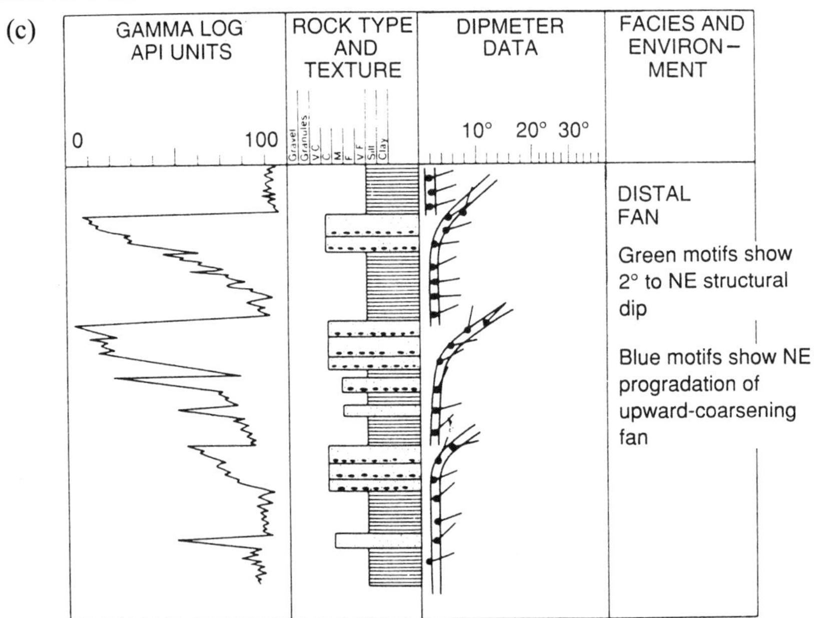

Figure 12.24b and c Deep sea sand dip motifs for proximal fan channel confined turbidites (b: upper) and for distal fan turbidites (c: lower). For explanations see text.

REFERENCES

The Jurassic case history from Sutherland, Scotland, and the North Sea was based on the author's own work and:

Bailey, Sir E. B. and Weir, J. (1932) Submarine faulting in Kimmeridgian times: East Sutherland, *Trans. Roy. Soc. Edinb.*, **47**, 431–67.

Cherry, S. T. J. (1993) The interaction of structure and sedimentary process controlling deposition of the Upper Jurassic Brae Formation Conglomerate, Block 16/7, North Sea. In J. R. Parker (ed.), *Petroleum Geology of Northwest Europe*, Geol. Soc. London, 387–400.

Crowell, J. C. (1961) Depositional Structures from Jurassic Boulder Beds, East Sutherland, *Trans. Edinb. Geol. Soc.*, **18**, 202–19.

Garland, C. R. (1993) Miller Field: reservoir stratigraphy and its impact on development. In J. R. Parker (ed.), *Petroleum Geology of Northwest Europe*, Geol. Soc. London, 401–14.

Harms, J. C. and McMichael, W. J. (1982) Sedimentology of the Brae oil field area, North Sea, *J. Petrol. Geol.*, **5**, 437–9.

Harms, J. C., Tackenberg, P., Pickles, E. and Pollock, R. E. (1981) The Brae oil field area. In L. V. Illing and G. D. Hobson (eds), *Petroleum Geology of the Continental Shelf of North-west Europe*, Heyden Press, London, 352–7.

Harris, J. P. and Fowler, R. M. (1986) Enhanced prospectivity of the Mid-Late Jurassic sediments of the South Viking Graben, northern North Sea. In J. Brooks and K. W. Glennie (eds), *Petroleum Geology of North West Europe*, Graham & Trotman, London, 879–98.

Lam, K. and Porter, R. (1977) The distribution of palynomorphs in the Jurassic rocks of the Brora Outlier. NE Scotland, *Jl. Geol. Soc. Lond.*, **134**, 45–55.

Neves, R. and Selley, R. C. (1975) A review of thc Jurassic rocks of North-east Scotland. In K. Finstad and R. C. Selley (eds), *Proceedings of the Northern North Sea Symposium*, Norwegian Petroleum Society, Oslo, Paper No. 5.

Pickering, K. T. (1984) The Upper Jurassic 'Boulder Beds' and related deposits: a fault controlled submarine slope, NE Scotland, *Jl. Geol. Soc. Lond.*, **141**, 357–74.

Stow, D. A. V., Bishop, C. D. and Mills, S. J. (1982) Sedimentology of the Brae Oil Field, North Sea: fan models and control, *J. Petrol. Geol.*, **5**, 129–58.

Turner, C. C., Cohen, J. M., Connell, E. R. and D. M. Cooper (1986) A depositional model for the South Brae oilfield. In J. Brooks and K. W. Glennie (eds), *Petroleum Geology of North West Europe*, Graham & Trotman, London, 853–64.

Other references cited in this chapter:

Bagnold, R. A. (1966) An approach to the sediment transport problem from general physics, US Geol. Surv. Prof. Paps., 422-I.

Bouma, A. H. (1962) *Sedimentology of some flysch deposits*, Elsevier, Amsterdam.

—— and Brouwer, A. (1964) *Turbidites*, Elsevier, Amsterdam.

—— , Normark, W. R. and Barnes, N. E. (1985) *Submarine Fans and related Turbidite Systems*, Springer-Verlag, Berlin.

Carozzi, A. V. and Frost, S. H. (1966) Turbidites in dolomitized flank beds of Niagaran (Silurian) Reefs, Lapel, Indiana, *J. Sediment. Petrol.*, **36**, 563–75.

Cummings, W. A. (1962) The greywacke problem, *Lpool. Manchr. Geol. J.*, **3**, 51–72.

Dill, R. F. (1964) Sedimentation and erosion in Scripps submarine canyon head. In R. L. Miller (ed.), *Marine Geology*, Macmillan, London, 23–41.

Dott, R. H. (ed.) (1974) *Modern and Ancient Geosynclinal sedimentation*, Soc. Econ. Pal. Min., Spec. Pub. **9**.

Duff, P. McL. D., Hallam, A. and Walton, E. K. (1967) *Cyclic Sedimentation*, Elsevier, Amsterdam.

Dzulinski, S. and Walton, E. K. (1965) *Sedimentary Features of Flysch and Greywackes*, Elsevier, Amsterdam.

Galloway, W. E. and Hobday, D. K. (1983) *Terrigenous Clastic Depositional Systems*, Springer-Verlag, Berlin.

Grover, N. C. and Howard, C. S. (1938) The passage of turbid water through Lake Mead, *Trans. Amer. Soc. Civ. Eng.*, **103**, 720–90.

Hand, B. M. and Emery, J. O. (1964) Turbidites and topography of North end of San Diego trough, California, *J. Geol.*, **72**, 526–42.

Harms, J. C. and Fahnestock, R. K. (1965) Stratification, bed forms, and flow phenomena (with an example from the Rio Grande). In G. V. Middleton (ed.), *Primary Sedimentary Structures and Their Hydrodynamic Interpretation*, Soc. Econ. Min. Pal., Sp. Pub., **13**, 84–115.

Heezen, B. C. (1963) Turbidity currents. In M. N. Hill (ed.), *The Sea*, vol. III, Interscience, NY, 742–75.

—— and Ewing, M. (1952) Turbidity currents and submarine slumps and the Grand Banks earthquake, *Am. J. Sc.*, **250**, 849–73.

Heritier, F. E., Lossel, P. and Wathne, E. (1981) The Frigg gas field. In L. V. Illing and G. D. Hobson (eds), *Petroleum Geology of the Continental Shelf of North-West Europe*, Heyden Press, London, 380–94.

Holtedahl, H. (1965) Recent turbidites in the Hardangerfjord, Norway. In W. F. Whittard and R. Bradshaw (eds), *Submarine Geology and Geophysics*, Butterworths, London, 107–42.

Horn, D. R., Ewing, M., Delach, M. N. and Horn, B. M. (1971) Turbidites of the northeast Pacific, *Sedimentology*, **16**, 55–69.

Horn, E. R., Ewing, J. I., and Ewing, M. (1972) Graded-bed sequences emplaced by turbidity currents north of 20° in the Pacific, Atlantic and Mediterranean, *Sedimentology*, **18**, 247–75.

Howell, D. G. and Normark, W. R. (1982) Submarine fans. In P. A. Scholle and D. Spearing (eds), *Sandstone Depositional Environments*, Amer. Assoc. Petrol. Geol. Tulsa, 365–404.

Hsu, K. J. (1970) The meaning of the word Flysch, a short historical search. In J. Lejoie (ed.), *Flysch Sedimentology in North America*, Geol. Soc. Can., Sp. Paper **7**, 1–11.

Hubert, J. F. (1964) Textural evidence for deposition for many western North Atlantic deep-sea sands by ocean bottom currents rather than turbidity currents, *J. Geol.*, **72**, 757–85.

—— (1967) Sedimentology of pre-Alpine flysch sequences, Switzerland, *J. Sediment. Petrol.*, **37**, 885–907.

Irvine, T. N. (1965) Sedimentary structures in Igneous intrusions with particular reference to Duke Island Ultramafic Complex. In G. V. Middleton (ed.), *Primary Sedimentary Structures and Their Hydrodynamic Interpretation*, Soc. Econ. Min. Pal., Sp. Pub., **13**, 220–32.

Jerzmanska, A. (1960) Ichthyo-fauna from the Jaslo shales at Sobniow (Poland), *Acta. Palaeontol. Polon.*, **5**, 367–419.

Kelling, G. (1964) The turbidite concept in Britain. In A. H. Bouma and A. Brouwer (eds), *Turbidites*, Elsevier, Amsterdam, 75–92.

Kelling, G. and Stanley, D. J. (1976) Sedimentation in canyon, slope, and base-of-slope environments. In D. J. Stanley and D. J. P. Swift (eds), *Marine Sediment Transport and Environment management*, J. Wiley, New York, 379–435.

Klein, G. de V. (1967) Paleocurrent analysis in relation to modern sediment dispersal patterns, *Bull. Amer. Assoc. Petrol. Geol.*, **51**, 366–82.

Kuenen, P. H. (1948) Turbidity currents of high density, *Int. Geol. Congr. 18th Session*, 44–52.
—— (1951) Mechanics of varve formation and the action of turbidity currents, *Geol. Foren, Stockholm Forh.*, **73**, 69–84.
—— (1965) Comment. In G. V. Middleton (ed.), *Primary Sedimentary Structures and Their Hydrodynamic Interpretation*, Soc. Econ. Min. Pal., Sp. Pub., **13**, 217–18.
—— (1967) Emplacement of flysch-type sand beds, *Sedimentology*, **9**, 203–43.
—— and Migliorini, C. I. (1950) Turbidity currents as a cause of graded bedding, *J. Geol.*, **58**, 91–127.
Lajoie, J. (1970) *Flysch Sedimentology in North America*, Geol. Soc. Can., Sp. Pap. 7.
Lowe, D. R. (1982) Sediment gravity flows: II. Depositional models with special reference to the deposits of high-density turbidity currents, *J. Sediment. Petrol.*, **52**, 279–97.
Mangin, J. P. (1962) Traces de pattes d'oiseaux et flute-casts associés dans un 'facies flysch' du Tertiare pyrénéen, *Sedimentology*, **1**, 163–6.
Martin, B. D. (1963) Rosedale Channel: evidence for Late Miocene submarine erosion in Great Valley of California, *Bull. Amer. Assoc. Petrol. Geol.*, **47**, 441–56.
Menard, H. W. (1952) Deep ripple marks in the sea, *J. Sediment. Petrol.*, **33**, 3–9.
Middleton, G. V. (1966a) Small-scale models of turbidity currents and the criterion for auto-suspension, *J. Sediment. Petrol.*, **36**, 202–8.
—— (1966b) Experiments on density and turbidity currents. I. Motion of the head, *Can. Jnl. Earth Sc.*, **3**, 523–46.
—— (1966c) Experiments on density and turbidity currents. II. Uniform flow of density currents, *Can. Jnl. Earth Sc.*, **3**, 627–37.
—— (1967) Experiments on density and turbidity currents. III, *Can. Jnl. Earth Sc.*, **4**, 475–505.
Mutti, E. and Lucchi, F. R. (1972) Le turbiditii dell'Appennino settentrionale: introduzione all'analisi di facies, *Mem. Soc. Geol. Ital.*, **11**, 161–99.
Natland, Migliorini, I. and Kuenen, P. H. (1951) *Sedimentary History of the Ventura Basin, California, and the Action of Turbidity Currents*, Soc. Econ. Min. Pal., Sp. Pub., **2**, 76–107.
Parker, J. R. (1975) Lower Tertiary sand development in the central North Sea. In A. W. Woodland (ed.), *Petroleum and the Continental Shelf Northwest Europe*, Applied Science Publishers, London, 447–56.
—— (1977) Deep-sea sands. In G. D. Hobson (ed.), *Developments in Petroleum Geology*, **1**, Applied Science Pubs, 225–42.
Pettijohn, F. J. and Potter, P. E. (1964) *Atlas and Glossary of Primary Sedimentary Structures*, Springer-Verlag, Berlin.
Rech-Frollo, M. (1962) Quelques aspects des conditions de depot du flysch, *Bull. Geol. Soc. France*, **7**, 41–8.
—— (1973) Flysch carbonates, *Bull. Centre Rech. Pau.*, **7**, 245–63.
Rusnak, G. A. and Nesteroff, W. D. (1964) Modern turbidites: terrigenous abyssal plain versus bioclastic basin. In R. L. Miller (ed.), *Marine Geology*, Macmillan, New York, 488-503.
Scott, K. M. (1966) Sedimentology and dispersal pattern of a Cretaceous flysch sequence, Patagonian Andes, southern Chile, *Bull. Amer. Assoc. Petrol. Geol.*, **50**, 72–107.
Selley, R. C. (1976) Sub-surface environmental analysis of North Sea sediments, *Bull. Amer. Assoc. Petrol. Geol.*, **60**, 184–95.
—— (1979) Dipmeter and log motifs in North Sea submarine-fans, *Bull. Amer. Assoc. Petrol. Geol.*, **63**, 905–17.
—— (1989) Deltaic reservoir prediction from rotational dipmeter patterns. In M.

Whateley and K. Pickering (eds), *Deltas, Traps and Sites for Fossil Fuels*, Geol. Soc. Lond. Sp. Pub., **41**.
Shanmugam, G. and Moiola, R. J. (1982) Eustatic control of turbidites and winnowed turbidites, *Geology*, **10**, 231–5.
—— and —— (1988) Submarine fans: characteristics, models, classification, and reservoir potential, *Earth Sci. Rev.*, **24**, 383–428.
Shelton, J. W. (1967) Stratigraphic models and general criteria for recognition of alluvial, barrier bar and turbidity current sand deposits, *Bull. Amer. Assoc. Petrol. Geol.*, **51**, 2441–60.
Shepard, F. (1971) *Submarine Canyons and other Sea Valleys*, Wiley, New York.
Stanley, D. J. (1968) *Graded Bedding – Sole Markings – Graywacke Assemblage and Related Sedimentary Structures in Some Carboniferous Flood Deposits, Eastern Massachusetts*, Geol. Soc. Amer. Spec. Pap., **106**, 211–39.
—— and Unrug, R. (1972) Submarine channel deposits, fluxoturbidites and other indicators of slope and base of slope environments in modern and ancient marine basins. In J. K. Rigby and W. K. Hamblin (eds), *Recognition of Ancient Sedimentary Environments*, Soc. Econ. Pal. Min., Sp. Pub, **16**, 287–340.
Stauffer, P. H. (1967) Grainflow deposits and their implications, Santa Ynez Mountains, California, *J. Sediment. Petrol.*, **37**, 487–508.
Stow, D. A. V. and Faugeres, J. C. (eds) (1993) *Contourites and Bottom Currents*, Sed. Geol. Sp. Vol. **82**.
Sullwold, H. H. (1961) Turbidites in oil exploration. In J. A. Peterson and J. C. Osmond (eds), *Geometry of Sandstone Bodies*, Amer. Assoc. Petrol. Geol., 63–81.
Thomson, A. F. and Thomasson, M. R. (1969) Shallow to deep water facies development in the Dimple limestone (Lower Pennsylvanian), Marathon Region, Texas. In G. M. Friedman (ed.), *Depositional Environments in Carbonate Rocks*, Soc. Econ. Min. Pal., Sp. Pub., **14**, 57–78.
Van de Kamp, P. C., Conniff, J. J. and Morris, D. A. (1974) Facies relations in the Eocene-Oligocene in the Santa Ynez Mountains California, *J. Geol. Soc. Lond.*, **130**, 554–66.
Van der Lingen, G. J. (1969) The turbidite problem, *N. Z. Jl. Geol. Geophys.*, **12**, 7–50.
Van Straaten, L. M. J. U. (1964) Turbidite sediments in the southeastern Adriatic sea. In A. H. Bouma and A. Brouwer (eds), *Turbidites*, Elsevier, Amsterdam, 142–7.
Walker, R. G. (1965) The origin and significance of the internal sedimentary structures of turbidites, *Proc. Yorks. Geol. Soc.*, **35**, 1–32.
—— (1967) Turbidite sedimentary structures and their relationship to proximal and distal depositional environments, *J. Sediment. Petrol.*, **37**, 25–43.
—— (1970) Review of the geometry and facies organisation of turbidites and turbidite bearing basins. In J. Lajoie (ed.), *Flysch Sedimentology of North America*, Geol. Soc. Can., Sp. Pub., 219–51.
Webb, F. W. (1965) The stratigraphy and sedimentary petrology of Miocene turbidites in San Joaquin Valley (Abs.), *Bull. Amer. Assoc. Petrol. Geol.*, **49**.
Whitaker, J. H. McD. (ed.) (1976) Submarine canyons and deep-sea fans, *Modern and Ancient*, Dowden, Hutchinson & Ross, Stroudsburg, Pa.

13 Pelagic deposits

INTRODUCTION: RECENT PELAGIC DEPOSITS

This chapter describes pelagic deposits. The term pelagic 'is generally applied to marine sediments in which the fraction derived from the continents indicates deposition from a dilute mineral suspension distributed throughout deep-ocean water' (Arrhenius, 1963, p. 655). To prove that an ancient sediment was deposited in a pelagic environment, as defined above, it is necessary to demonstrate both great depth and an absence of terrestrial influence. As this chapter shows, it is not easy to prove both these points in ancient sediments. It may be safer therefore to define pelagic deposits as 'of the open sea' (Riedel, 1963, p. 866). The need to make this distinction between depth and 'oceanicity' in analysing the environment of supposed ancient pelagic deposits has been stressed by Hallam (1967, p. 330).

The Recent seas can be divided into those of the continental shelves which are separated by the shelf break, at about 200 m from the oceans. Deposits of the Recent oceans have been studied intensively. Particularly important accounts of the modern ocean floor sediments are given in the Deep Sea Drilling Project (DSDP) reports. More concise reviews occur in volumes edited by Lisitzin (1972), Inderbitzen (1974), Hsu and Jenkyns (1974), Cook and Enos (1977), Ross (1982), Scrutton and Talwani (1982), and Stow and Piper (1984). Recent oceanic sediments can be broadly classified into these types:

1. terrigenous sediments,
2. calcareous oozes,
3. siliceous oozes,
4. red clays,
5. manganiferous deposits.

These will now be briefly described and their distribution discussed. Terrigenous sediments are present adjacent to the continents. They consist of argillaceous deposits and the deep-sea sands, of possible turbidite origin, discussed in the previous chapter. Calcareous oozes, or muds, are composed largely of the shells of microfossils (Lucas, 1973). Two main types may be distinguished: Pteropod ooze made up largely of the aragonitic shells of this mollusc, and foraminiferal oozes, composed largely of calcitic foraminiferal tests, often of *Globigerina*.

Siliceous oozes are made of the skeletons.of Diatoms and Radiolaria. Red clays are red and dark brown muds which are believed to form from the finest of wind-blown dust from continental deserts, together with extremely fine particles of volcanic ash and cosmic detritus. The last type of pelagic deposit is diagenetic rather than depositional in origin; these are scoured sea floor surfaces termed 'hard grounds' impregnated and overlain by nodules rich in manganese. It has been estimated that this is the most extensive type of hard rock surface on the lithosphere (Mero, in Shepard, 1963).

The distribution of the various Recent pelagic deposits is roughly correlated with depth. Thus over much of the Atlantic and Pacific oceans Red clays occur in the deepest parts, Radiolarian oozes form in shallower waters deeper than about 4500 m, calcitic foraminiferal oozes lie on the ocean bed between 4500 m and 3500 m, above which point aragonitic pteropod and foraminiferal oozes are found (Figure 13.1).

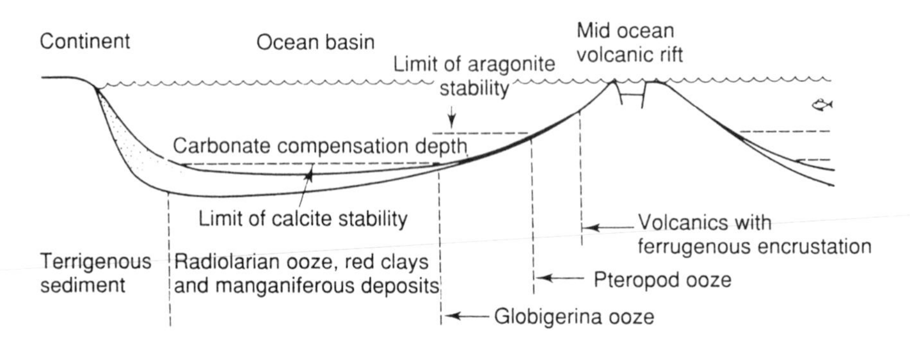

Figure 13.1 Diagrammatic cross-section through a modern ocean basin, complete with mid-ocean rift, to show the relationship between deposits and water depth. Note that the carbonate compensation depth varies with latitude.

Though the distribution of pelagic deposits is loosely correlated with depth it is actually controlled by a variety of factors such as rate of sedimentation and rate of solution. Thus the sequence, with increasing depth, of aragonitic, calcitic, siliceous, and argillaceous deposits largely reflects their increasing chemical stability. The rate of solution of these minerals is a function of their rate of burial, the water temperature and its state of saturation by the various chemicals, as well as the hydrostatic pressure. Only the last of these is actually depth-dependent.

This point is illustrated by the fact that, regardless of depth, calcareous oozes are poorly developed in polar regions. This is due to the low temperature of the bottom water which causes the solution of aragonite and calcite at faster rates than in the more equable equatorial oceans. This fact, that the distribution of Recent pelagic deposits is not a direct reflection of depth, is critical to attempts to determine absolute depths of ancient pelagic deposits.

At the present time the oceans cover much more of the earth's surface than the continental regions. There does not appear to be a corresponding dominance of deep over shallow water sediments in the geological column; on the contrary, ancient deep-sea deposits appear to be very rare. This supports the concept of isostasy, since it suggests that the continents have never been submerged to great depths below the oceans. Not only are ancient deep-sea deposits very rare but their depth is often debatable. This is because it is very hard to separate criteria of depth of deposition from those which indicate distance from the land. The two are not synonymous, as the wide continental shelves of the Atlantic indicate today. This dilemma will become apparent in the discussion of the case history which follows.

TRIASSIC-JURASSIC DEPOSITS OF THE MEDITERRANEAN

Description

Lengthy geological studies have revealed that during Mesozoic and early Tertiary time marine sediments were laid down in a trough stretching from southern France, northern Italy, and beyond to the east. In mid-Tertiary time these deposits were folded and uplifted to form the Alps and associated mountain chains. This trough is called the Tethys ocean. In general the following sequence of events seems to have taken place in this region:

1. Deposition of marine carbonates and fine-grained clastics (Triassic Jurassic) (thought to be shelf sediments).
2. Deposition of thin sequences of clays, cherts, and limestones, sometimes associated with vulcanicity (Triassic-Lower Cretaceous) (thought to be pelagic in origin).
3. Deposition of thick flysch sand: shale sequences (Cretaceous-Oligocene) (thought to be turbidites).
4. Phase of folding, thrusting and uplift.
5. Deposition of clastic sediments in troughs marginal to the mountains, e.g. the Swiss plain and the north Italian plain. This, the molasse facies, is generally continental, at its base is locally marine and sometimes calcareous (Oligocene-Recent).

The timing of these events varies considerably from place to place, two phases often occurring simultaneously in adjacent areas. The clays, cherts and limestones of the second phase are often attributed to deposition in pelagic and deep water environments. They will now be described and the reasons for this interpretation will be discussed. These deposits may be divided into three sub-facies defined as follows:

1. clay shales, marls, and micritic limestones;
2. radiolarian cherts;
3. nodular red limestones.

Clay shales, marls, and micritic limestones

This sub-facies consists of clay shales, sometimes dark and pyritic, sometimes calcareous, marls, and fine-grained micro-crystalline limestones. These beds are generally massive or laminated. The various lithologies sometimes occur rhythmically interbedded with one another. Fossils are generally rare but often well-preserved and sometimes crowded on bedding surfaces. They include thin-shelled lamellibranchs termed *Posidonia* or *Bositra*, belemnites, ammonites, and brachiopods. Often ammonite shells are absent but the lids of their shells (aptychi) occur.

A particularly characteristic brachiopod of this facies is *Pygope*, an unusual form with a hole in the middle. Beds of this sub-facies are named according to their fauna, e.g. Aptychus marls, Pygope limestone, and so on. It is widely distributed in the Upper Jurassic strata of the Alps, the Apennines and Greece.

Radiolarian cherts

Radiolarian cherts are often associated with the previous sub-facies, and share a similar geographic and stratigraphic distribution. They are generally thin-bedded and dark grey, black, or red in colour. Bedding is often rhythmic with thin shale partings. Siliceous limestones are sometimes present transitional between the cherts and the micrites of the previous sub-facies. The cherts are often largely composed of the microscopic siliceous tests of Radiolaria. In addition they contain minor quantities of silicified sponge spicules and foraminifera. The macrofauna is similar to that of the previous sub-facies but is generally much rarer. This chert facies has been considered as an insoluble residue from which all calcium carbonate has been dissolved.

Nodular red limestones

Red nodular limestones, termed 'ammonitico rosso', occur associated with the two previous sub-facies and are widespread over the Alps, Apennines, and parts of Greece. They range in age from Triassic to Jurassic. Lithologically this sub-facies consists of layers of thin pink micrite nodules in a dark red marl matrix. The nodules are often coated with iron and manganese (Figure 13.2).

Thin beds of manganese nodules are sometimes present. The characteristic fossils of this facies are the ammonites from which it gets its name. These are locally abundant. Also present are the aptychi of ammonites, belemnites, crinoids, echinoids, lammellibranchs, gastropods, foraminifera

Figure 13.2 Polished slab of Ammonitico Rosso, Adnet Limestone, Lower Lias, Adnet, Austria. Pale pink micritic limestone fragments set in a matrix of red clay rich in iron and manganese. From Hallam (1967, Plate 2, Figure 1), by courtesy of the Editors of the Scottish Journal of Geology.

and sponge spicules. These fossils often show signs of corrosion and mineral impregnation. Scoured bedding surfaces, termed 'hard grounds', are sometimes present. Hard ground substrates are generally very rich in iron and manganese and are occasionally phosphatized. They are sometimes pierced by *Chondrites* burrows and overlain by a thin layer of manganese or phosphate nodules.

Interpretation

The three sub-facies just described have a number of points in common. They all show an absence of land-derived quartzose sediment and have obvious marine faunas. Due to complex tectonics, their stratigraphic relations are often uncertain, but palaeontology indicates that they were deposited at the same time as adjacent thicker shelf sediments whose depth must be relatively shallow since they are sometimes reefs. These rocks are often overlain by flysch turbidites.

All the sub-facies are fine-grained and devoid of sedimentary structures such as cross-bedding and channelling which would suggest strong current action. The 'hard grounds' may be due to gentle scouring of early cemented sea beds. The fauna is characterized by an abundance of pelagic fossils and a scarcity of benthos. Though the shells are generally preserved as calcite, as are most fossils, there is a curious absence of shells of original aragonitic composition. Thus the calcitic aptychi of ammonites often occur, while their aragonitic shells do not. Chemically the rocks are unusual due to the abundance of organic silica and manganese.

The fine grainsize and absence of cross-bedding and channelling indicate that these are low-energy deposits which must have formed below wave base, though as the 'hard grounds' show, subjected to intermittent scouring by currents.

According to the deductions just made these deposits are of pelagic origin, in the sense that they originated in the open sea far from continental influence. There is, though, some controversy as to the precise depth of their deposition. The evidence cited in favour of great depth hinges on the analogy of these sediments with Recent deep-sea deposits. Similarities are to be found in the absence of land-derived clastics, the scarcity of current structures, and scarcity of benthos, despite the frequent oxygenation of the sea bed, indicated by the presence of red ferric iron. The limestones invite comparison with Recent Globigerina oozes, the cherts with Recent radiolarian oozes, the manganiferous beds with the red clays and manganiferous nodules.

The lack of fossils of aragonitic shells can be attributed to the greater solubility of aragonite over calcite. At the present day aragonite goes into solution at depths of about 3500 m (Friedman, 1965) and calcite at about 4500 m (Berner, 1965). Hard ground horizons occur between 200 and 3500 m (Fischer and Garrison, 1967). All these analogies suggest a deep water origin for the Alpine beds described above. On the other hand it can be argued that many of these criteria are either dubious on chemical grounds, or that they indicate distance from the land, not great depth. This is particularly valid in the case of absence of land-derived quartz and the abundance of fine-grained clay, and organic calcareous and siliceous sediment.

Geochemical studies of Recent deep-sea nodules have laid stress on their oxygen, iron and manganese ratios and on their trace element content. Recent manganese nodules occur today in widely varying depths and even in some lakes. Their chemical variation may be not just depth-controlled but also due to proximity of land-derived chemicals (Price, 1967). Similarly the solution rates of aragonite and calcite are controlled not just by depth, but also by temperature, as pointed out in the introduction to this chapter. Sedimentation rate will also control solution, slow burial increasing the time available for a shell to corrode on the sea bed (Hudson, 1967).

The facies relationships of the supposed deep water deposits are sometimes curious too. For example in the Unken syncline of Austria there is a 300 m sequence of Jurassic pelagic limestones, radiolarian cherts, and Ammonitico Rosso for which Garrison and Fischer (1969) have discussed depths in the order of up to 4000–5000 m. This sequence overlies a karstic topography of Upper Triassic limestones and at the top is unconformably overlain by Lower Cretaceous (Berriasian) limestones. There must have been quite a tectonic flutter between the end of the Triassic and the Early Cretaceous.

In conclusion the fine-grained limestones, Ammonitico Rosso, and radiolarian cherts of the Mediterranean are clearly marine deposits. They were laid down in a low-energy environment below the level of strong current activity and, as the absence of algae show, below the photic zone. The condensed stratigraphy shows that sedimentation was slow, with little influx of land-derived material (Figure 13.3).

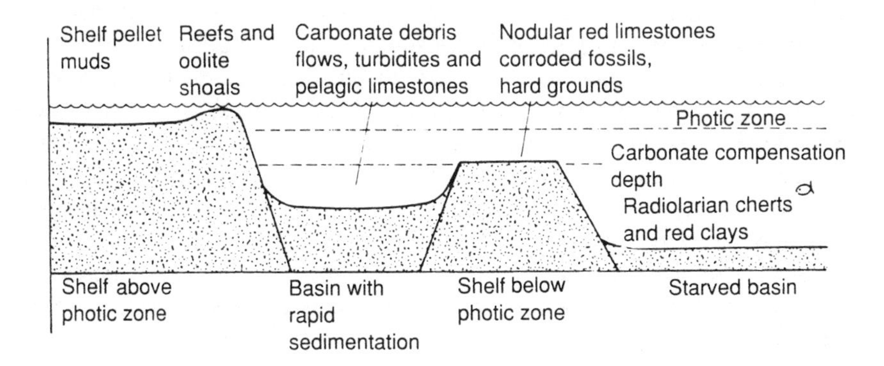

Figure 13.3
Geophantasmogram to show the formation of Ammonitico Rosso and related facies of the Tethys during the Triassic-Jurassic periods. Note that fault blocks may yoyo from the photic zone to below the carbonate compensation depth in response to tectonic and eustatic events. (Simplified from Bernoulli and Jenkyns, 1974, and Santantonio, 1994.)

They may be attributed to a pelagic environment with confidence. Comparison with Recent deep sea deposits suggests a deep-water origin. However the distribution of these is not directly depth-dependent, many of the similarities may be controlled, not necessarily by depth, but by distance from the land. This comparison may not therefore be valid.

GENERAL DISCUSSION OF PELAGIC DEPOSITS

Whether the deposits described above are deep water or not one very important point must be noted. Pelagic facies are often present in a similar position in other ancient geosynclinal sequences. This can be seen in Upper Palaeozoic sediments of the Variscan geosyncline which extended from southern Ireland and southwestern England eastwards through Germany. Devonian and Carboniferous rocks of these regions often consist of laminated black shales with ostracods and trilobites, radiolarian cherts, and red nodular cephalopod bearing limestones not unlike the Ammonitico Rosso (Tucker, 1974).

Goldring (1962) has produced an elegant study combining absolute age determinations with the thicknesses of different facies in various regions. This enables an estimation of sedimentation rates to be made. The data suggest that deposition slowed markedly when the Old Red Sandstone fluviatile sedimentation was followed by the 'bathyal' pelagic facies in the axis of the geosyncline. Sedimentation speeded up again at different times in different regions when the bathyal phase was succeeded by flysch of the Culm Series.

A further common characteristic of ancient early geosynclinal pelagic sediments is their association with volcanic activity (Garrison and Fischer, 1974). This typically consists of diverse pillow lavas, spilites, basalts, and serpentinites. This association, generally termed the Ophiolitic Suite, is found in the Jurassic and Cretaceous rocks of Greece, the Apennines and the Alps. It occurs in Upper Palaeozoic rocks of the Variscan geosyncline in south-west England and Germany, and is present in the Lower Palaeozoic Caledonian geosyncline, notably in the southern uplands of Scotland.

A further notable example occurs associated with large-scale gravity tectonics in the Oman Mountains of Arabia (Wilson, 1969). There has been considerable speculation that sub-marine volcanic eruptions locally increased the silica content of sea water, encouraging population explosions of radiolaria. Aubouin (1965) has shown, however, that radiolarian chert formation was independent of vulcanicity in the Jurassic and Lower Cretaceous sediments of the Pindus furrow in Greece. It might be interesting to relate the present distributions of radiolarian oozes to sub-marine vulcanism.

Finally one more example of possible ancient deep-water sediments deserves mention. This occurs in the island of Portuguese Timor between Australia and south-east Asia. Here some curious beds of Late Cretaceous age lie beneath a thick Tertiary flysch sequence. In western Timor these contain red clays with manganese nodules, sharks teeth and rare fish bones. In eastern Timor there is a diverse assemblage of radiolarian cherts, calcilutites, marls, turbidites, and shales rich in iron and manganese and containing manganese nodules. These sediments have been the subject of a detailed study by Audley-Charles (1965; 1966).

The nodules of western Timor are geochemically similar to Recent deep sea nodules while the eastern Timor examples show characteristics midway between Recent deep sea and shelf nodules. Here again, however, the direct correlation of depth and geochemistry has been questioned by Price (1967) who, as previously mentioned, has suggested that the chemical variation of modern manganese nodules is controlled by oceanicity rather than depth.

To conclude this discussion of ancient pelagic sediments the following points are important. Turbidite flysch sequences often overlie a distinctive facies which can be recognized in various parts of the world in rocks of varying ages. This is composed of fine-grained limestones, shales and marls. The limestones are sometimes red and nodular. They are associated with radiolarian cherts and, sometimes, with sub-marine vulcanicity. The fauna of this facies is typically pelagic; benthonic fossils are generally extremely rare. Sedimentary structures suggesting strong current action are lacking. Sequences of this facies generally show few stratigraphic breaks and are thinner than nearby time-equivalent facies.

It is clear that such rocks were slowly deposited in marine environments below the photic zone and away from strong current action. In many ways these sediments are comparable to Recent deep-sea sands. Is this similarity due to the fact that they both formed in deep water, or due to distance from the land?

ECONOMIC SIGNIFICANCE OF DEEP-SEA DEPOSITS

Pelagic deposits are of interest for a number of reasons. They may be source beds from which petroleum may be expelled. They also host several types of metallic ore bodies. The oceans are generally oxygen deficient between about 200 and 1500 m (Emery, 1963). The upper limit corresponds to the average depth of the continental shelf margin. Thus barred basins whose floors lie between 200–1500 m may be areas where organic rich sediments can accumulate and ultimately become petroleum source beds.

Detailed analysis of recent DSDP data has shown that there is an optimum sedimentation rate for the preservation of organic matter. This optimum rate varies with sediment type (Johnson-Ibach, 1982). It is about 14 m/m.y. for calcareous muds, 21 m/m.y. for siliceous oozes, and about 40 m/m.y. for clays. Clays are the sediments with the highest preserved organic content. Johnson-Ibach suggests that when sedimentation rates are too slow then organic matter is oxidized, when they are too high the organic content is diluted.

At the present day the optimum situation for organic-rich sedimentation lies below the present depth of the continental shelf. At times of globally high sea level however, the anoxic zone may extend across the continental shelf (Figure 13.4).

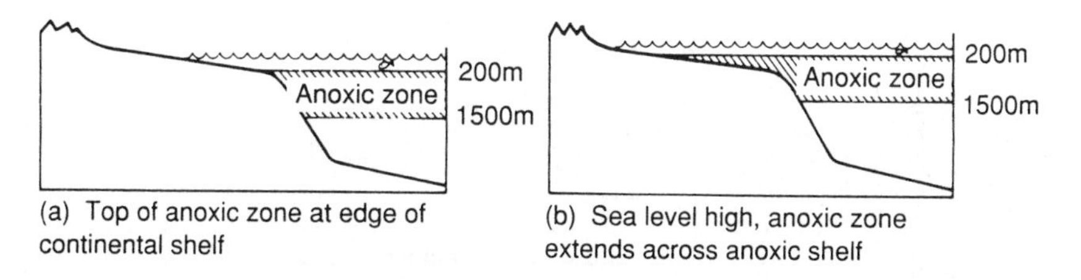

Figure 13.4 Cross-sections to show how the oceanic anoxic zone may extend across continental shelves during periods of high sea level. This will favour the deposition of transgressive petroleum source beds.

This may explain the occasional occurrence of widespread organic rich petroleum source beds, such as those of the Upper Jurassic and Lower Cretaceous (Schlanger and Cita, 1982). Indeed, it has been argued that the only good thing to come out of sequence stratigraphy is an appreciation that episodes of source rock deposition correlate with transgressions, and deep sea sand deposition with regressions ('high stands' and 'low stands' that is).

The pelagic environment also favours the development of some mineral deposits, notably manganese and sulphide ores (Cronan, 1980). The occurrence of manganese nodules and areas of sea floor encrustations has been already noted. Feasibility studies have been carried out on the commercial exploitation of these deposits. Nodules recovered from the floor of the Pacific Ocean contain up to 20% manganese and traces of copper, nickel and other metals.

There are several ancient bedded manganese deposits which appear to have formed in ancient deep sea environments, notably in Cyprus and Oman (Hutchison, 1983, Robertson and Fleet, 1986). These deposits generally overly pillow lavas and/or ophiolites and are interbedded and overlain by cherts. These are in turn succeeded by pelagic fine-grained limestones or turbidites. The sequence appears to indicate a gradual shallowing and increase in sedimentation rate from the time of formation of the manganese beds. The igneous rocks below the ore were formed as old ocean bed, on which the manganese nodules and encrustations developed. The cherts formed as siliceous oozes, and the pelagic micrites were deposited in shallower conditions above the carbonate compensation level (Figure 13.5).

Studies of recent mid-oceanic ridges, notably in the Red Sea, have revealed 'hot holes' where hot metal-rich brine springs emerge onto the sea

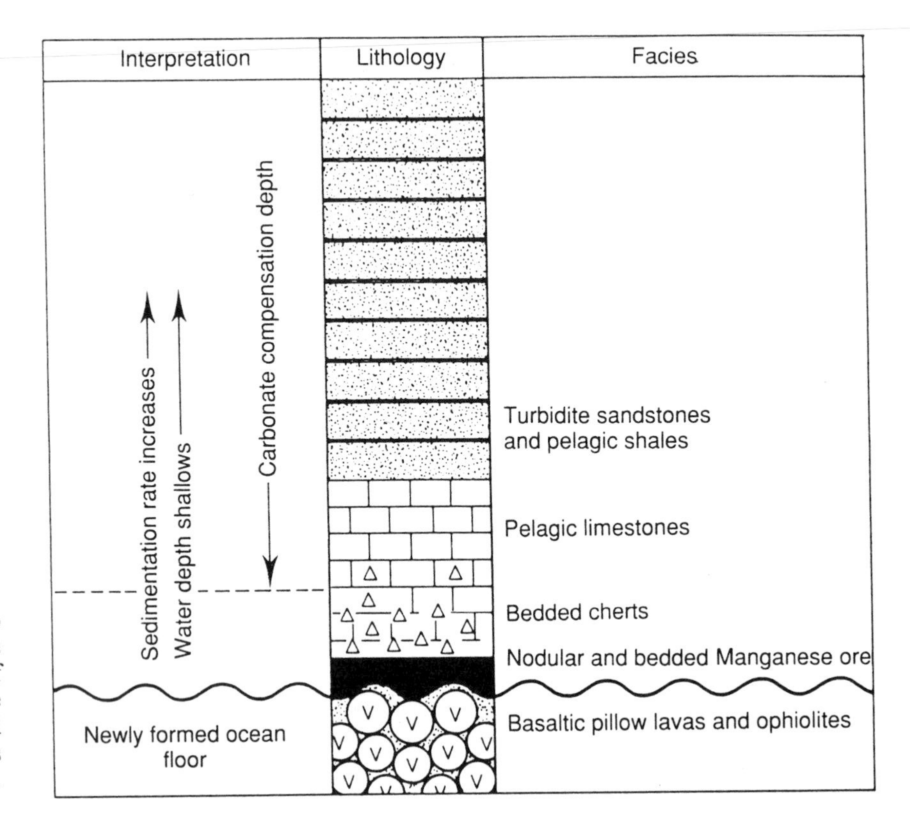

Figure 13.5 Figure to show the lithology, facies and environments of ancient deep marine manganese ores and associated deposits, such as those of Cyprus and Oman.

floor (Degens and Ross, 1969; Rona 1984). These attain temperatures of 50–60° C, and salinities of over 200 pp thousand. Hydrothermal sulphide ores are forming on the sea floor near the springs, aided by specialized thermaphilic bacteria. These recent deposits include sulphides of iron, copper, lead and zinc, together with traces of native copper, silver and gold. Ancient analogues of these deposits include the sulphide ore bodies of Rammelsberg, Germany; and Semail, Oman (see Jensen and Bateman, 1981, Chapter 8, for a review).

Observations on the rifted crests of the spreading centres in the East Pacific and other oceans have revealed the existence of dramatic underwater springs which are termed 'smokers'. Black smokers expel metallic sulphide particles at temperatures of some 400° C. White smokers expel barite, silica and pyrite at somewhat lower temperatures. It is believed that sea water is drawn into convective flow cells in the cooling basalts of the mid-ocean ridges, leaches metals out of the magma, and emerges to precipitate out the metallic sulphides (Haymon, 1983, Rona, 1984). Smokers precipitate out chimneys that grow at speeds of up to 8–30 cm per day, but these chimneys are constantly collapsing, so the smokers gradually build up mounds of a rubble of sulphide ore material that is encrusted by ore (Figure 13.6).

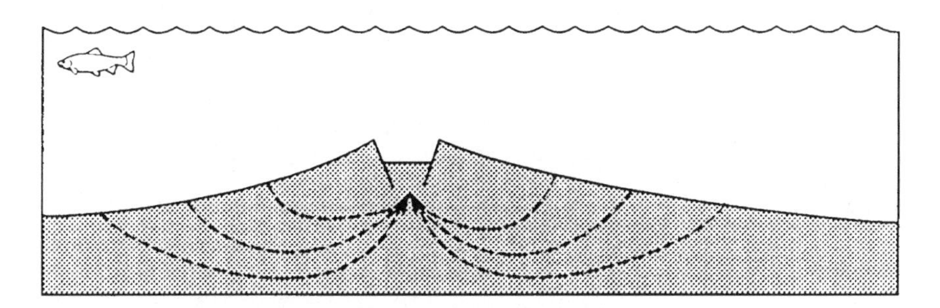

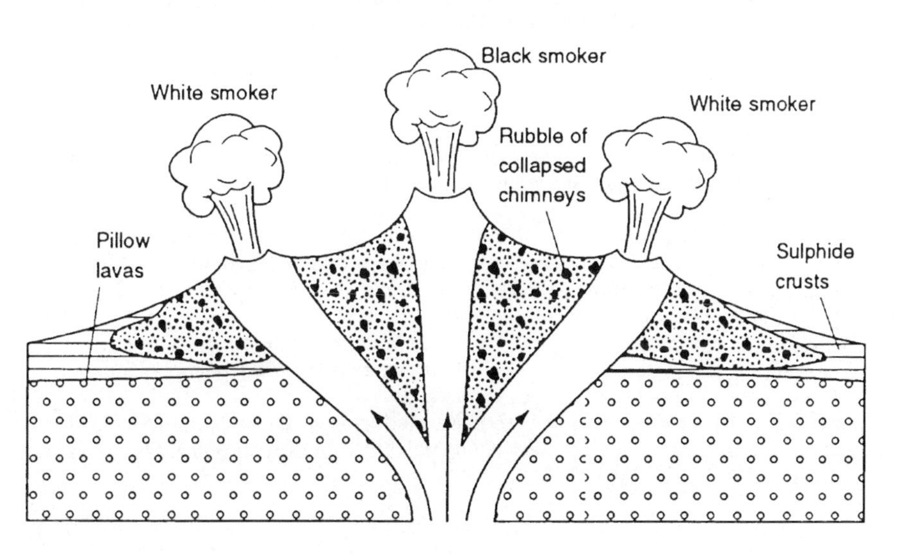

Figure 13.6 Upper: Cross-section of mid-ocean spreading centre, to show convection cells of sea water moving through new basaltic crust, emerging as hot metalliferous springs within the median rift. Lower: Cartoon of sulphide ore mound and peripheral crust apron produced by the development and collapse of 'smoker' chimneys found on modern mid-ocean ridge rifts. Ancient analogues have been described from Japan, the Oman, and other areas of preserved ancient oceanic sea floor.

These crusts become more regular, but thinner, away from the smoker mounds. Smokers are also noted for their bizarre ecosystem, based on thiophyllic bacteria, and culminating in gigantic tube worms and clams.

Analogues of these smoker mounds have now been identified in many ancient suphide ore bodies that overly pillow lavas of ancient oceanic crust (e.g. Robertson and Fleet, 1986; Ijma *et al.*, 1990).

Some bedded barite also seem to have formed in deep marine conditions. The Monitor Valley deposits of Nevada are an example. These occur in an Ordovician geosynclinal sequence of radiolarian cherts and shales. Barytes is interbedded with these, and forms thin conglomerates mixed with phosphate nodules. The presence of re-worked barytes fragments strongly supports its primary origin (Shawe, Poole and Brobst, 1969).

REFERENCES

The description of Alpine deep-sea deposits was based on

Aubouin, J. (1965) *Geosynclines*, Elsevier, Amsterdam.

Barrett, T. J. (1982) Stratigraphy and sedimentology of Jurassic bedded chert overlying ophiolites in the North Appennines, Italy, *Sedimentology*, **29**, 353–73.

Bernoulli, D. and Jenkyns, H. C. (1974) Alpine, Mediterranean, and Central Atlantic Mesozoic facies in relation to the early evolution of the Tethys. In R. E. Dott and R. H. Shaver (eds), *Modern and Ancient Geosynclinal Sedimentation*, Spec. Pub. Soc. Econ. Pal. Min., **19**, Tulsa, 129–60.

Farinacci, A. and Elmi, S. (1981) *Rosso Ammonitico Symposium Proceedings*, Ed. Tec. Roma.

Garrison, R. E. and Fischer, A. G. (1969) Deep water limestones and radiolarites of the Alpine Jurassic. In G. M. Friedman (ed.), *Depositional environments in carbonate rocks*, Soc. Econ. Min. Pal., Sp. Pub., **14**, 20–56.

—— and —— (1974) Radiolarian cherts, pelagic limestones and igneous rocks in eugeosynclinal assemblages. In K. J. Hsu and H. C. Jenkyns (eds), *Pelagic Sediments: On Land and Sea*, Blackwell Scientific Publications, Oxford, 367–99.

Hallam, A. (1967) Sedimentology and palaeographic significance of certain red limestones and associated beds in the Lias of the Alpine regions, *Scott. J. Geol.*, **3**, 195–220.

Hudson, J. D. and Jenkyns, H. C. (1969) Conglomerates in the Adnet Limestones of Adnet (Austria) and the origin of 'Scheck', *N. Jb. Geol. Palaeont. Mh., Jg.* H.9, 552–8.

Jenkyns, H. C. (1974) Origin of red nodular limestones (Ammonitico Rosso, Knollenkalke) in the Mediterranean Jurassic: a diagenetic model. In K. J. Hsu and H. C. Jenkyns (eds), *Pelagic Sediments: On Land and Sea*, Blackwell Scientific Publications, Oxford, 249–71.

Nocera, D. S. and Scandone, P. (1977) Triassic nannoplankton limestones of deep basinal origin in the central Mediterranean region, *Palaeogeog. Palaeoclim. Palaeoecol.*, **21**, 101–11.

Santantonio, M. (1994) Pelagic carbonate platforms in the geologic record: their classification, and sedimentary and paleotectonic evolution, *Amer. Assoc. Petrol. Geol. Bull.*, **78**, 122–41.

Other references cited in this chapter:

Arrhenius, G. (1963) Pelagic sediments. In M. N. Hill (ed.), *The Sea*, vol. III, Interscience, NY, 655–727.

Audley-Charles, M. G. (1965) A geochemical study of Cretaceous ferromanganiferous sedimentary rocks from Timor. *Geochem. Cosmochim. Acta*, 29, 1153–73.

—— (1966) Mesozoic palaeography of Australasia, *Palaeogeography, Palaeoclimatology, Palaeoecology*, **2**, 1–25.

Berner, R. A. (1965) Activity coefficients of bicarbonate, carbonate and calcium ions in sea water, *Geochem. Cosmochim. Acta*, **29**, 947–65.

Cook, H. E. and Enos, P. (eds) (1977) *Deep Water Carbonate Environments*, Soc. Econ. Pal. Min., Sp. Pub., **25**.

Cronan, D. S. (1980) *Underwater Minerals*, Academic Press, London.

Degens, E. T. and Ross, D. A. (eds) (1969) *Hot Brines and Recent Heavy Metal Deposits in the Red Sea*, Springer-Verlag, Berlin.

Emery, K. O. (1963) Oceanic factors in accumulation of petroleum, *Proc. 6th World Petrol. Cong. Frankfurt*, Section 1, Paper 42, PD2, 483–91.

Fischer, A. G. and Garrison, R. E. (1967) Carbonate lithifaction on the sea floor, *J. Geol.*, **75**, 488–96.

Friedman, G. M. (1965) Occurrence and stability relationships of aragonite, high magnesium calcite and low-manganesium calcite under deep sea conditions, *Bull. Geol. Soc. Am.*, **76**, 1191–5.

Goldring, R. (1962) The Bathyal lull: Upper Devonian and Lower Carboniferous sedimentation in the Variscan geosyncline. In K. Coe (ed.), *Some Aspects of the Variscan Fold Belt*, Exeter Univ. Press, 75–91.

Haymon, R. M. (1983) Growth history of hydrothermal black smokers, *Nature*, London, **301**, 695–98.

Hsu, K. J. and Jenkyns, H. C. (eds) (1974) Pelagic sediments: on land and under the sea, *Internat. Assn. Sedimentol. Spec. Pub.*, No. 1, Blackwell Scientific Publications, Oxford.

Hudson, J. D. (1967) Speculations on the depth relations of calcium carbonate solution in recent and ancient seas. In A. Hallam (ed.), *Depth Indicators in Marine Sedimentary Environments*, Marine Geology Sp. Issue 5, **5**(6), 473–80.

Hutchison, C. S. (1983) *Economic Deposits and their Tectonic Setting*, Macmillan, London.

Ijma, A., Watanabo, Y., Ogihara, S. and Yamazaki, K. (1990) Mineoka Umber: a submarine hydrothermal deposit on an Eocene arc volcanic ridge in Central Japan. In *Sp. Pub. Internat. Assn. Sedol.*, **11**, 73–88

Inderbitzen, A. L. (ed.) (1974) *Deep-sea Sediments*, Plenum Press, New York.

Jensen, M. L. and Bateman, A. M. (1981) Economic Mineral Deposits, 3rd edn, J. Wiley & Sons, Chichester.

Johnson-Ibach, L. E. (1982) Relationship between sedimentation rate and total organic carbon content in ancient marine sediments, *Bull. Amer. Assoc. Petrol. Geol.*, **66**, 170–88.

Lisitzin, A. P. (1972) Sedimentation in the world ocean, *Soc. Econ. Pal. Min. Sp. Pub.*, **17**.

Lucas, G. (1973) Deep sea carbonate facies, *Bull. Centre Rech. Pau.*, **7**, 193–206.

Morris, D. A. (1976) Organic diagenesis of Miocene sediments from Site 341, Voring Plateau, Norway. In M. Talwani, G. Udintsev *et al.* (eds), *Initial Rept. Deep Sea Drilling Project*, **38**, Washington (US Govt Printing Office), 809–14.

Price, N. B. (1967) Some geochemical observations on manganese-iron oxide nodules from different depth environments. In A. Hallam (ed.), Depth Indicators in Marine Sedimentary Environments, *Marine Geology, Sp. Issue*, 5, No. 5/6, pp. 511-38.

Riedel, W. R. (1963) The preserved record: palaeontology of pelagic sediments. In M. N. Hill (ed.), *The Sea*, vol. III, Interscience, NY, 866–87.
Robertson, A. H. F. and Fleet, A. J. (1986) Geochemistry and palaeooceanography of metalliferous and pelagic sediments from the Late Cretaceous Oman ophiolites, *Mar. Pet. Geol.*, **3**, 315–37.
Rona, G. P. (1984) Hydrothermal mineralization and sea floor spreading centres, *Earth Sci. Rev.*, **20**, 1–40.
Ross, D. A. (1982) *Introduction to Oceanography*, Prentice-Hall, New Jersey.
Schlanger, S. O. and Cita, M. B. (1982) *Nature and Origin of Cretaceous Carbon-Rich Facies*, Academic Press, London.
Scrutton, R. A. and Talwani, M. (1982) *The Ocean Floor*, J. Wiley, Chichester.
Shawe, D. R., Poole, F. G. and Brobst, D. A. (1969) Newly discovered bedded Barite deposits in East Northumberland Canyon, Nye County, Nevada, *Econ. Geol.*, **64**, 245–54.
Shepard, F. P. (1963) *Submarine Geology*, Harper & Row, NY.
Stow, A. V. and Piper, D. J. W. (1984) Fine-grained Sediments, *Deep Water Processes and Facies*, Blackwell, Oxford.
Tucker, M. E. (1974) Sedimentology of Palaeozoic pelagic limestones: the Devonian Griotte (Southern France) and Cephalopodenkalk (Germany). In K. J. Hsu and H. C. Jenkyns (eds), *Pelagic Sediments: On Land and under the Sea*, Sp. Pub. Internat. Assn. Sedol., **1**, 71–92.
Wilson, H. H. (1969) Late Cretaceous and Eugeosynclinal sedimentation, gravity tectonics and ophiolite emplacement in Oman Mountains, southeast Arabia, *Bull. Amer. Assoc. Petrol. Geol.*, **53**, 626–71.

Index